American Book Company
THE STANDARDS EXPERTS

Georgia Online Testing is now available!

On your computer: 4 Easy Steps!
1. Visit americanbookcompany.com/takeatest
2. Select your state and grade
3. Select GA CCGPS Analytic Geometry
4. Enter the password "arc"

On your mobile device: 4 Easy Steps!
1. Scan the QR code below
2. Select your state and grade
3. Select GA CCGPS Analytic Geometry
4. Enter the password "arc"

Analytic Geometry
CCGPS
2014-2015 Edition

Scan this QR code with your smart device to jump to the online testing page.

Online testing available through:
August 1, 2015

CCGPS

Analytic Geometry

2014-2015 Edition

Bryan Portnoy

Clive Sombe

Colleen Pintozzi

AMERICAN BOOK COMPANY

P. O. BOX 2638

WOODSTOCK, GEORGIA 30188-1383

TOLL FREE PHONE 1 (888) 264-5877

TOLL FREE FAX 1 (866) 827-3240

WEB SITE: www.americanbookcompany.com

Acknowledgements

In preparing this book, we would like to acknowledge Eric Field for his contribution in developing graphics, Mary Reagan for her contribution in compiling this book, and Erica Day for some of the content creation. We would also like to thank our many students whose needs and questions inspired us to write this text.

Copyright © 2014
by American Book Company
P.O. Box 2638
Woodstock, GA 30188-1383

ALL RIGHTS RESERVED

The text of this publication, or any part thereof, may not be reproduced or transmitted in any form or by any means, electronic or mechanical, including photocopying, recording, storage in an information retrieval system, or otherwise, without the prior permission of the publisher.

Printed in the United States of America
06/14

Chart of Standards

Standard	Chapter Number(s)
G.SRT.1	3
G.SRT.2	3
G.SRT.3	3
G.SRT.4	1, 3
G.SRT.5	3
G.SRT.6	5
G.SRT.7	5
G.SRT.8	5
G.CO.6	4
G.CO.7	4
G.CO.8	4
G.CO.9	1
G.CO.10	1
G.CO.11	1
G.CO.12	2
G.CO.13	2
G.C.1	6
G.C.2	6
G.C.3	6
G.C.4	6
G.C.5	6
G.GMD.1	7
G.GMD.2	7
N.RN.1	7, 12
N.RN.2	8, 9, 12
N.RN.3	8, 9
N.CN.1	10
N.CN.2	10
N.CN.3	10

Standard	Chapter Number(s)
N.CN.7	13
A.APR.1	11, 12, 15
A.SSE.1	11
A.SSE.2	11
A.SSE.3	11
A.CED.1-2	13, 14, 15
A.CED.4	13, 14
A.REI.4	13
A.REI.7	15
F.IF.4	15
F.IF.5	15
F.IF.6	16
F.IF.7	15
F.IF.8	13, 15
F.IF.9	16
F.BF.1	16
F.BF.3	15, 16
F.LE.3	15
S.ID.6	16
G-GPE.1	6
G-GPE.2	16
G-GPE.4	4
S.CP.1	17, 18
S.CP.2	18
S.CP.3	18
S.CP.4	18
S.CP.5	18
S.CP.6	18
S.CP.7	18

Contents

Acknowledgements		ii
Preface		xi
Pretest		1
1 Logic and Geometric Proofs		**10**
1.1	Geometry Logic (DOK 2)	10
1.2	Definitions, Properties, Postulates, and Theorems (DOK 2	12
1.3	Two-Column Proofs (DOK 2 & 3)	20
1.4	Flowcharts (DOK 2 & 3)	22
	Chapter 1 Review	24
	Chapter 1 Test	26
2 Geometric Constructions		**28**
2.1	Copying a Line Segment (DOK 2)	29
2.2	Copying an Angle (DOK 2)	30
2.3	Constructing a Bisector of an Angle (DOK 2)	32
2.4	Constructing the Perpendicular Bisector of a Line Segment (DOK 2)	35
2.5	Points on a Perpendicular Bisector (DOK 2)	36
2.6	Constructing Perpendicular Lines (DOK 2)	37
2.7	Constructing Parallel Lines (DOK 2)	38
2.8	Constructing Polygons Inscribed in a Circle (DOK 2)	40
	Chapter 2 Review	41
	Chapter 2 Test	42
3 Similarity		**45**
3.1	Similar and Congruent (DOK 2)	45
3.2	Solving Proportions (DOK 1)	46
3.3	Similar Triangles (DOK 2)	47
3.4	Similar Plane Figures (DOK 2)	50
3.5	Geometric Relationships of Plane Figures (DOK 3)	51
3.6	Applying Similar Plane Figures (DOK 2)	53
	Chapter 3 Review	54
	Chapter 3 Test	55

Contents

4 Transformations — 56
- 4.1 Reflections (DOK 2) — 56
- 4.2 Translations (DOK 2) — 60
- 4.3 Rotations (DOK 2) — 62
- 4.4 Transformation Practice (DOK 2) — 63
- 4.5 Algebraic Transformation Notation (DOK 2) — 64
- 4.6 Transforming a Shape onto Itself (DOK 2) — 66
- 4.7 Using Transformations to Prove Congruency (DOK 3) — 68
- 4.8 Comparing Transformations (DOK 3) — 74
- 4.9 Going Deeper with Transformations (DOK 3) — 75
- 4.10 Dilations (DOK 2) — 76
- 4.11 Using Dilations to Prove Similarity (DOK 2) — 78
- Chapter 4 Review — 79
- Chapter 4 Test — 80

5 Triangle Trigonometry — 82
- 5.1 Pythagorean Theorem (DOK 2 & 3) — 82
- 5.2 Finding the Missing Leg of a Right Triangle (DOK 2) — 84
- 5.3 Applications of the Pythagorean Theorem (DOK 3) — 85
- 5.4 Special Right Triangles (DOK 2) — 87
- 5.5 Introduction to Trigonometric Ratios (DOK 2 & 3) — 89
- 5.6 More Similar Triangles (DOK 3) — 94
- 5.7 More Trigonometric Ratios (DOK 2) — 95
- 5.8 Trigonometric Ratios Applied to Other Shapes (DOK 2) — 97
- Chapter 5 Review — 98
- Chapter 5 Test — 100

6 Circles — 102
- 6.1 Parts of a Circle (DOK 2) — 102
- 6.2 Circumference (DOK 2) — 104
- 6.3 Area of a Circle (DOK 2) — 105
- 6.4 Proving Similarity of Circles (DOK 2) — 108
- 6.5 Constructing a Tangent Line To The Circle (DOK 2) — 109
- 6.6 Constructing an Inscribed Circle of a Triangle (DOK 2) — 111
- 6.7 Constructing a Circumscribed Circle of a Triangle (DOK 2) — 112
- 6.8 Arc Measures (DOK 2) — 113
- 6.9 Properties of Angles of a Quadrilateral (DOK 2) — 114

6.10	More Arc Lengths (DOK 3)	115
6.11	More Circle Properties (DOK 3)	116
6.12	Area of Sectors (DOK 2)	119
6.13	Finding Equations of Circles Using the Pythagorean Theorem (DOK 2)	120
	Chapter 6 Review	121
	Chapter 6 Test	123

7 Solid Geometry — 125

7.1	Types of 3-Dimensional Figures (DOK 1)	125
7.2	Cross Sections (DOK 2)	127
7.3	Formation of a Cube (DOK 2)	129
7.4	Formation of a Rectangular Prism (DOK 2)	129
7.5	Formation of a Cone (DOK 2)	130
7.6	Formation of a Sphere (DOK 2)	130
7.7	Formation of a Cylinder (DOK 2)	131
7.8	Volume of Spheres, Cones, Cylinders, and Pyramids (DOK 2)	132
7.9	Volume Word Problems (DOK 3)	134
7.10	Geometric Relationships of Solids (DOK 3)	135
7.11	Applying Similar Solid Figures (DOK 3)	137
	Chapter 7 Review	138
	Chapter 7 Test	142

8 Exponents — 147

8.1	Understanding Exponents (DOK 1)	147
8.2	Multiplying Exponents with the Same Base (DOK 1)	148
8.3	Multiplying Fractional Exponents with the Same Base (DOK 1)	148
8.4	Multiplying Exponents Raised to an Exponent (DOK 1)	149
8.5	Expressions Raised to a Power (DOK 1)	149
8.6	Fractions Raised to a Power (DOK 1)	150
8.7	More Multiplying Exponents (DOK 1)	150
8.8	More Multiplying Fractional Exponents (DOK 2)	151
8.9	Negative Exponents (DOK 1)	152
8.10	Multiplying with Negative Exponents (DOK 2)	152
8.11	Dividing with Exponents (DOK 1)	153
8.12	Dividing with Fractional Exponents (DOK 2)	154
8.13	Order of Operations (DOK 2)	155

Contents

        Chapter 8 Review   156
        Chapter 8 Test   157

9 Roots   **159**
    9.1   Square Root (DOK 1)   159
    9.2   Simplifying Square Roots Using Factors (DOK 1)   159
    9.3   Adding, Subtracting, and Simplifying Square Roots (DOK 2)   160
    9.4   Multiplying and Simplifying Square Roots (DOK 2)   161
    9.5   Dividing and Simplifying Square Roots (DOK 2)   162
    9.6   Cube Roots (DOK 1)   163
    9.7   Rational Exponents (DOK 1)   164
    9.8   Real Numbers (DOK 1)   165
    9.9   Computing with Real Numbers (DOK 1)   166
    9.10   Going Deeper into Roots (DOK 3)   166
        Chapter 9 Review   167
        Chapter 9 Test   168

10 Complex Numbers   **170**
    10.1   Complex Numbers Defined (DOK 1)   170
    10.2   Imaginary Numbers (DOK 1)   171
    10.3   Adding and Subtracting Complex Numbers (DOK 2)   172
    10.4   Multiplying Complex Numbers (DOK 2)   173
    10.5   Dividing Complex Numbers (DOK 2)   174
    10.6   Simplifying Complex Numbers (DOK 2)   175
        Chapter 10 Review   176
        Chapter 10 Test   177

11 Polynomials   **179**
    11.1   Parts of an Expression (DOK 1)   179
    11.2   Rewriting Expressions (DOK 2)   181
    11.3   Adding Polynomials (DOK 2)   183
    11.4   Subtracting Polynomials (DOK 2)   183
    11.5   Multiplying Monomials (DOK 2)   185
    11.6   Multiplying Monomials with Different Variables (DOK 2)   185
    11.7   Multiplying Monomials by Polynomials (DOK 2)   186
    11.8   Dividing Polynomials by Monomials (DOK 2)   187
    11.9   Dividing Polynomials Using Synthetic Division(DOK 2)   188
    11.10   Removing Parentheses and Simplifying (DOK 2)   189

	Chapter 11 Review	190
	Chapter 11 Test	191

12 Solving Multi-Step Equations and Inequalities — 194

12.1	Two-Step Algebra Problems (DOK 2)	194
12.2	Multi-Step Algebra Problems (DOK 2)	195
12.3	Solving Radical Equations (DOK 2)	197
12.4	More Equations (DOK 2)	198
12.5	Equation Word Problems (DOK 3)	199
12.6	Multi-Step Inequalities (DOK 2)	200
12.7	Solving Equations and Inequalities with Absolute Values (DOK 2)	202
12.8	More Solving Equations and Inequalities with Absolute Values (DOK 2)	203
12.9	Inequality Word Problems (DOK 3)	205
12.10	Recognizing Errors in Problems (DOK 3)	206
	Chapter 12 Review	207
	Chapter 12 Test	208

13 Solving Quadratic Equations — 209

13.1	Fundamental Theorem of Algebra (DOK 2)	209
13.2	Using the Quadratic Formula (DOK 2)	211
13.3	Solving Quadratic Equations with Complex Roots (DOK 2)	212
13.4	Solving the Difference of Two Squares (DOK 2)	213
13.5	Solving Perfect Squares (DOK 2)	214
13.6	Completing the Square (DOK 2)	215
13.7	Real-World Quadratic Equations (DOK 3)	216
	Chapter 13 Review	220
	Chapter 13 Test	221

14 Systems of Equations and Inequalities — 224

14.1	Solving Systems of Linear Equations by Substitution (DOK 2)	224
14.2	Graphing Systems of Linear Inequalities (DOK 2)	226
14.3	Graphing Systems of Three Linear Inequalities (DOK 3)	228
14.4	Manipulating Formulas and Equations (DOK 2)	230
	Chapter 14 Review	231
	Chapter 14 Test	232

15 Polynomial Functions — 235

15.1	Solving Polynomial Equations Analytically (DOK 3)	235

Contents

15.2	Solving Polynomial Equations Graphically (DOK 2)	242
15.3	Solving Polynomial Inequalities Analytically (DOK 3)	247
15.4	Graph Transformation of $f(x) = ax^n$ (DOK 2)	253
15.5	Multiplicity of Graphs of Polynomial Functions (DOK 3)	256
15.6	Real-World Polynomial Functions (DOK 3)	261
	Chapter 15 Review	263
	Chapter 15 Test	266

16 Combining and Comparing Functions — **270**

16.1	Parabolas (DOK 3)	270
16.2	Solutions of Equations (DOK 2)	272
16.3	Finding Common Solutions of Functions (DOK 2)	274
16.4	Finding Rate of Change of a Function (DOK 2, 3)	275
16.5	Comparing Functions (DOK 3)	277
16.6	Comparing Real-Life Functions (DOK 3)	279
16.7	Combining Functions (DOK 2)	280
16.8	Function Symmetry (DOK 2)	281
16.9	Symmetry of Graphs of Functions (DOK 2)	283
16.10	Solutions of Equations (DOK 2)	286
16.11	Modeling Data with Quadratic Functions (DOK 2)	287
	Chapter 16 Review	290
	Chapter 16 Test	293

17 Sets — **299**

17.1	Set Notation (DOK 2)	299
17.2	Subsets (DOK 2)	300
17.3	Intersection of Sets (DOK 2 & 3)	301
17.4	Union of Sets (DOK 2 & 3)	302
17.5	Reading Venn Diagrams (DOK 3)	303
	Chapter 17 Review	304
	Chapter 17 Test	305

18 Probability — **307**

18.1	Probability Terms (DOK 2)	307
18.2	Probability (DOK 2)	308
18.3	More Probability (DOK 2)	310
18.4	Independent and Dependent Events (DOK 3)	311
18.5	Mutually Exclusive Events (DOK 3)	313

18.6	Conditional Probability (DOK 3)	314
18.7	The Multiplication Rule (DOK 2)	316
18.8	Two-Way Frequency Tables (DOK 3)	317
	Chapter 18 Review	319
	Chapter 18 Test	322

Preface

CCGPS Analytic Geometry will help you review and learn important concepts and skills related to high school geometry. **The materials in this book are based on the CCGPS in mathematics coordinated by the National Governors Association Center for Best Practices and the Council of Chief State School Officers. The complete list of standards is located at the beginning of the Answer Key. Each chapter is referenced to the standards.**

This book contains chapters that teach the concepts and skills for Geometry. Answers to the pretest and exercises are in a separate manual.

CCGPS Analytic Geometry includes Depth of Knowledge levels for four content areas based on Norman Webb's Model of interpreting and assigning depth of knowledge levels to both objectives within standards and assessment items for alignment analysis.

The four levels of Depth of Knowledge:

Level 1: Recall and Reproduction (DOK 1)

Questions at this level (DOK 1) may ask you to recall information such as facts, definitions, or simple procedures, as well as perform a simple algorithm or carry out a one-step, well-defined, and straightforward procedure. A few example DOK level 1 questions are listed below.

- State the associative property of multiplication.
- Identify the divisor in a problem.
- Measure the perimeter of a figure.
- Calculate $4.3 + 8.5$.

Level 2: Skills and Concepts/Basic Reasoning (DOK 2)

DOK level 2 questions involve some mental processing beyond a habitual response. They require students to make some decisions as to how to approach the problem, as well as to classify, organize, estimate, make observations, collect, display, and compare data. A few example DOK level 2 questions are listed below.

- Interpret the bar graph to answer questions about a given population.
- Classify the different types of polygons based upon their characteristics.
- Compare the two sets of data using their measures of central tendency.
- Extend the algebraic pattern.

Level 3: Strategic Thinking/Complex Reasoning (DOK 3)

This level (DOK 3) includes problems that require reasoning, planning, using evidence, and higher levels of thinking beyond what was required in DOK levels 1 and 2. This level requires students to explain their thinking, and its cognitive demands are more complex and abstract. DOK 3 demands that students use reasoning skills to draw conclusions from observations and make conjectures. Some examples of level 3 DOK questions are listed below.

- Explain how to determine if two triangles are similar.
- Formulate an expression to determine the next few terms in a pattern.

Preface

- Construct and conduct a survey and analyze the results to determine the most popular movie genre.

Level 4: Extended Thinking/Reasoning (DOK 4)

DOK level 4 questions include things such as complex reasoning, planning, and developing. Students' thinking will most likely take place over an extended period of time and will include taking into consideration a number of variables. Students should make several connections and relate ideas within the content area or among other content areas. Students should select one approach among many alternatives to solve problems. At this level students will be expected to design and conduct their own experiments, make connections between findings, and relate them to other concepts and phenomena. A few example problems are presented below.

- Explore real-world phenomena of Cartesian planes and create a report to present your fndings.

- Connect your knowledge of integers to the plate tectonics of Earth.

- Design, carry out, and analyze the results of your own experiment to determine the fairness of common game pieces (i.e., dice, spinners, etc.) using what you know about the probability of events.

ABOUT THE AUTHORS

Bryan Portnoy has spent over 15 years in education, most of them teaching mathematics at the high school level. He has a Bachelor's of Arts Degree in Mathematics from SUNY Albany and a Master's in Educational Leadership from Georgia State University. Bryan's focus in the classroom was helping all students achieve by meeting the needs of each individual student through their own learning style. Bryan is currently serving as the Math Curriculum Director.

Colleen Pintozzi has taught mathematics at the middle school, junior high, senior high, and adult level for 22 years. She holds a B.S. degree from Wright State University in Dayton, Ohio and has done graduate work at Wright State University, Duke University, and the University of North Carolina at Chapel Hill. She is the author of many mathematics books.

Clive Sombe received most of his education from his home country, Zambia, and from The United Kingdom. Clive holds an Advanced Diploma in Management Accounting as well as A-Levels in Mathematics, Chemistry, and Physics from Cambridge University. Clive has most recently earned a Bachelor of Science Degree in Mathematics with honors from Kennesaw State University. He is Green Belt-Six Sigma certified and enjoys tutoring, writing and spends some of his spare time exploring mathematical trivia.

Pretest

1 When the scale factor of a dilation is greater than 1, the image is a(n)...

A reduction

B enlargement

C rotation

D reflection

G.SRT.1 DOK 1

2 Suppose a chord of a circle is 50 cm long, and the distance from its midpoint to the center of the circle is 60 cm. Find the radius of the circle.

A 105 cm

B 43 cm

C 65 cm

D 58 cm

G.C.2 DOK 2

3 Simplify:

$$\frac{5 + 4i}{3 - 2i}$$

A $\dfrac{7 + 22i}{13}$

B $\dfrac{13}{23 + 22i}$

C $\dfrac{9i}{i}$

D $\dfrac{8}{2i}$

G.CN.3 DOK 2

4 In a soccer team of 22 players, 7 are strikers and the rest are defenders. During a practice session, 4 strikers each scored 2 or more goals, and 5 of the defenders each scored 2 or more goals. If a player is chosen at random from the team, what is the probability of choosing a striker or a player that scored 2 or more goals?

A $\dfrac{5}{22}$

B $\dfrac{15}{22}$

C $\dfrac{7}{22}$

D $\dfrac{7}{15}$

S.CP.7 DOK 3

5 The table shows the summer jobs taken up by students at a local high school.

	Lawns	Cars	Dogs	Total
Girls	96	122	56	274
Boys	184	58	30	272
Total	280	180	86	546

If a student is selected at random, what is the probability that this student got a car washing job (Cars) in the summer?

A 0.50

B 0.330

C 0.88

D 0.72

S.CP.4 DOK 3

6 Find the roots: $2x^2 - 26x - 336 = 0$

A $x = \dfrac{4}{7} \pm \dfrac{\sqrt{5}}{7}$

B $x = -8, 21$

C $x = \dfrac{7}{4} \pm \dfrac{\sqrt{7}}{5}$

D $x = 7, 3$

N.CN.7 DOK 2

7 Which statement best describes the behavior of the function within the interval $(-3, 3)$ to $(0, 0)$?

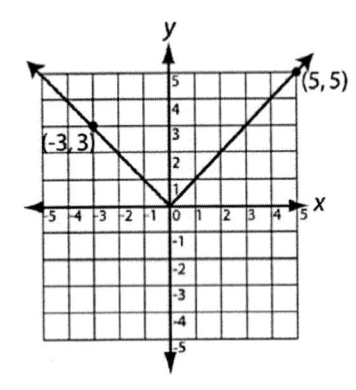

A from left to right, the function falls and then rises

B from left to right, the function rises and then falls

C from left to right, the function falls only

D from left to right, the function rises only

F.IF.4 DOK 2

8 $(1 - 6i) - (3 - 2i)$

A $-2 - 4i$

B $4 - 2i$

C $5 - 5i$

D $-7i + 8i$

N.CN.2 DOK 1

9 Which of the following types of distributions best describe the graph below?

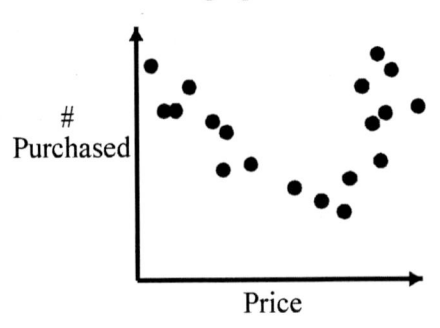

A quadratic

B linear

C exponential

D piecewise

S.ID.6a DOK 2

10 Rewrite the expression $(5x^3y^4z^5) \times (-4x^{-3}y^8z)$, without negative exponents.

A $20xy^2z^6$

B $-20x^6y^2z^6$

C $20xy^{12}z^6$

D $-20y^{12}z^6$

A.APR.1 DOK 2

11 What is the length of the missing side?

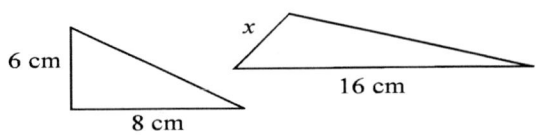

A 5

B 2

C 8

D These triangles are not similar, so it cannot be determined.

G.SRT.5 DOK 2

12 Find the points of intersection of the graphs of the system:

$$x^2 + y^2 = 5$$
$$y = -2x$$

A $(1,-2), (-1,2)$

B $(1,2), (1,2)$

C $(-1,-2), (1,2)$

D $(1,2), (1,-2)$

A.REI.7 DOK 2

13 Write the quadratic function in vertex form. Identify the vertex.

$$x^2 + 6x - 7 = 0$$

A $(x+6)^2 - 7$, vertex: $(-6,7)$

B $x^2 + 1$, vertex: $(0,1)$

C $(x-3)^2 + 49$, vertex: $(3,-49)$

D $(x+3)^2 + 16$, vertex: $(-3,-16)$

F.IF.8a DOK 3

14 The figures below are similar. What is the measure of x?

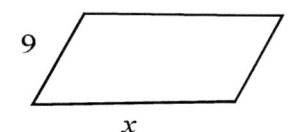

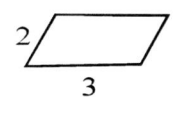

A 10

B 13.5

C 26

D cannot be determined

G.SRT.2 DOK 2

15 If the figure below is translated in the direction described by the arrow, what will be the new coordinates of point D after the transformation?

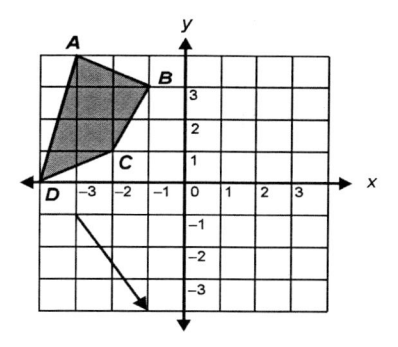

A $(-2,-4)$

B $(-1,-3)$

C $(-1,-4)$

D $(-2,-3)$

G.CO.6 DOK 2

16 You deposit $\$1,500$ into an account that pays 5% interest that compounds yearly. Find the balance after 6 years.

A $\$7,500$

B $\$1,950$

C $\$4,333.33$

D $\$2,010.14$

F.BF.1b DOK 3

17 Which point lies on a circle with center $(-7,8)$ and radius 3?

A $(-8,4)$

B $(-4,8)$

C $(1,6)$

D $(-4,7)$

G.GPE.4 DOK 2

18 What is the probability of getting 3 heads in a row when tossing a coin?

  A 0.125

  B 0.133

  C 0.5

  D 1

S.CP.2 DOK 2

19 A segment that touches a circle at two points is called a

  A tangent

  B chord

  C secant

  D inscribed line

G.C.4 DOK 1

20 Simplify:
$$\left(\frac{14a^3b^4}{7a^6b^2}\right)^4$$

  A $\dfrac{2a^3}{b^2}$

  B $\dfrac{b^4}{7a^6}$

  C $\dfrac{16a^{12}}{b^8}$

  D $\dfrac{16b^8}{a^{12}}$

A.SSE.1 DOK 2

21 How many roots does the quadratic equation $x^2 - 1 = 0$, have?

  A 1

  B 2

  C none

  D 3

A.REI.4a DOK 1

22 From what point will the function $y = 16^x$ surpass $y = 16x$?

  A $x = 1$

  B $x > 1$

  C $x = 16$

  D $x > 16$

F.LE.3 DOK 2

23 What is the equation of the parabola with focus at $(-4, 9)$ and directrix $y = 6$.

  A $y = \frac{3}{4}x^2 + \frac{1}{6}x + 6$

  B $y = 6x^2 + 4x + 61$

  C $y = \frac{1}{6}x^2 + \frac{4}{3}x + \frac{61}{6}$

  D $y = \frac{5}{6}x^2 + \frac{4}{3}x + \frac{31}{3}$

G.GPE.2 DOK 3

24 What are the center and radius of a circle described by the equation
$$(x+7)^2 + (y-8)^2 = 9$$

  A Center $= (0, 0)$, radius $= 9$ units

  B Center $= (7, -8)$, radius $= 9$ units

  C Center $= (0, 0)$, radius $= 3$ units

  D Center $= (-7, 8)$, radius $= 3$ units

G.GPE.1 DOK 2

25 Find the measure of the missing side.

  A 6

  B 8

  C 12

  D 5.33

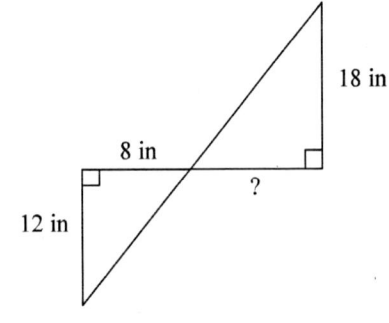

G.SRT.6 DOK 2

26 What, to the nearest mile, is the distance of Carmen's house from the police station?

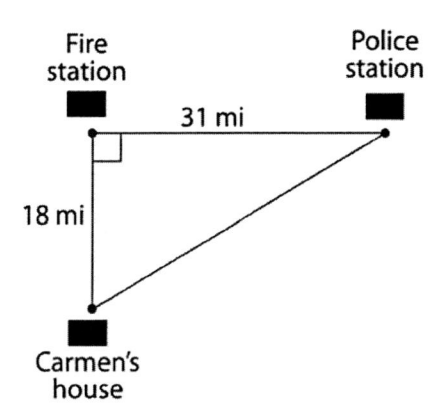

- **A** 35 miles
- **B** 36 miles
- **C** 37 miles
- **D** 25 miles

G.SRT.8 DOK 2

27 A job pays $300 per week. For every 3 months of continued work, a worker gets $10 per week raise. Write a formula for a workers weekly pay after n 3-month periods.

- **A** $300n + 10$
- **B** $300 + 10n$
- **C** $3n(310)^n$
- **D** $(310n)^3$

F.BF.1a DOK 2

28 Describe the transformation made to the parent graph of $y = (x+4)^2 + 5$

- **A** 4 units to the right, 5 units down
- **B** 5 units to the right, 4 units down
- **C** 4 units to the left, 5 units up
- **D** 25 units to the left, 4 units down

F.BF.3 DOK 2

29 You are bored while shopping with your mother. You watch 20 people walk by. Of those, 12 are female. From those 12 females, 7 have blonde hair. How many females do not have blonde hair?

- **A** 3
- **B** 7
- **C** 5
- **D** 1

S.CP.1 DOK 1

30 What are the maxima and minima of the graph?

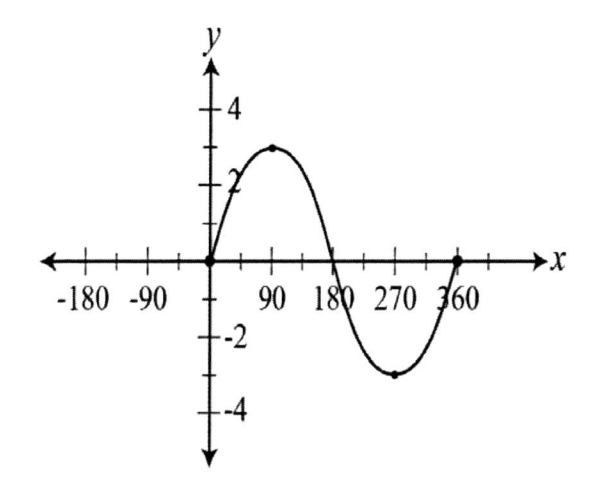

- **A** maxima: $(3, 90)$, minima: $(-3, 270)$
- **B** maxima: $(90, 3)$, minima: $(270, -3)$
- **C** maxima: $(0, 0)$, minima: $(-180, 0)$
- **D** maxima: $(180, 0)$, minima: $(0, 0)$

F.IF.7a DOK 2

31 If $\angle X \cong \angle Y$, what similarity postulate explains why triangles XQW and YQZ are similar?

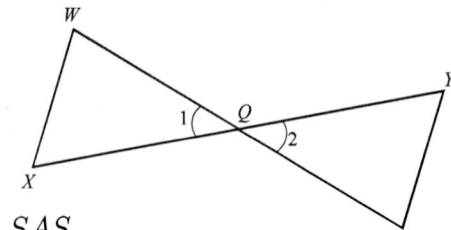

A SAS
B SSS
C AA
D SSA

G.SRT.3 DOK 1

32 $(3+4i)+(2+5i) =$

A $1+3i$
B $5+9i$
C $3-i$
D $4+8i$

N.CN.1 DOK 1

33 Consider the function $y = x^2 + 6x + 4$. Find the average rate of change from $x = 1$ to $x = 4$.

A 11
B 3
C 21
D 16

F.IF.6 DOK 2

34 $64^{\frac{2}{3}} + 81^{\frac{1}{4}} - 125^{\frac{1}{3}} =$

A 12
B 13
C 14
D 15

N.RN.2 DOK 2

35 Which function has the greater slope?

A:
x	0	100	200	300	400
y	0	10	20	30	40

B: $y = 0.04x + 18$

A Function B, Slope $= 0.1$
B Function A, Slope $= 0.1$
C Function B, Slope $= 18$
D Function A: Slope $= 0.04$

F.IF.9 DOK 2

36 A basketball has a diameter of 28 inches. Which expression uses the formula for volume of a sphere correctly?

A $\dfrac{3}{4}\pi (14)^3$

B $\dfrac{4}{3}(\pi)^3$

C $\dfrac{4}{3}\pi (28)^2$

D $\dfrac{4}{3}\pi (14)^3$

G.GMD.3 DOK 1

37 The sum of 2 numbers is 126 and their difference is 42. What are the two numbers?

A 100, and 26
B 108, and 18
C 64, and 62
D 84, and 42

A.CED.1 DOK 3

38 Solve $x^2 - 22x = -117$ by completing the square.

- **A** $x = 13, x = 9$
- **B** $x = 5 \pm 3\sqrt{5}$
- **C** $x = 22 \pm \sqrt{117}$
- **D** $x = -13, x = -9$

A.SSE.3 DOK 2

39 A rectangle has an area of 32 m². If you are told that its width is 4 m less than its length, find the length of the rectangle.

- **A** 4 m
- **B** 7 m
- **C** 8 m
- **D** 5 m

A.REI.4b DOK 3

40 Find the missing side from the following similar triangles.

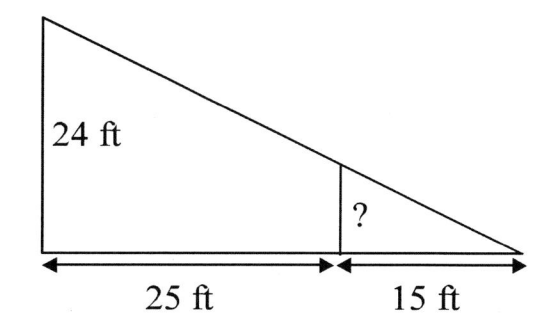

- **A** 12 ft
- **B** 14.4 ft
- **C** 15.625 ft
- **D** 9 ft

G.SRT.4 DOK 2

41 Solve the following radical equation?

$$2\sqrt{x - 8} + 18 = 34$$

- **A** $x = 72$
- **B** $x = 8$
- **C** $x = 36$
- **D** $x = 16$

N.RN.1 DOK 3

42 Today there is a 60% chance of rain, 15% of lightning, and 20% chance of lightning if it is raining. What is the chance of both rain and lightning today?

- **A** 20%
- **B** 45%
- **C** 12%
- **D** 35%

S.CP.5 DOK 3

43 $P(A) = 0.17$
$P(B) = 0.4$

Find $P(A \text{ and } B)$ that would show $P(A)$ and $P(B)$ are independent.

- **A** 0.425
- **B** 0.23
- **C** 0.57
- **D** 0.068

S.CP.3 DOK 2

44 Find the circumference of the a circle with radius 15 cm.

- **A** 30π cm
- **B** 7.5 cm
- **C** 225π cm²
- **D** 45π

C.GMD.1 DOK 1

45 What is the domain for the following graph?

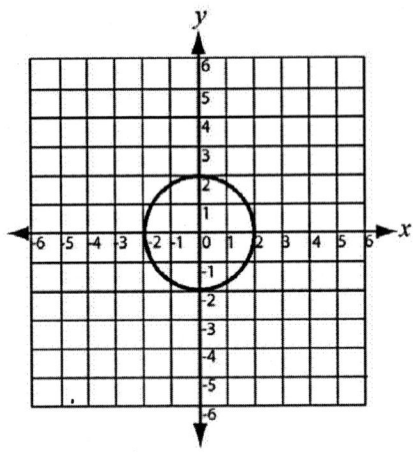

- **A** $-2 \leq y \geq 2$
- **B** $-4 \leq x \leq 2$
- **C** $-2 \leq y \leq 2$
- **D** $-2 \leq x \leq 2$

F.IF.5 DOK 1

46 Use the circle to find the measure of $m\widehat{AD}$. It is given that $m\angle BDE = 60°$, $m\angle ABD = 35°$, and that $m\widehat{AB} = 100$.

- **A** 70
- **B** 95
- **C** 85
- **D** 35

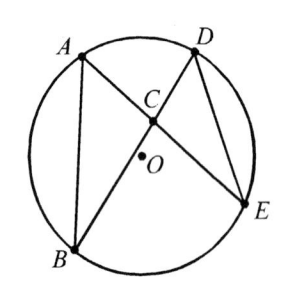

G.C.5 DOK 2

47 Solve:
$$2x - y = 1$$
$$5 - 3x = 2y$$

- **A** $(1, 1)$
- **B** $(2, 1)$
- **C** $(-8, 6)$
- **D** $(3, 4)$

A.CED.4 DOK 2

48 The product of two consecutive integers is 7656. What are these two integers?

- **A** 76 and 56
- **B** 76 and 77
- **C** 87 and 88
- **D** 66 and 67

A.CED.2 DOK 1

49 The interior angles of a quadrilateral inscribed in a circle are 110°, 70°, $2x$, and x. What is the value of x?

- **A** 60°
- **B** 120°
- **C** 70°
- **D** 110°

G.C.3 DOK 2

50 Find the area of the shaded region.

- **A** 50.27 in^2
- **B** 12.57 in^2
- **C** 62.83 in^2
- **D** 113.10 in^2

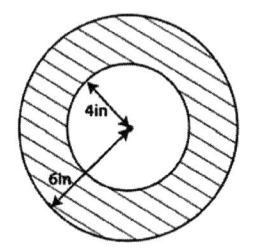

G.C1 DOK 2

51 Find a.

- **A** $2\sqrt{2}$
- **B** $3\sqrt{2}$
- **C** $\sqrt{6}$
- **D** $2\sqrt{3}$

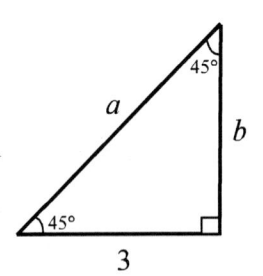

G.SRT.7 DOK 2

Evaluation Chart for the Mathematics Pretest

Directions: On the following chart, circle the question numbers that you answered incorrectly. Then turn to the appropriate topics (listed by chapters), read the explanations, and complete the exercises. Review the other chapters as needed.

	Pretest	Pages
Chapter 1:	40	10 – 27
Chapter 2:		28 – 44
Chapter 3:	1, 11, 14, 31	45 – 55
Chapter 4:	17, 15	56 – 81
Chapter 5:	25, 26, 51	82 – 101
Chapter 6:	2, 19, 24, 46, 49, 50	102 – 124
Chapter 7:	36, 41, 44	125 – 146
Chapter 8:	34	147 – 158
Chapter 9:	34	159 – 169
Chapter 10:	3, 8, 32	170 – 178
Chapter 11:	10, 20, 38	179 – 193
Chapter 12:	10, 34, 41	194 – 208
Chapter 13:	6, 13, 21, 37, 39, 47, 48	209 – 223
Chapter 14:	37, 47, 48	224 – 234
Chapter 15:	7, 10, 12, 13, 22, 28, 30, 37, 45, 48	235 – 269
Chapter 16:	9, 16, 23, 27, 33, 35	270 – 298
Chapter 17:	29	299 – 306
Chapter 18:	4, 5, 18, 29, 42, 43	307 – 325

Chapter 1
Logic and Geometric Proofs

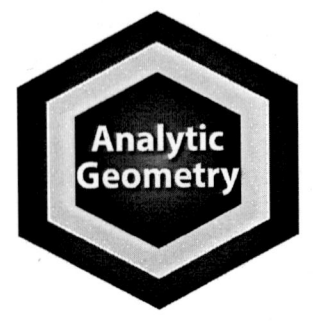

This chapter covers the following CCGPS standard(s):

	Content Standards
Congruence	G.CO.9, G.CO.10, G.CO.11
Similarity, Right Triangles, and Trigonometry	G-SRT.4

1.1 Geometry Logic (DOK 2)

A **conditional statement** is a type of logical statement that has two parts, a **hypothesis** and a **conclusion**. The statement is written in "if-then" form, where the "if" part contains the hypothesis and the "then" part contains the conclusion. For example, let's start with the statement, "Two lines intersect at exactly one point." We can rewrite this as a conditional statement in "if-then" form as follows:

$$\underbrace{\text{If two lines intersect}}_{\text{hypothesis}}, \text{ then } \underbrace{\text{their intersection is at exactly one point}}_{\text{conclusion}}.$$

Conditional statements may be true or false. To show that a statement is false, you need only to provide a single **counterexample** which shows that the statement is not always true. To show that a statement is true, on the other hand, you must show that the conclusion is true for all occasions in which the hypothesis occurs. This is often much more difficult.

Example 1: Provide a counterexample to show that the following conditional statement is false:
If $x^2 = 4$, then $x = 2$.
To begin with, let $x = -2$.
The hypothesis is true, because $(-2)^2 = 4$.
For $x = -2$, however, the conclusion is false even though the hypothesis is true. Therefore, we have provided a counterexample to show that the conditional statement is false.

The **converse** of a conditional statement is an "if-then" statement written by switching the hypothesis and the conclusion. For example, for the conditional statement "If a figure is a quadrilateral, then it is a rectangle," the converse is "If a figure is a rectangle, then it is a quadrilateral."

The **inverse** of a conditional statement is written by negating the hypothesis and conclusion of the original "if-then" conditional statement. Negating means to change the meaning so it is the negative, or opposite, of its original meaning. The inverse of the conditional statement "If a figure is a quadrilateral, then it is a rectangle" is "If a figure is **not** a quadrilateral, then it is **not** a rectangle."

The **contrapositive** of a conditional statement is written by negating the converse. That is, switch the hypothesis and conclusion of the original statement, and make them both negative. The contrapositive of the conditional statement "If a figure is a quadrilateral, then it is a rectangle" is "If a figure is not a rectangle, then it is not a quadrilateral."

1.1 Geometry Logic (DOK 2)

Example 2: Given the conditional statement, "If $m\angle F = 60°$, then $\angle F$ is acute." Write the converse, inverse, and contrapositive.

Step 1: The converse is constructed by switching the hypothesis and the conclusion: If $\angle F$ is acute, then $m\angle F = 60°$.

Step 2: The inverse is constructed by negating the original statement: If $m\angle F \neq 60°$, then $\angle F$ is not acute.

Step 3: The contrapositive is the negation of the converse: If $\angle F$ is not acute, then $m\angle F \neq 60°$.

Answer the following problems about geometry logic.

1. Rewrite the following as a conditional statement in "if-then" form: A number divisible by 8 is also divisible by 4.

2. Write the converse of the following conditional statement: If two circles have equal radii, then the circles are congruent.

3. Given the conditional statement: If $x^4 = 81$, then $x = 3$. Is the statement true? Provide a counterexample if it is false.

4. Given the statement: A line contains at least two points. Write as a conditional statement in "if-then" form, then write the converse, inverse, and contrapositive of the conditional statement.

5. "If a parallelogram has four congruent sides, then it is a rhombus." Write the converse, inverse, and contrapositive for the conditional statement. Which are true? Which are false?

6. "If a triangle has one right angle, then the acute angles are complementary." Write the converse, inverse, and contrapositive for the conditional statement. Indicate whether each is true or false.

7. "If a rectangle has four congruent sides, then it is a square." Write the contrapositive for the conditional statement and indicate whether it is true or false. Give a counterexample if it is false.

1.2 Definitions, Properties, Postulates, and Theorems (DOK 2)

Proofs use definitions, properties, postulates, and theorems to provide convincing arguments about geometric relationships. **Postulates** are statements that are assumed to be true, especially in arguments. **Theorems** are propositions that need to be proven first before they can be used in a proof.

Example 3: Given: $\triangle ABC$, $\angle B$ is a right triangle
Prove: $\angle A$ and $\angle C$ are complementary

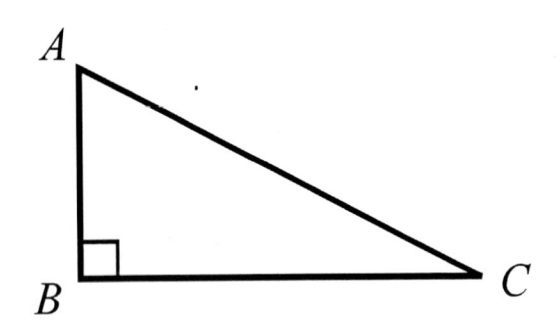

Statement	Reason
$\angle B$ is a right angle	Given (facts are provided)
$m\angle B = 90°$	Defintion of a right angle
$\triangle ABC$	Given
$m\angle A + m\angle B + m\angle C = 180°$	Triangle angle - sum theorem
$m\angle A + 90° + m\angle C = 180°$	Substitution property
$m\angle A + m\angle C = 90°$	Subtraction property
$\angle A$ and $\angle C$ are complementary	Definition of complementary angles

All theorems that are presented in this book have already been proven and may be used in proofs.

1.2 Definitions, Properties, Postulates, and Theorems (DOK 2)

The following tables contain geometric definitions, properties, postulates, and theorems. This list does not represent all of the concepts used in geometric proofs, but does provide us with a solid foundation as we learn the process of proving geometric relationships.

DEFINITIONS

WORD	DEFINITION	ILLUSTRATION/EXAMPLE
Congruent Segments	Segments are congruent if they have the same length.	$\overline{AB} \cong \overline{CD}$
Midpoint of a Segment	The point on a segment that determines two congruent segments	$\overline{AM} \cong \overline{MB}$
Congruent Angles	Angles are congruent if they have the same measure.	$\angle A \cong \angle B$
Complementary Angles	Two angles whose sum of their measures is 90°	$\angle A$ and $\angle B$ are complementary.
Supplementary Angles	Two angles whose sum of their measures is 180°	$\angle A$ and $\angle B$ are supplementary.
Vertical Angles	Two angles whose sides form pairs of opposite rays	$\angle A$ and $\angle B$ are vertical angles.
Transversal	A line that intersects two or more lines at different points (line *t* in the diagram at right)	
Corresponding Angles	Two angles that occupy corresponding positions (such as $\angle 1$ and $\angle 5$)	
Alternate Interior Angles	Two angles that lie between *l* and *m* on opposite sides of *t* (such as $\angle 2$ and $\angle 8$)	
Alternate Exterior Angles	Two angles that lie outside *l* and *m* on opposite sides of *t* (such as $\angle 1$ and $\angle 7$)	
Consecutive Interior Angles	Angles that lie between *l* and *m* on the same side of *t* (such as $\angle 2$ and $\angle 5$)	
Consecutive Exterior Angles	Angles that lie outside of *l* and *m* on the same side of *t* (such as $\angle 1$ and $\angle 6$)	

Chapter 1   Logic and Geometric Proofs

PROPERTIES

PROPERTY	DESCRIPTION	ILLUSTRATION/EXAMPLE
Addition Property of Equality	If two expressions are equal, you can add the same constant to both sides and the two sides remain equal.	If $a = b$, then $a + c = b + c$.
Subtraction Property of Equality	If two expressions are equal, you can subtract the same constant from both sides and the two sides remain equal.	If $a = b$, then $a - c = b - c$.
Multiplication Property of Equality	If two expressions are equal, you can multiply both sides by the same constant and the two sides remain equal.	If $a = b$, then $a \times c = b \times c$.
Division Property of Equality	If two expressions are equal, you can divide both sides by the same constant as long as the constant does not equal zero and the two sides remain equal.	If $a = b$, then $\frac{a}{c} = \frac{b}{c}$ ($c \neq 0$).
Reflexive Property of Equality	A number is equal to itself.	$a = a$
Symmetric Property of Equality	If the first number is equal to the second, then the second is equal to the first.	If $a = b$, then $b = a$.
Transitive Property of Equality	If the first number is equal to the second and the second is equal to the third, then the first must be equal to the third.	If $a = b$ and $b = c$, then $a = c$.
Substitution Property of Equality	If two expressions are equal, then you can substitute one for the other.	If $a = b$, then a can be substituted for b.
Reflexive Property of Congruence	Any geometric object is congruent to itself.	$\triangle ABC \cong \triangle ABC$
Symmetric Property of Congruence	If one geometric object is congruent to a second, then the second is congruent to the first.	If $\triangle ABC \cong \triangle XYZ$, then $\triangle XYZ \cong \triangle ABC$.
Transitive Property of Congruence	If one geometric object is congruent to a second and the second is congruent to a third, then the first is congruent to the third.	If $\triangle ABC \cong \triangle MNO$ and $\triangle MNO \cong \triangle XYZ$, then $\triangle ABC \cong \triangle XYZ$.

Copyright © American Book Company

1.2 Definitions, Properties, Postulates, and Theorems (DOK 2)

POSTULATES

POSTULATE	DESCRIPTION	ILLUSTRATION/EXAMPLE
Segment Addition Postulate	If B is between A and C, then $\overline{AB} + \overline{BC} = \overline{AC}$.	$A \quad B \quad\quad\quad C$
Angle Addition Postulate	If C is in the interior of $\angle AOD$, then $m\angle AOC + m\angle COD = m\angle AOD$.	
Corresponding Angles Postulate	If two parallel lines are cut by a transversal, then the pairs of corresponding angles are congruent.	$\angle 1 \cong \angle 3$ $\angle 2 \cong \angle 4$ $\angle 5 \cong \angle 7$ $\angle 6 \cong \angle 8$ $l \parallel m$
Corresponding Angles Converse Postulate	If two lines are cut by a transversal so that corresponding angles are congruent, then the lines are parallel.	If you can show that $\angle 1 \cong \angle 3$, $\angle 2 \cong \angle 4$, $\angle 5 \cong \angle 7$, or $\angle 6 \cong \angle 8$, then $l \parallel m$.
Side-Side-Side (SSS) Congruence Postulate	If three sides of one triangle are congruent to three sides of another triangle, then the two triangles are congruent.	If $\overline{AB} \cong \overline{xy}$, $\overline{BC} \cong \overline{yz}$, and $\overline{AC} \cong \overline{xz}$, then $\triangle ABC \cong \triangle xyz$.
Side-Angle-Side (SAS) Congruence Postulate	If two sides and the included angle of one triangle are congruent to two sides and the included angle of a second triangle, then the two triangles are congruent.	If $\overline{AB} \cong \overline{xy}$, $\angle A \cong \angle x$, and $\overline{AC} \cong \overline{xz}$, then $\triangle ABC \cong \triangle xyz$.
Angle-Side-Angle (ASA) Congruence Postulate	If two angles and the included side of one triangle are congruent to two sides and the included angle of a second triangle, then the two triangles are congruent.	If $\angle B \cong \angle y$, $\angle C \cong \angle z$, and $\overline{BC} \cong \overline{yz}$, then $\triangle ABC \cong \triangle xyz$.
Supplementary Angles Postulate	If two adjacent angles form a linear pair, then they are supplementary.	$m\angle 1 + m\angle 2 = 180°$

Chapter 1   Logic and Geometric Proofs

THEOREMS

THEOREM	DESCRIPTION	ILLUSTRATION/EXAMPLE
Congruent Supplements Theorem	If two angles are supplementary to the same angle or to congruent angles, then they are congruent.	If $\angle A \cong \angle B$, $\angle A$ is supplementary to $\angle x$, and $\angle B$ is supplementary to $\angle y$, then $\angle x \cong \angle y$.
Congruent Complements Theorem	If two angles are complementary to the same angle or to congruent angles, then they are congruent.	If $\angle A \cong \angle B$, $\angle A$ is complementary to $\angle x$, and $\angle B$ is complementary to $\angle y$, then $\angle x \cong \angle y$.
Vertical Angles Theorem	If two angles are vertical angles, then they are congruent.	$\angle A \cong \angle B$ and $\angle x \cong \angle y$
Transitivity of Parallel Lines Theorem	If two lines are parallel to the same line, then they are parallel to each other.	If $l \parallel m$ and $m \parallel n$, then $l \parallel n$.
Property of Perpendicular Lines Theorem	If two coplanar lines are perpendicular to the same line, then they are parallel to each other.	If $l \perp s$ and $m \perp s$, then $l \parallel m$.
Right Angles Theorem	All right angles are congruent.	All right angles are 90°. Therefore, all right angles are congruent.
Adjacent Congruent Angles Theorem	If two lines intersect to form a pair of adjacent congruent angles, then the lines are perpendicular.	If $\angle A \cong \angle B$, then $l \perp s$.
Alternate Interior Angles Theorem	If two parallel lines are cut by a transversal, then the pairs of alternate interior angles are congruent.	If $l \parallel m$, then $\angle 1 \cong \angle 4$ and $\angle 3 \cong \angle 2$.
Consecutive Interior Angles Theorem	If two parallel lines are cut by a transversal, then the pairs of consecutive interior angles are supplementary.	If $l \parallel m$, then $m\angle 3 + m\angle 4 = 180°$ and $m\angle 1 + m\angle 2 = 180°$.
Alternate Exterior Angles Theorem	If two parallel lines are cut by a transversal, then the pairs of alternate exterior angles are congruent.	If $l \parallel m$, then $\angle 3 \cong \angle 2$ and $\angle 1 \cong \angle 4$.

1.2   Definitions, Properties, Postulates, and Theorems (DOK 2)

THEOREMS (CONTINUED)

THEOREM	DESCRIPTION	ILLUSTRATION/EXAMPLE
Perpendicular Transversal Theorem	If a transversal is perpendicular to one of two parallel lines, then it is perpendicular to the second.	If $l \parallel m$ and $t \perp l$, then $t \perp m$.
Properties of Congruent Triangles Theorem	1. Every triangle is congruent to itself. 2. If $\triangle ABC$ is congruent to $\triangle PQR$, then $\triangle PQR$ is congruent to $\triangle ABC$. 3. If $\triangle ABC$ is congruent to $\triangle PQR$ and $\triangle PQR$ is congruent to $\triangle TUV$, then $\triangle ABC$ is congruent to $\triangle TUV$.	
Triangles Sum Theorem	The sum of the measures of the interior angles of a triangle is 180°.	$m\angle A + m\angle B + m\angle C = 180°$
Third Angles Theorem	If two angles of one triangle are congruent to two angles of a second triangle, then the third angles are also congruent.	If $\angle A \cong \angle x$ and $\angle B \cong \angle y$, then $\angle C \cong \angle z$.
Acute Angles Theorem	The acute angles of a right triangle are complementary.	$m\angle A + m\angle B = 90°$
Angle-Angle-Side (AAS) Congruence Theorem	If two angles and a nonincluded side of one triangle are congruent to two angles and the corresponding side of a second triangle, then the two triangles are congruent.	If $\angle A \cong \angle x$, $\angle C \cong \angle z$, and $\overline{BC} \cong \overline{yz}$, then $\triangle ABC \cong \triangle xyz$.
Base Angles Theorem	If two sides of a triangle are congruent, then the angles opposite them are congruent.	If $\overline{AC} \cong \overline{AB}$, then $\angle B \cong \angle C$. ($\angle B$ is opposite $\overline{AC}$, and $\angle C$ is opposite $\overline{AB}$.)

Chapter 1   Logic and Geometric Proofs

A **corollary** is a statement that follows a theorem. For example, Corollary to Base Angles Theorem (below) is an additional statement that is made as a result of the definition of the Base Angles Theorem.

THEOREMS (CONTINUED)

THEOREM	DESCRIPTION	ILLUSTRATION/EXAMPLE
Corollary to Base Angles Theorem	If a triangle is equilateral, then it is also equiangular.	If $\overline{AB} \cong \overline{BC} \cong \overline{AC}$, then $\angle A \cong \angle B \cong \angle C$.
Angle-Side Theorem	If two angles of a triangle are congruent, then the sides opposite them are congruent.	If $\angle C \cong \angle B$, then $\overline{AB} \cong \overline{AC}$.
Corollary to Angle-Side Theorem	If a triangle is equiangular, then it is also equilateral.	If $\angle A \cong \angle B \cong \angle C$, then $\overline{CB} \cong \overline{AC} \cong \overline{AB}$.
Hypotenuse-Leg (HL) Congruence Theorem	If the hypotenuse and a leg of a right triangle are congruent to the hypotenuse and the leg of a second right triangle, then the two triangles are congruent.	If $\overline{AB} \cong \overline{xy}$ and $\overline{AC} \cong \overline{xz}$, then $\triangle ABC \cong \triangle xyz$.
CPCTC Theorem	Corresponding parts of congruent triangles are congruent.	If $\triangle ABC \cong \triangle xyz$, then $\angle A \cong \angle x$, $\angle B \cong \angle y$, $\angle C \cong \angle z$, $\overline{AB} \cong \overline{xy}$, $\overline{BC} \cong \overline{yz}$, and $\overline{AC} \cong \overline{xz}$.

1.2 Definitions, Properties, Postulates, and Theorems (DOK 2)

If each statement in the first column is given information, what is the reason for each corresponding conclusion? Remember that your reasons will be definitions, properties, postulates, and/or theorems. The first two problems are done for you. A conclusion may require more than one reason.

	Given	**Conclusion**	**Reason**
1.	$\angle XYZ \cong \angle ABC$ and $\angle ABC \cong \angle MNO$	$\angle XYZ \cong \angle MNO$	transitive property
2.	$m\angle QRS = 20°$ and $m\angle DEF = 28°$	$m\angle QRS + m\angle DEF = 48°$	addition property
3.	$\overline{PT} = 30$	$\dfrac{\overline{PT}}{2} = 15$	_____
4.	$m\angle KLM = m\angle TUV$ and $m\angle TUV = 85°$	$m\angle KLM = 85°$	_____
5.	$\overline{AB} + \overline{CD} = 34$ and $\overline{CD} = 14$	$\overline{AB} = 20$	_____
6.	$\angle GHJ$ and $\angle PQR$ are complementary.	$\angle GHJ + \angle PQR = 90°$	_____
7.	$\angle BCD$ and $\angle FGH$ are congruent, and $m\angle BCD = 65°$.	$m\angle FGH = 65°$	_____
8.	$\overline{OT}$ is in the interior of $\angle NOP$.	$m\angle NOT + m\angle TOP = m\angle NOP$	_____
9.	$\triangle ABC$ and $\triangle XYZ$, $\overline{AB} \cong \overline{XY}, \overline{BC} \cong \overline{YZ}$, and $\angle B \cong \angle Y$	$\triangle ABC \cong \triangle XYZ$	_____
10.	$\triangle RST, m\angle S = 35°$, and $m\angle T = 55°$	$m\angle R = 90°$	_____
11.	$\overleftrightarrow{JK} \perp \overleftrightarrow{XY}$, and $\overleftrightarrow{MN} \perp \overleftrightarrow{XY}$	$\overleftrightarrow{JK} \parallel \overleftrightarrow{MN}$	_____
12.	$\triangle JKL$ $\overline{KL} \cong \overline{JK}$	$\angle J \cong \angle L$	_____
13.	$\angle S \cong \angle T$ $m\angle A + m\angle S = 180°$ $m\angle B + m\angle T = 180°$	$\angle A \cong \angle B$	_____

Chapter 1   Logic and Geometric Proofs

1.3   Two-Column Proofs (DOK 2 & 3)

A proof provides supporting evidence that a conclusion is true. Two-column proofs are organized into two columns: a logical sequence of statements is written in the column on the left, and the corresponding reasons are written in the column on the right.

Example 4:

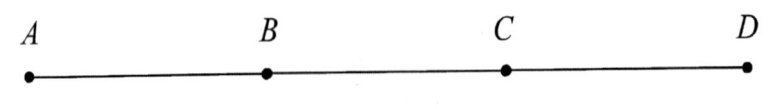

A   B   C   D

Given: $\overline{AB} + \overline{BC} = 22$
$\overline{AB} \cong \overline{CD}$
Prove: $\overline{CD} + \overline{BC} = 22$

Statement	Reasons
1. $\overline{AB} \cong \overline{CD}$	Given
2. $\overline{AB} = \overline{CD}$	Definition of congruent segments
3. $\overline{AB} + \overline{BC} = 22$	Given
4. $\overline{CD} + \overline{BC} = 22$	Substitution property

Example 5:

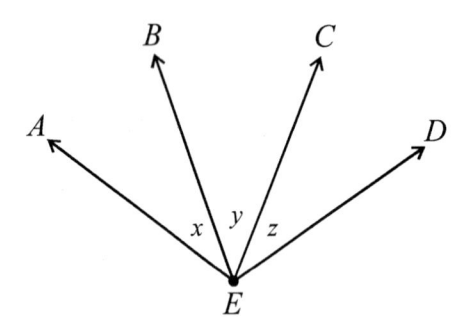

Given: $m\angle x = m\angle z$
Prove: $m\angle AEC = m\angle BED$

Statement	Reasons
1. $m\angle x = m\angle z$	Given
2. $m\angle x + m\angle y = m\angle z + m\angle y$	Addition property
3. $m\angle AEC = m\angle x + m\angle y$	Angle addition postulate
4. $m\angle BED = m\angle z + m\angle y$	Angle addition postulate
5. $m\angle AEC = m\angle BED$	Substitution property

1.3 Two-Column Proofs (DOK 2 & 3)

Write the missing reasons for the following proof. Use the definitions, properties, postulates, and theorems listed in the previous sections.

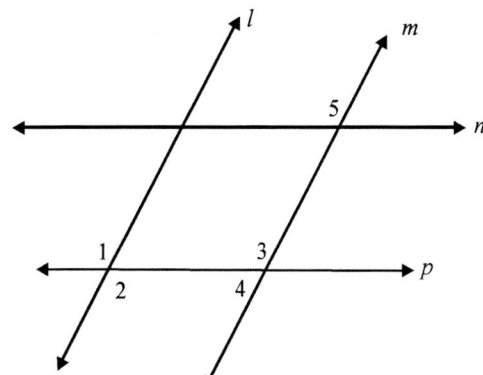

1.

Given: ∠1 ≅ ∠3
∠2 ≅ ∠5
Prove: n ∥ p

Statement   Reasons

1. ∠2 ≅ ∠1   _____

2. ∠1 ≅ ∠3   _____

3. ∠2 ≅ ∠3   _____

4. ∠2 ≅ ∠5   _____

5. ∠3 ≅ ∠5   _____

6. n ∥ p   _____

Chapter 1 Logic and Geometric Proofs

1.4 Flowcharts (DOK 2 & 3)

Example 6: Given: $\triangle MJK \cong \triangle KLM$
Prove: $MJKL$ is a parallelogram

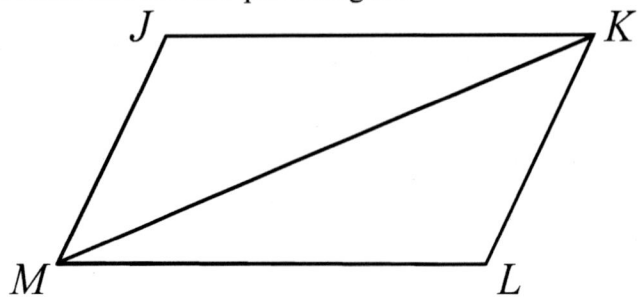

Step 1: First, draw boxes for everything that is given on the left hand side of the page. The statement goes in top of the box, and the reason goes in the bottom of the box.

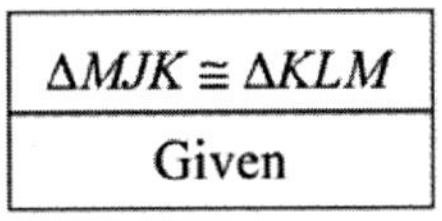

$\triangle MJK \cong \triangle KLM$
Given

Step 2: Next, draw arrows to the boxes of the conclusions that can be drawn directly from the statements that are given. In this case, since the triangles are congruent, we can conclude that the corresponding sides are congruent by the congruence theorem.

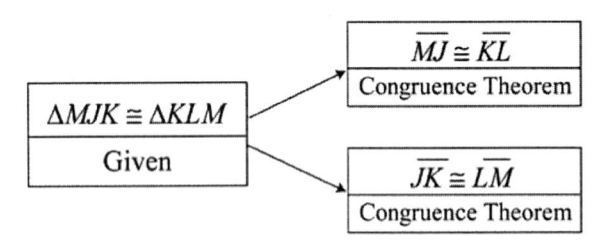

$\triangle MJK \cong \triangle KLM$ / Given
→ $\overline{MJ} \cong \overline{KL}$ / Congruence Theorem
→ $\overline{JK} \cong \overline{LM}$ / Congruence Theorem

Step 3: Using these two statements, we can conclude our proof, so we connect these boxes to the last single box containing the proof statement. In this example, because both pairs of opposite sides are congruent, it allows us to conclude that $MJKL$ is indeed a parallelogram.

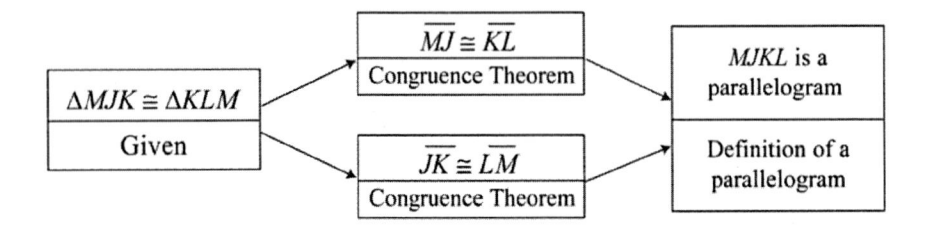

$\triangle MJK \cong \triangle KLM$ / Given
→ $\overline{MJ} \cong \overline{KL}$ / Congruence Theorem → $MJKL$ is a parallelogram
→ $\overline{JK} \cong \overline{LM}$ / Congruence Theorem → Definition of a parallelogram

1.4 Flowcharts (DOK 2 & 3)

Solve the following problems.

1. Given: ∠1 and ∠2 are a linear pair. ∠2 and ∠3 are a linear pair. Prove: ∠1 ≅ ∠3.

2. Given: $l \parallel m$. Write a flowchart proof to show that ∠1 ≅ ∠3.

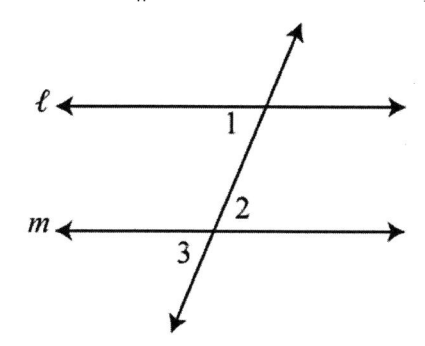

3. $l \perp p$, $m \perp p$. Write a flowchart proof to show that $l \parallel m$.

Chapter 1 Review

Match the items on the left with the descriptions or illustrations on the right. (DOK 2)

1. Addition Property of Equality _____

2. Multiplication Property of Equality _____

3. Transitive Property of Equality _____

4. Corresponding Angles Postulate _____

5. Angle-Side-Angle (ASA) Congruence Postulate _____

6. Side-Angle-Side (SAS) Congruence Postulate _____

7. Vertical Angles Theorem _____

8. Consecutive Interior Angles Theorem _____

9. Triangles Sum Theorem _____

10. Acute Angles Theorem _____

11. Base Angles Theorem _____

12. CPCTC Theorem _____

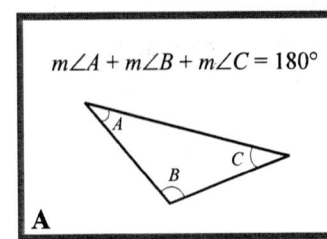

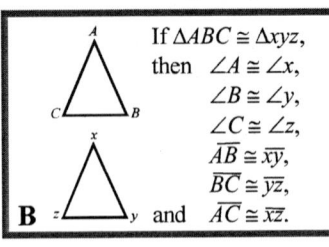

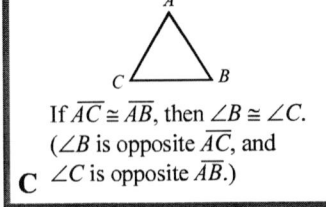

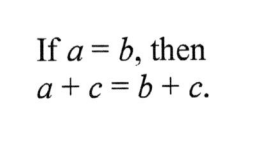

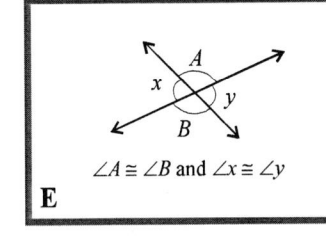

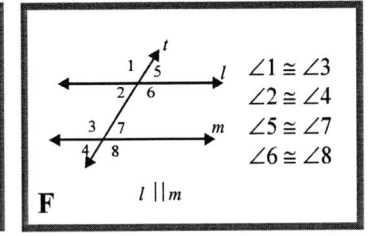

A. $m\angle A + m\angle B + m\angle C = 180°$

B. If $\triangle ABC \cong \triangle xyz$, then $\angle A \cong \angle x$, $\angle B \cong \angle y$, $\angle C \cong \angle z$, $\overline{AB} \cong \overline{xy}$, $\overline{BC} \cong \overline{yz}$, and $\overline{AC} \cong \overline{xz}$.

C. If $\overline{AC} \cong \overline{AB}$, then $\angle B \cong \angle C$. ($\angle B$ is opposite $\overline{AC}$, and $\angle C$ is opposite $\overline{AB}$.)

D. If $a = b$, then $a + c = b + c$.

E. $\angle A \cong \angle B$ and $\angle x \cong \angle y$

F. $\angle 1 \cong \angle 3$, $\angle 2 \cong \angle 4$, $\angle 5 \cong \angle 7$, $\angle 6 \cong \angle 8$; $l \parallel m$

G. If $a = b$, then $a \times c = b \times c$.

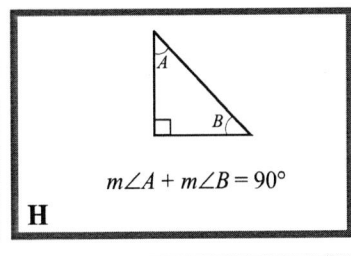

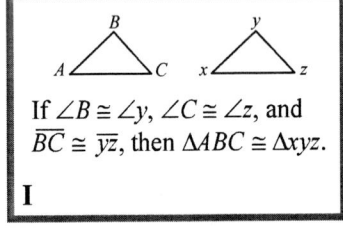

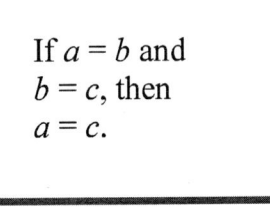

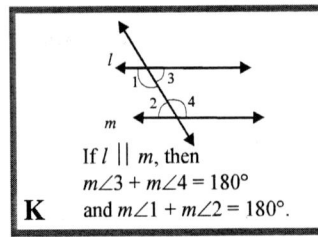

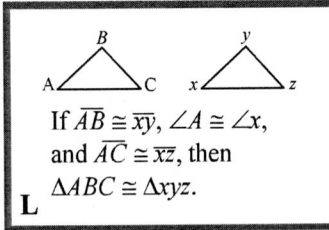

H. $m\angle A + m\angle B = 90°$

I. If $\angle B \cong \angle y$, $\angle C \cong \angle z$, and $\overline{BC} \cong \overline{yz}$, then $\triangle ABC \cong \triangle xyz$.

J. If $a = b$ and $b = c$, then $a = c$.

K. If $l \parallel m$, then $m\angle 3 + m\angle 4 = 180°$ and $m\angle 1 + m\angle 2 = 180°$.

L. If $\overline{AB} \cong \overline{xy}$, $\angle A \cong \angle x$, and $\overline{AC} \cong \overline{xz}$, then $\triangle ABC \cong \triangle xyz$.

Chapter 1 Review

Solve the following problems. (DOK 3)

13. "If today is Tuesday, then it is raining." Write the converse, inverse, and contrapositive for the conditional statement.

Fill in the two-column proof for each of the following exercises. (DOK 2 & 3)

14. Given: $l \parallel m$, $\angle 2 \cong \angle 3$
    Prove: $\angle 1 \cong \angle 3$

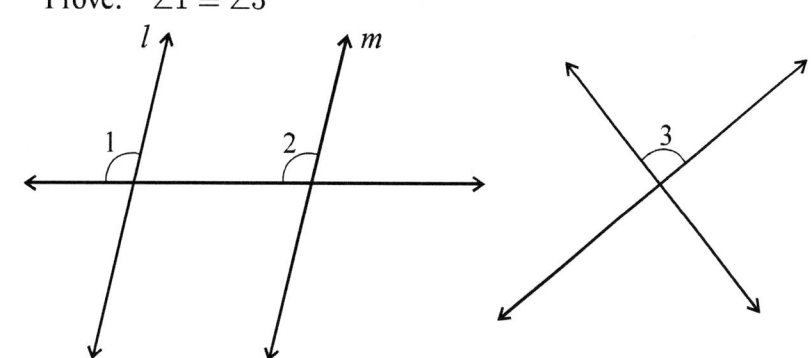

Statement	Reasons
1. _____	Given
2. $\angle 1 \cong \angle 2$	_____
3. $\angle 2 \cong \angle 3$	_____
4. _____	Transitive Property

Chapter 1 Test

1 Which is the contrapositive of the following conditional statement: "If a trapezoid is isosceles, then its base angles are congruent?"

  A If a trapezoid has congruent base angles, then it is isosceles.
  B If a trapezoid is not isosceles, then its base angles are not congruent.
  C If a trapezoid's base angles are not congruent, then it is not isosceles.
  D A trapezoid is isosceles if and only if its base angles are congruent.   (DOK 2)

2 Below is an incomplete two-column proof. Reading through the proof, what statement is missing to make the proof complete?

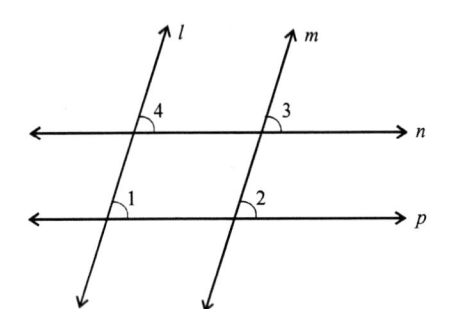

Given: $l \parallel m$ and $\angle 4 \cong \angle 2$
Prove: $n \parallel p$

Statement	Reasons
1. $l \parallel m$	Given
2. $\angle 1 \cong \angle 2$	Corresponding Angles Postulate
3. $\angle 4 \cong \angle 2$	Given
4. _____	Transitive property
5. $n \parallel p$	Corresponding Angles Converse Postulate

  A $\angle 1 \cong \angle 4$
  B $\angle 3 \cong \angle 4$
  C $\angle 2 \cong \angle 3$
  D $\angle 1 \cong \angle 3$ (DOK 2)

3 Below is an incomplete two-column proof. Reading through the proof, what reason is missing to make the proof complete?

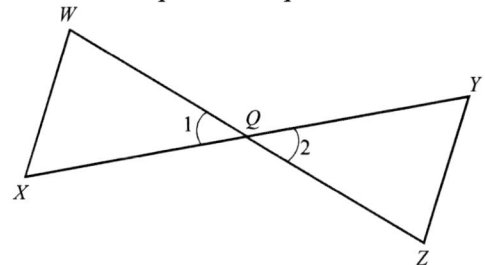

Given: Q is the midpoint of $\overline{WZ}$ and of $\overline{XY}$
Prove: $\triangle WXQ \cong \triangle ZYQ$

Statement	Reasons
1. $\angle 1 \cong \angle 2$	Vertical angles
2. Q is the midpoint of $\overline{WZ}$	Given
3. $\overline{WQ} \cong \overline{QZ}$	Definition of midpoint
4. $\overline{XQ} \cong \overline{QY}$	Definition of midpoint
5. $\triangle WXQ \cong \triangle ZYQ$	

A Definition of midpoint
B Symmetric Property of Congruence
C SAS theorem
D Vertical Angles theorem

(DOK 2)

4 For the conditional statement "If a figure is a rectangle, then its diagonals are congruent," which of the following are correct?

A The conditional statement and the inverse are both true.
B The converse and the inverse are both true.
C The conditional statement and the contrapositive are both true.
D The conditional statement, converse, inverse, and contrapositive are all true.

(DOK 3)

Chapter 2
Geometric Constructions

This chapter covers the following CCGPS standard(s):

	Content Standards
Congruence	G.CO.12, G.CO.13

This chapter focuses on constructing angles and lines. We will copy a line segment and an angle. We will construct a bisector of an angle, the perpendicular bisector of a line segment, parallel lines, perpendicular lines, and polygons inscribed in a circle.

The constructions are created by using a compass and a straight edge. A **straight edge** can be anything that has a straight side on it. Some examples are books, notebooks (not the spiral side), boxes, and paper. The best example of a straight edge is a ruler. A **compass** is a tool for constructing angles. It has a point on one end and a pencil on the other. Below is a picture of one type of compass.

* Remember: When constructing, do NOT erase construction marks.

2.1 Copying a Line Segment (DOK 2)

Example 1: Copy the following line segment.

$A \bullet \longrightarrow \bullet B$

Step 1: Make a point separate from the line segment AB and name it.

$A \bullet \longrightarrow \bullet B \quad C \bullet$

Step 2: Measure $\overline{AB}$ with the compass by putting the pointed end on A stretching the pencil end to reach point B.

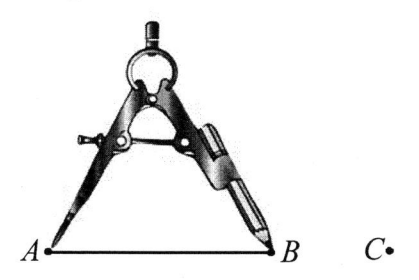

$A \bullet \longrightarrow \bullet B \quad C \bullet$

Step 3: Without closing the compass, move the compass, so the pointed end is on point C. With the pencil draw a little line.

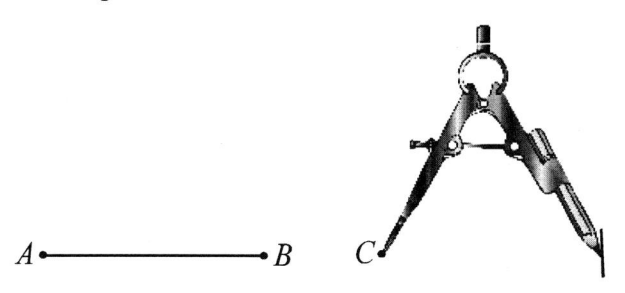

$A \bullet \longrightarrow \bullet B \quad C \bullet$

Step 4: With a straight edge, draw a straight line from C to the vertical line drawn in step 3. Label this end point.

$A \bullet \longrightarrow \bullet B \quad C \bullet \longrightarrow \bullet D$

Copy the following line segments using a compass and straight edge.

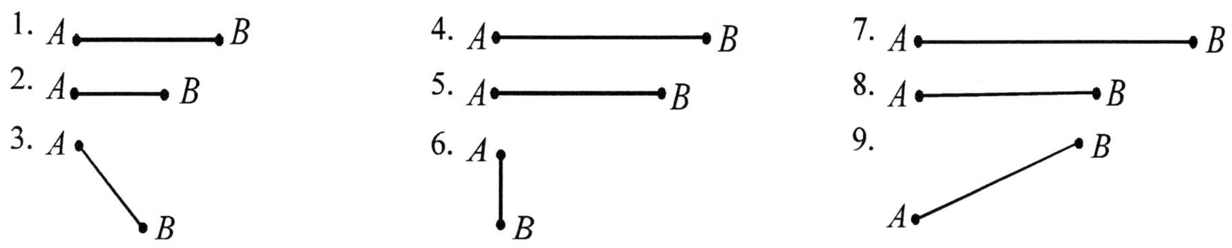

1. $A \bullet \longrightarrow \bullet B$
2. $A \bullet \longrightarrow B$
3. $A \bullet \searrow \bullet B$
4. $A \bullet \longrightarrow \bullet B$
5. $A \bullet \longrightarrow \bullet B$
6. $A \bullet \downarrow \bullet B$
7. $A \bullet \longrightarrow \bullet B$
8. $A \bullet \longrightarrow \bullet B$
9. $A \bullet \nearrow \bullet B$

Chapter 2 Geometric Constructions

2.2 Copying an Angle (DOK 2)

Example 2: Copy the following angle.

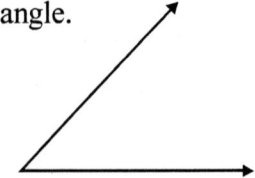

Step 1: Pick a point and draw a straight line using a straight edge.

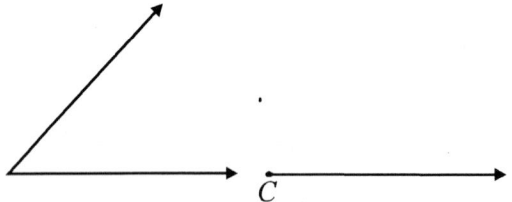

Step 2: Construct an arc through the angle using the compass. Label the points where they cross over the angle.

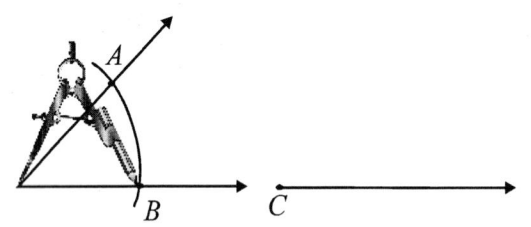

Step 3: Draw an arc with the same measure by putting the pointed end of the compass at endpoint C. Label the point D where the arc goes through the ray.

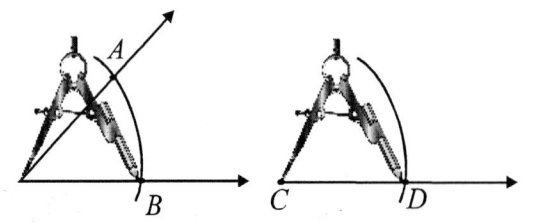

Step 4: Measure the distance from point A to point B with the compass by putting the pointed end on B and the pencil on A. Use the same measure, move the compass to point D, and make a small arc where the pencil crosses the bigger arc. Label the point of intersection E.

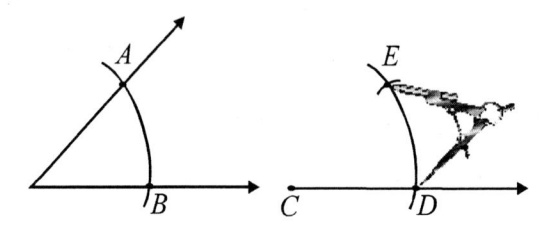

2.2 Copying an Angle (DOK 2)

Step 5: With a straight edge draw a line from point C to point E. You have copied the angle.

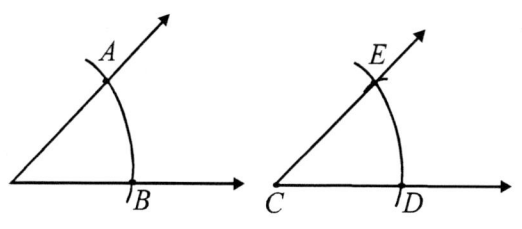

Copy the following angles using a compass and straight edge.

1.

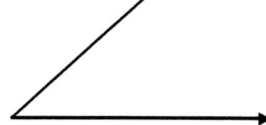

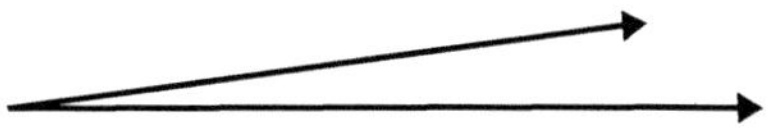

2.

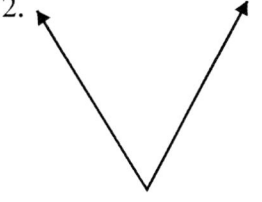

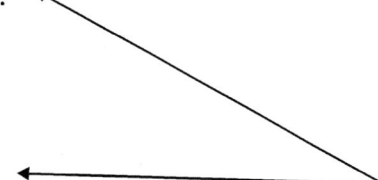

3.

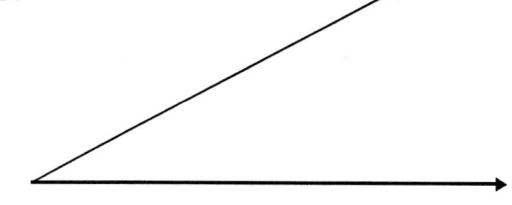

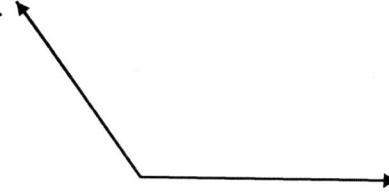

4.

5.

6.

7.

8.

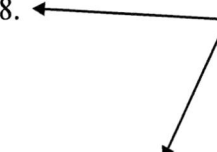

Copyright © American Book Company   31

Chapter 2 Geometric Constructions

2.3 Constructing a Bisector of an Angle (DOK 2)

A **bisector** is a line or ray that intersects something, like an angle or another line, directly through the middle to divide it into two equal parts.

Example 3: Construct the bisector of ∠ACB.

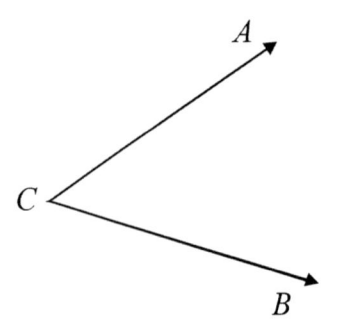

Step 1: By placing the pointed end of a compass at vertex C of the angle, draw an arc crossing about $\frac{2}{3}$ of $\overrightarrow{CA}$.

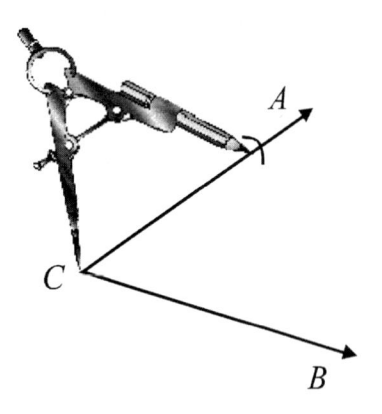

Step 2: Repeat step 1 for $\overrightarrow{CB}$. Make sure that the measure of the compass is the same for both $\overrightarrow{CA}$ and $\overrightarrow{CB}$.

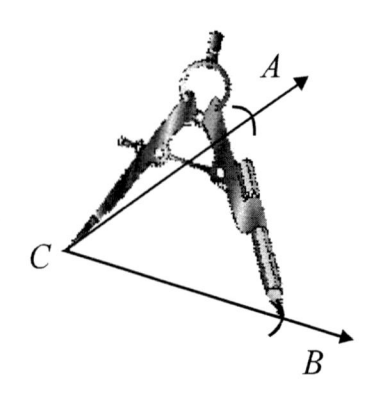

2.3 Constructing a Bisector of an Angle (DOK 2)

Step 3: Using the same distance for steps 1 and 2, put the pointed side of the compass on the small line on $\overrightarrow{CA}$, construct an arc with the pencil.

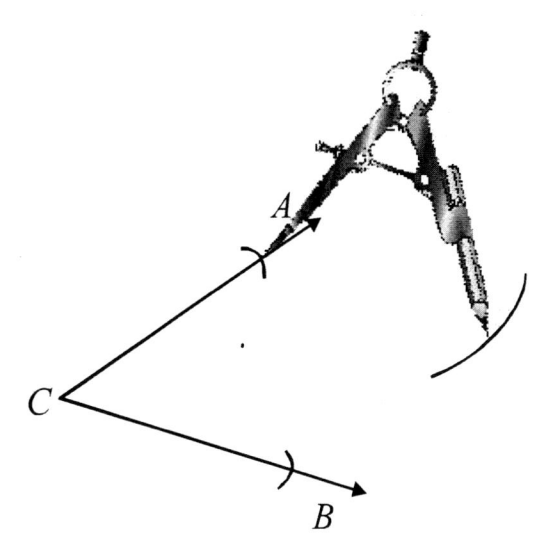

Step 4: Repeat step 3 for $\overrightarrow{CB}$. Make sure that the measure of the compass is the same for both $\overrightarrow{CA}$ and $\overrightarrow{CB}$.

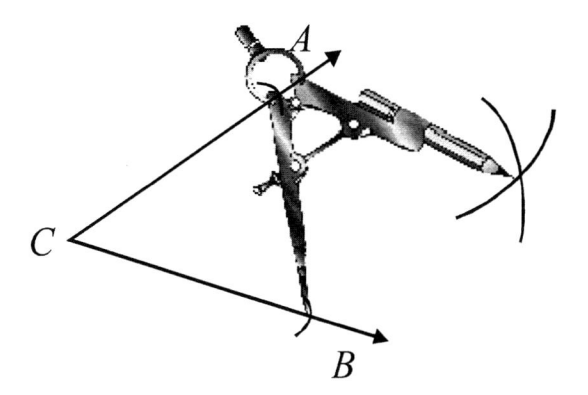

Step 5: Find the point where the two arcs intersect. Construct a ray using a straight edge from the vertex of $\angle ACB$ through that point.

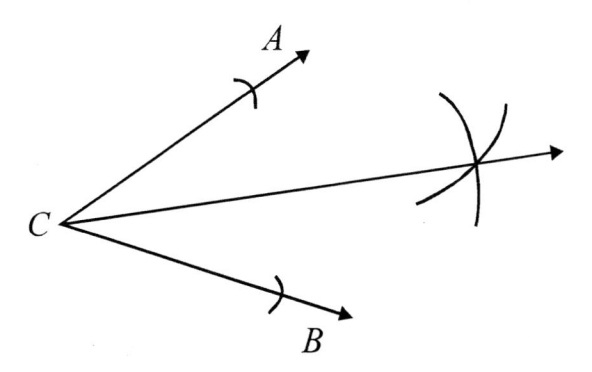

Chapter 2 Geometric Constructions

Construct the bisector of the following angles.

1.

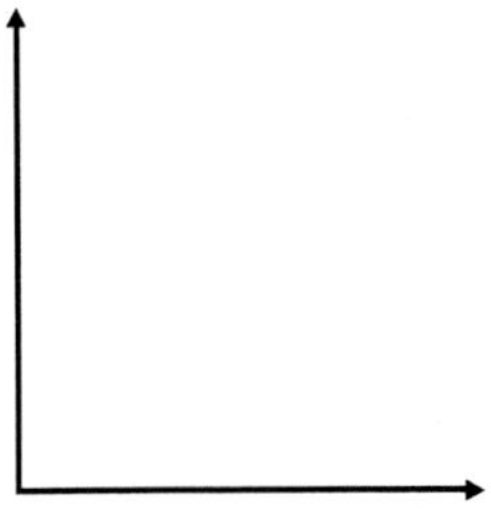

2.

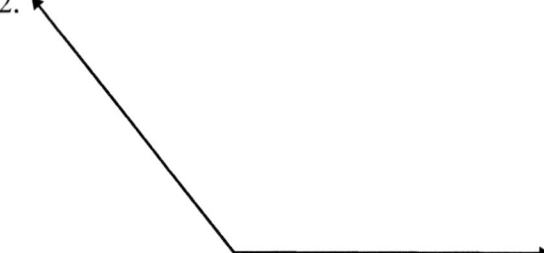

3.

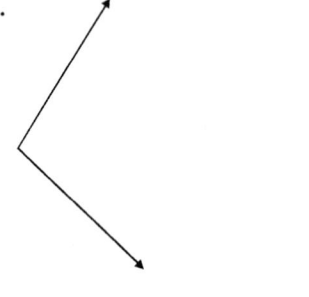

4.

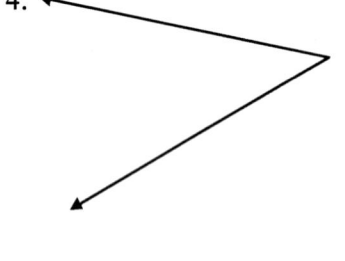

5.

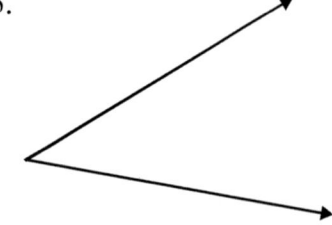

6.

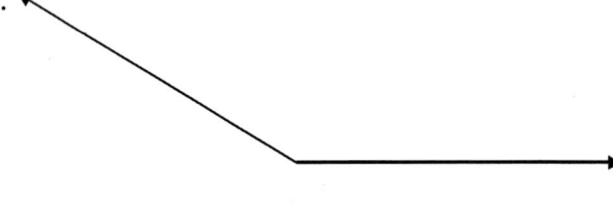

7.

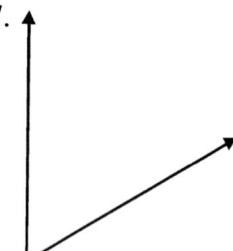

8.

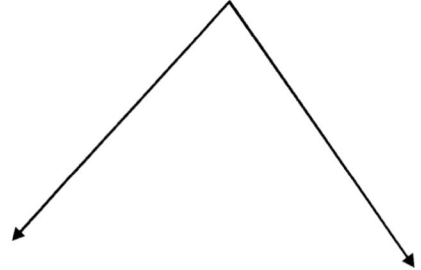

2.4 Constructing the Perpendicular Bisector of a Line Segment (DOK 2)

A **perpendicular bisector** is a line perpendicular to a segment passing through the segment's midpoint. Perpendicular lines meet at a 90° angle.

Example 4: Construct a perpendicular bisector of line segment AB.

$A \bullet \longrightarrow \bullet B$

Step 1: Put the pointed end of the compass on point A. Set the pencil more than $\frac{1}{2}$ way across the segment and draw an arc from above the line segment to below the line segment.

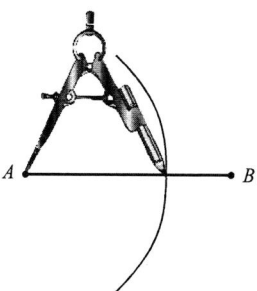

Step 2: Repeat step 1 for point B by putting the pointed end on point B.

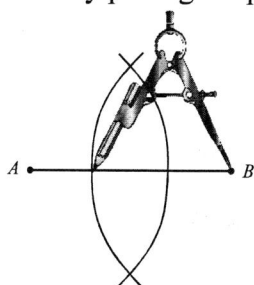

Step 3: There are two points where the arcs intersect. Using a straight edge, draw a straight line between the two points. You have bisected the line segment.

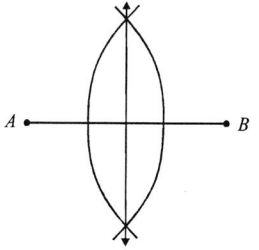

Construct a perpendicular bisector of $\overline{AB}$.

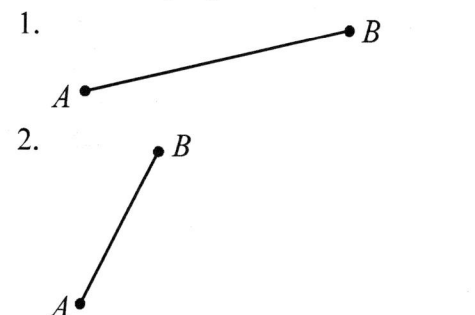

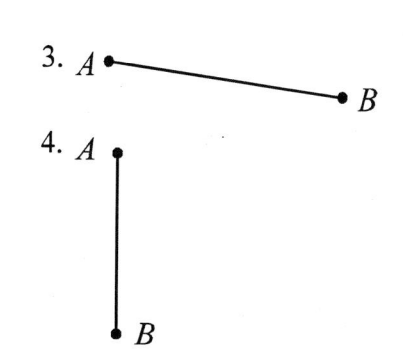

1.
2.
3.
4.

2.5 Points on a Perpendicular Bisector (DOK 2)

This section will show that points that lie on a perpendicular bisector of a line segment are exactly equidistant from the line segment's end points.

We will use a "question-answer" type of proof method to arrive at the stated conclusion.

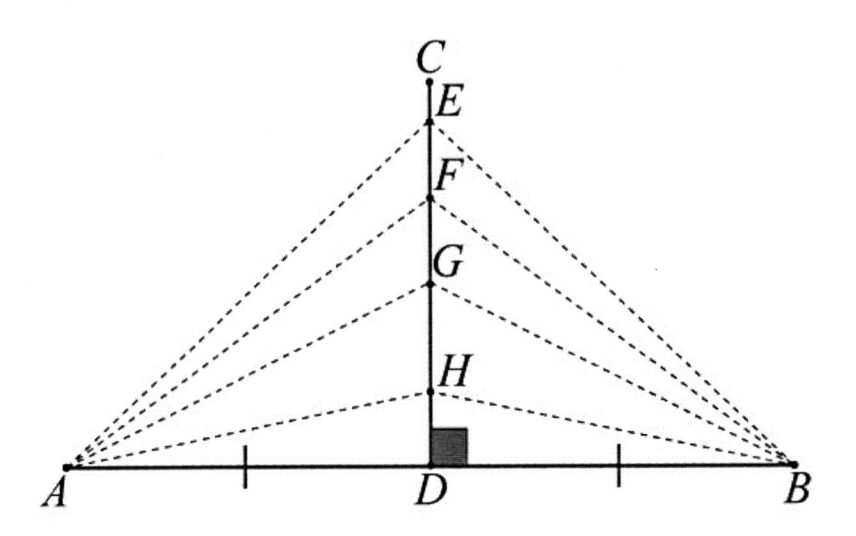

$\overline{CD}$ is the perpendicular bisector of the line segment AB and on it are the points H, G, F, and E. For this example, we will only consider point E on the line segment. The proof of this point will represent every other point that lies on the perpendicular bisector $\overline{CD}$, of $\overline{AB}$.

Claim: $\overline{AE} = \overline{BE}$

Proof:

Question	Answer
1. Is $\overline{ED}$ a perpendicular bisector to $\overline{AB}$?	Yes (Given)
2. Is $\overline{ED}$ perpendicular to $\overline{AB}$?	Yes (Definition of perpendicular bisector)
3. Is $\overline{AD} \cong \overline{DB}$?	Yes (Perpendicular bisector)
4. Is $\overline{ED} \cong \overline{ED}$?	Yes (Reflexive property and congruency)
5. Is $\angle ADE \cong \angle BDE$?	Yes (Definition of a right angle)
6. Is $\triangle ADE \cong \triangle BDE$?	Yes (2 $\triangle$'s are congruent if any pair of corresponding sides and their included angles are equal in both $\triangle$'s.)
7. Is $\overline{AE} \cong \overline{BE}$?	Yes (Corresponding parts of congruent $\triangle$'s are congruent (CPCTC))

2.6 Constructing Perpendicular Lines (DOK 2)

Example 5: Construct two perpendicular line segments.

Step 1: The method used to construct a perpendicular bisector only requires a compass and a ruler. It is similar to constructing a bisector of a line segment. Find a point C on the line segment AB.

$A \bullet \text{———————} \bullet B$

Step 2: Set the compass width to a reasonable width (shorter than the distance from point C to either point A or B). Set it on the point C and mark short arcs on the line segment $\overline{AB}$ on either side of point C. Label these points of arc and line segment intersection, as x and y.

Step 3: Now increase the compass width slightly, set it on either x or y and mark off an arc on one side of the line segment.

Step 4: Keeping the compass width unchanged, repeat Step 4 with the compass point on the other point, so that the two arcs intersect.

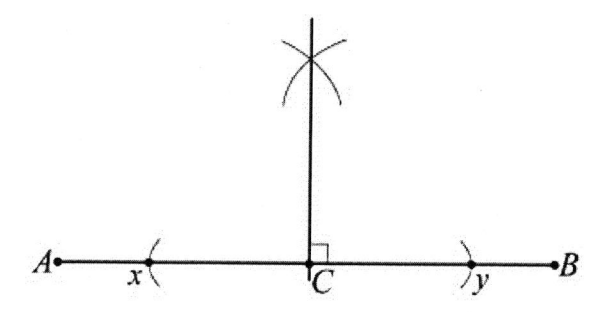

As shown above, construct a line through point C and the intersection of the two arcs. This line is perpendicular to the line segment $\overline{AB}$.

Chapter 2 Geometric Constructions

2.7 Constructing Parallel Lines (DOK 2)

Example 6: Construct a line that is parallel to line l.

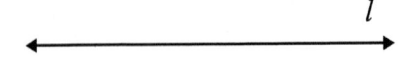

Step 1: Pick two points on line l and label them A and B. Also, pick a point somewhere above the line and label it C. Using a straight edge, draw a straight edge from point A to point C.

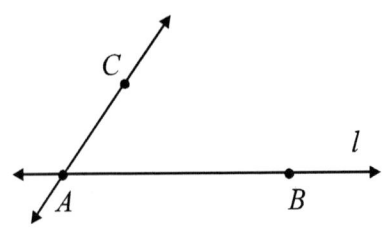

Step 2: Put the pointed side of the compass on A, and with the pencil, draw an arc that crosses $\overrightarrow{AB}$ and $\overrightarrow{AC}$.

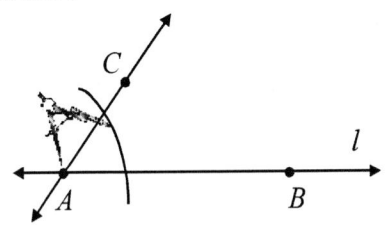

Step 3: Using the same distance on the compass, put the pointed end of the compass on point C, and draw a similar arc.

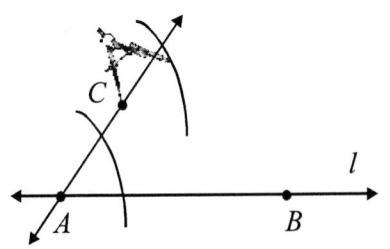

Step 4: Put the pointed end of the compass on the point where the first arc drawn intersects with $\overrightarrow{AC}$. Stretch the compass so the pencil touches the point where that same arc crosses $\overrightarrow{AB}$.

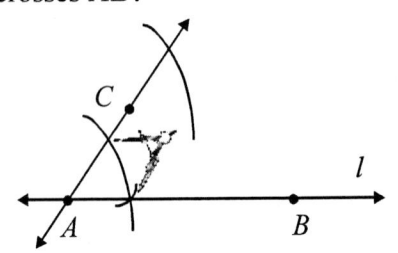

2.7 Constructing Parallel Lines (DOK 2)

Step 5: Using the same measure on the compass from step 4, put the pointed end of the compass on the point where the second arc crosses $\overrightarrow{AC}$. Draw a similar arc to the one drawn in step 4 on the second arc.

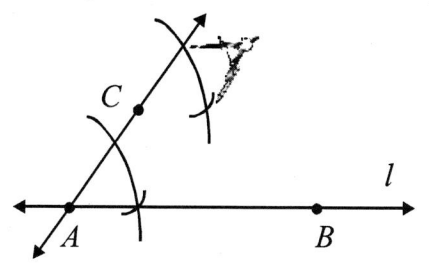

Step 6: Using a straight edge draw a line through point C and the point where the two arcs intersect.

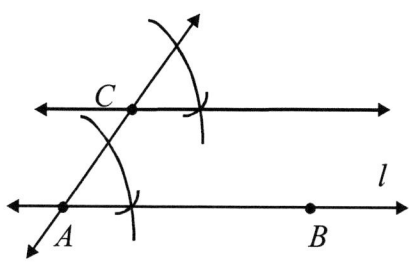

Construct lines parallel to the lines below.

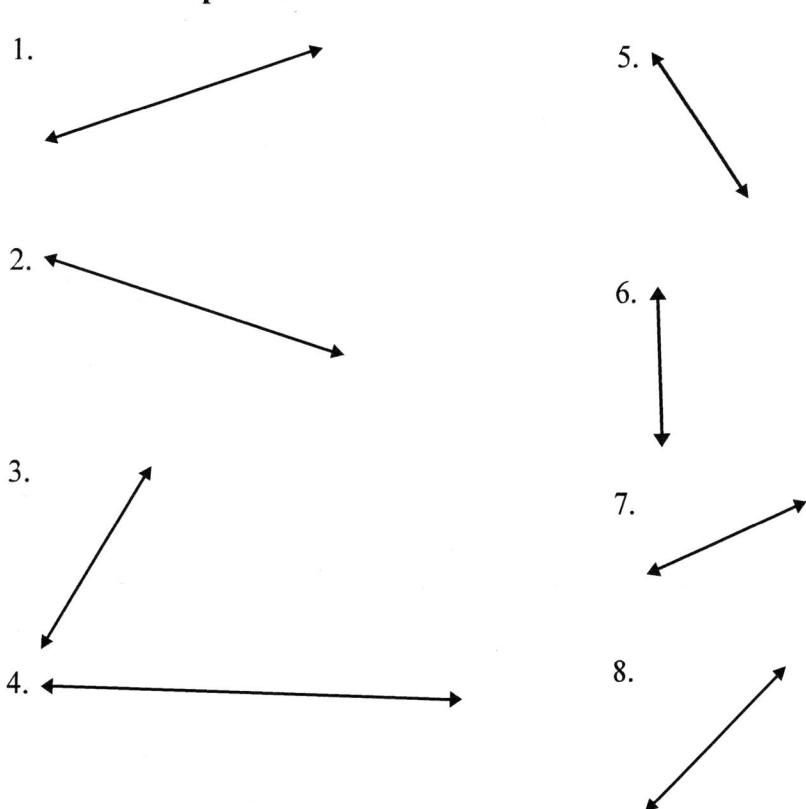

1.

2.

3.

4.

5.

6.

7.

8.

Chapter 2 Geometric Constructions

2.8 Constructing Polygons Inscribed in a Circle (DOK 2)

A polygon inscribed in a circle is one that is "nested" in a circle. To construct an inscribed regular polygon, the circumference of the circle has to be divided into equal sectors. The number of sectors to be drawn is equal to the number of sides of the polygon. Once the sectors are drawn, all that remains is to join the points where the sector boundaries touch the circles circumference.

Example 7: Inscribe an equilateral triangle and a regular hexagon in a circle.

For any circle, its radius can be struck exactly six times around its circumference, which can yield a hexagon.

To create a hexagon, begin by opening a compass to the radius of the circle. Next, without adjusting the width of the compass place the endpoint anywhere on the circumference of the circle. Then using the compass create an arc that intersects with the circumference of the circle. Finally, without changing the width of the compass move the point of the compass to the newly created arc. Repeat this process until you have moved all the way around the circle.

To create an equilateral triangle, connect the intersection of every other arc, and to create a hexagon connect each successive interception.

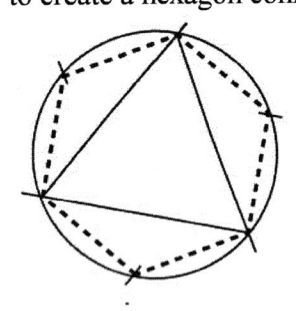

Example 8: Inscribe a square in a circle.

Step 1: Construct a horizontal diameter of the circle and use your knowledge of perpendicular bisectors to construct the vertical diameter.

Step 2: Construct the diagonal diameters by bisecting the right angles created by the intersection of the vertical and horizontal diameters.

Step 3: Connect the diagonal diameters of the circle to get the inscribed square.

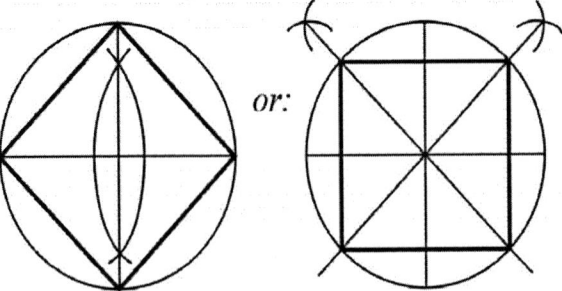

Both diagrams above show a square inscribed in a circle. The one on the left shows a tilted square and was created by simply joining the vertical and horizontal diameters.

Chapter 2 Review

Copy the following angles and line segments using a compass and straight edge. (DOK 2)

1.

2.

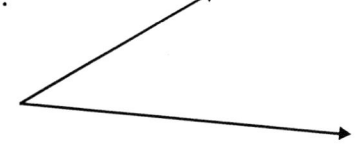

3.

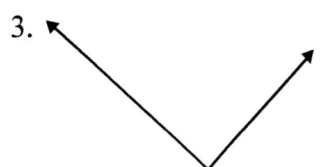

4.

Construct a bisector for each of the following angles. (DOK 2)

5.

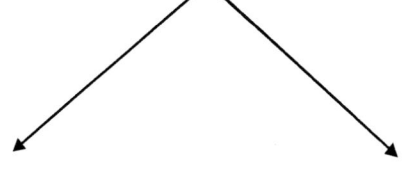

6.

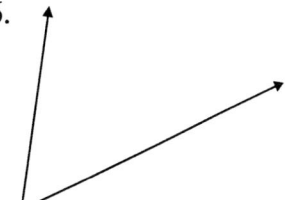

Construct a perpendicular bisector and a parallel line for each of the following line segments. (DOK 2)

7.

8.

9.

10.

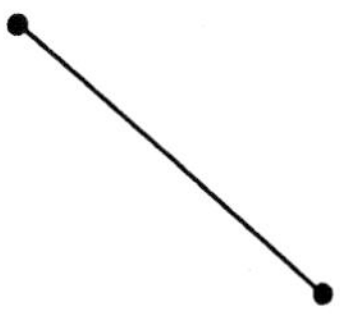

Chapter 2 Test

1 To copy a line segment which two objects do you need to use?

   A compass and straight edge

   B protractor and straight edge

   C compass and protractor

   D None of the above. (DOK 1)

2 Steve is attempting to copy the angle shown below. So far, this is what he has done.

Original Image   Steve's Image

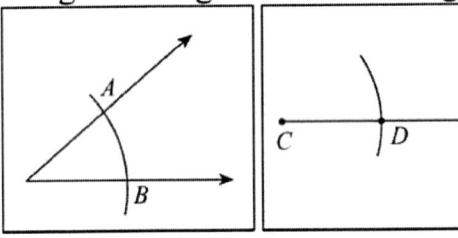

What should be the next step that Steve takes?

   A Measure the distance from point A to point B using a ruler.

   B Measure the distance from point A to point B using the compass.

   C Without moving the compass place the pointed end on point D, and sketch an arc so that the new arc intersects with the previously drawn arc.

   D Draw a line connecting point C to the end of the unknown arc, and label the intersection as point E.

(DOK 2)

3 The image shown below is the final image for what construction?

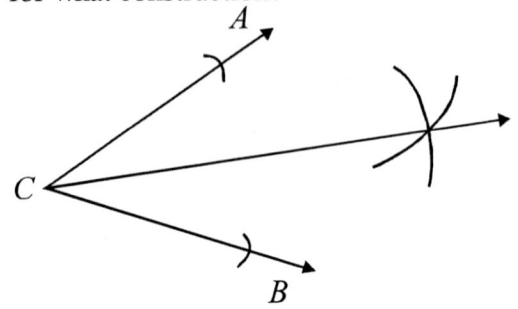

   A Construction of a Perpendicular Line

   B Copying an Angle

   C Perpendicular Bisector

   D Angle Bisector (DOK 1)

4 The diagram seen below shows construction marks that are labeled I, II, III, and IV. If you were creating a perpendicular bisector of $\overline{AB}$, which two construction marks would you use next?

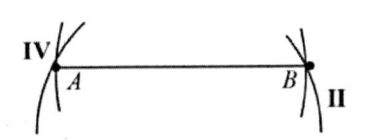

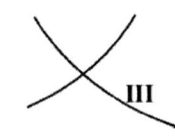

   A I and II

   B I and III

   C II and IV

   D II and III (DOK 1)

Chapter 2 Test

5 Given the image shown below what is the relationship between $\overline{AE}$ and $\overline{BE}$?

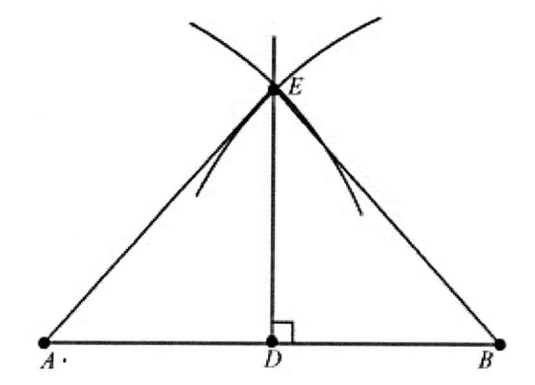

A $m\overline{AE}$ is equal to $m\overline{BE}$.

B $\overline{AE}$ is similar to $\overline{BE}$.

C $\overline{AE}$ is perpendicular to $\overline{BE}$.

D There is no relationship between $\overline{AE}$ and $\overline{BE}$.

(DOK 1)

6 Shawn's teacher drew the construction below on the board. Based upon this construction which of the following statements does Shawn know to be true?

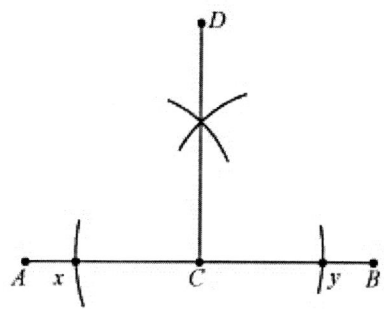

A Point C is the midpoint of $\overline{AB}$.

B $\overline{CD}$ is the perpendicular bisector of $\overline{AB}$.

C $\overline{AC}$ is equal to $\overline{CB}$.

D $\overline{CD}$ is perpendicular to $\overline{AB}$.

(DOK 1)

7 What do you use to copy an angle?

A thermometer
B calculator
C graph paper
D compass

(DOK 1)

8 Danielle is constructing a non-tilted square inscribed inside a circle. The image below shows where she is in her construction so far. What should the next step be in Danielle's construction?

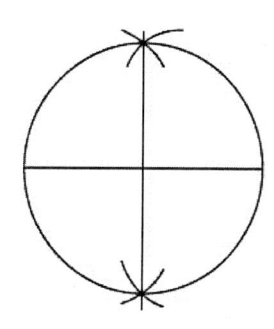

A Create 4 perpendicular lines at each of the end points of diameters.

B Connect the intersection of the two arcs with the end points of the horizontal diameter of the circle.

C Create 4 angle bisectors to bisect the 4 right angles.

D Make 4 chords inside the circle by connecting all the end points of the diameter together.

(DOK 2)

9 Which object would you need to use to construct a perfect circle?

A protractor
B compass
C straight edge
D thermometer

(DOK 1)

Chapter 2 Geometric Constructions

10 Lilly is using constructions to draw a parallel line to line AB. On which step did Lilly make her first mistake?

A Begin by creating points A and B on line l. Then create a point C above line l. Next, create a new line that goes through points A and C.

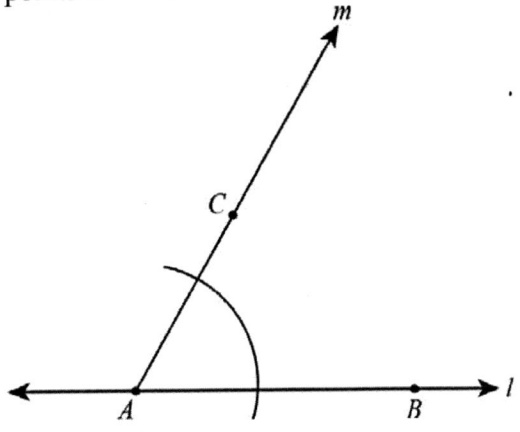

B Next, place the pointed end of the compass on point A. Create an arc that goes through line l and m. Without changing the width of the compass, place the pointed end of the compass on point C and create a similar arc.

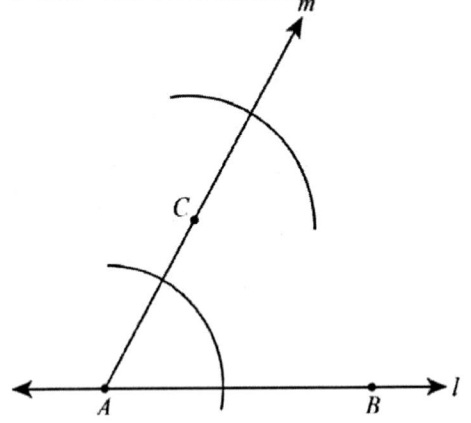

C Using the compass, place the point end on point B and extend the compass so it is the length of $\overline{AB}$. Using this length, create an arc that crosses through the previous arc without moving the compass off of point B.

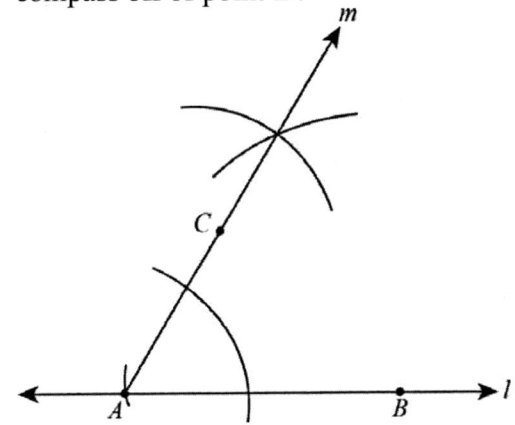

D Using a straight edge, draw a line that goes through the intersection of the arcs.

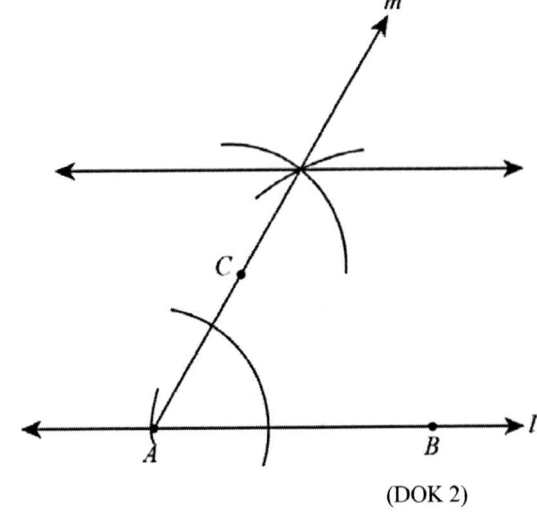

(DOK 2)

Chapter 3
Similarity

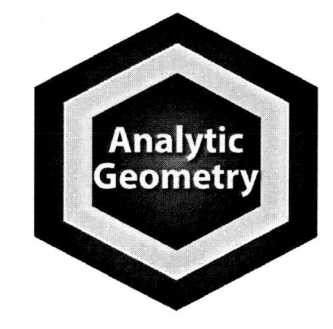

This chapter covers the following CCGPS standard(s):

	Content Standards
Similarity, Right Triangles, and Trigonometry	G.SRT.1, G.SRT.2, G.SRT.3, G.SRT.4 G.SRT.5

3.1 Similar and Congruent (DOK 2)

Similar figures have the same shape, but are two different sizes. Their corresponding sides are proportional. **Congruent figures** are exactly alike in size and shape and their corresponding sides and angles are equal. See the examples below.

SIMILAR

CONGRUENT

Label each pair of figures below as either S if they are similar, C if they are congruent, or N if they are neither.

1.
2.
3.
4.
5.
6.
7.
8.
9.
10.
11.
12.
13.
14.

Chapter 3 Similarity

3.2 Solving Proportions (DOK 1)

Two **ratios (fractions)** that are equal to each other are called **proportions**. For example, $\frac{1}{4} = \frac{2}{8}$. Read the following example to see how to find a number missing from a proportion.

Example 1: $\qquad \dfrac{5}{15} = \dfrac{8}{x}$

Step 1: To find x, you first multiply the two numbers that are diagonal to each other.

$$\dfrac{5}{\{15\}} = \dfrac{\{8\}}{x}$$

$$15 \times 8 = 5 \times x$$
$$120 = 5x$$

Step 2: Then solve for x.

$$120 = 5x \Rightarrow \dfrac{120}{5} = \dfrac{5x}{5}$$
$$\dfrac{120}{5} = x$$

Therefore, $x = 24$ and $\dfrac{5}{15} = \dfrac{8}{24}$

Practice finding the number missing from the following proportions. First, multiply the two numbers that are diagonal from each other. Then divide by the other number.

1. $\dfrac{2}{5} = \dfrac{6}{x}$

2. $\dfrac{9}{3} = \dfrac{x}{5}$

3. $\dfrac{x}{12} = \dfrac{3}{4}$

4. $\dfrac{7}{x} = \dfrac{3}{9}$

5. $\dfrac{12}{x} = \dfrac{2}{5}$

6. $\dfrac{12}{x} = \dfrac{4}{3}$

7. $\dfrac{27}{3} = \dfrac{x}{2}$

8. $\dfrac{1}{x} = \dfrac{3}{12}$

9. $\dfrac{15}{2} = \dfrac{x}{4}$

10. $\dfrac{7}{14} = \dfrac{x}{6}$

11. $\dfrac{5}{6} = \dfrac{10}{x}$

12. $\dfrac{4}{x} = \dfrac{3}{6}$

13. $\dfrac{x}{5} = \dfrac{9}{15}$

14. $\dfrac{9}{18} = \dfrac{x}{2}$

15. $\dfrac{5}{7} = \dfrac{35}{x}$

16. $\dfrac{x}{2} = \dfrac{8}{4}$

17. $\dfrac{15}{20} = \dfrac{x}{8}$

18. $\dfrac{x}{40} = \dfrac{5}{100}$

3.3 Similar Triangles (DOK 2)

Two triangles are similar if there are three pairs of congruent angles. The **AA theorem** states when 2 angles of one triangle are equal to 2 corresponding angles of the other triangle, the two triangles must be similar. The corresponding sides of similar triangles are proportional. Also, a line parallel to one side of a triangle divides the other two sides proportionally. If the line dividing the triangle is parallel to one of the sides, it creates corresponding angles, and thus a smaller, similar triangle.

Claim: A line parallel to one side of a triangle divides the other two proportionally.

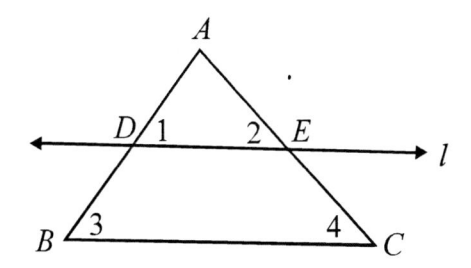

Given: $\triangle ABC$ and $\overline{DE}$ is parallel to $\overline{BC}$

Prove: $\dfrac{\overline{BD}}{\overline{DA}} = \dfrac{\overline{CE}}{\overline{EA}}$

Statement	Reason
1. $\overline{DE}$ is parallel to $\overline{BC}$.	1. Given
2. $\angle 3 \cong \angle 1$ and $\angle 4 \cong \angle 2$	2. If lines are parallel, then corresponding angles are congruent.
3. $\triangle ABC \cong \triangle ADE$	3. AA theorem
4. $\dfrac{\overline{BA}}{\overline{DA}} = \dfrac{\overline{CA}}{\overline{EA}}$	4. Corresponding sides of similar triangles are proportional.
5. $\overline{BA} = \overline{BD} + \overline{DA}$ and $\overline{CA} = \overline{CE} + \overline{EA}$	5. Segment Addition Postulate
6. $\dfrac{\overline{BD} + \overline{DA}}{\overline{DA}} = \dfrac{\overline{CE} + \overline{EA}}{\overline{EA}}$	6. Substitution
7. $\dfrac{\overline{BD}}{\overline{DA}} + \dfrac{\overline{DA}}{\overline{DA}} = \dfrac{\overline{CE}}{\overline{EA}} + \dfrac{\overline{EA}}{\overline{EA}}$	7. Property of Fractions
8. $\dfrac{\overline{BD}}{\overline{DA}} + 1 = \dfrac{\overline{CE}}{\overline{EA}} + 1$	8. Simplify.
9. $\dfrac{\overline{BD}}{\overline{DA}} = \dfrac{\overline{CE}}{\overline{EA}}$	9. Addition Property of Equality

Chapter 3   Similarity

Corresponding Sides - The triangles below are similar. Therefore, the two shortest sides from each triangle, c and f, are corresponding. The two longest sides from each triangle, a and d, are corresponding. The two medium length sides, b and e, are corresponding.

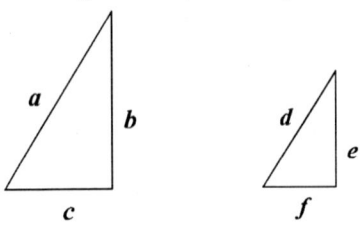

Proportional - The corresponding sides of similar triangles are proportional to each other. This means if all the measurements of one triangle are known, and only one measurement of the other triangle is given, then the measurements of the two other sides can be determined by using proportions. The two triangles below are similar.

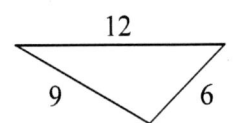

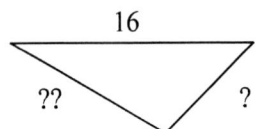

Note: To set up the proportion correctly, it is important to keep the measurements of each triangle on opposite sides of the equal sign.

To find the short side:	To find the medium length side:
Step 1: Set up the proportion	**Step 1:** Set up the proportion
$\dfrac{\text{long side}}{\text{short side}} \quad \dfrac{12}{6} = \dfrac{16}{?}$	$\dfrac{\text{long side}}{\text{medium}} \quad \dfrac{12}{9} = \dfrac{16}{??}$
Step 2: Solve the proportion. Multiply the two numbers diagonal to each other, and then divide by the other number. $16 \times 6 = 96$ $96 \div 12 = 8$	**Step 2:** Solve the proportion. Multiply the two numbers diagonal to each other, and then divide by the other number. $16 \times 9 = 144$ $144 \div 12 = 12$

To find the **scale factor**, divide a value from the second triangle by the corresponding value from the first triangle. The value 16 is from the second triangle, and the corresponding value from the first triangle is 12. $k = \dfrac{16}{12} = \dfrac{4}{3}$ The scale factor in this problem is $\frac{4}{3}$.

To check this answer, multiply every length in the first triangle by the scale factor, and you will get every corresponding length in the second triangle.

$$12 \times \frac{4}{3} = 16 \qquad 9 \times \frac{4}{3} = 12 \qquad 6 \times \frac{4}{3} = 8$$

3.3   Similar Triangles (DOK 2)

Find the missing side from the following similar triangles.

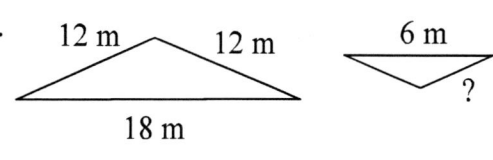

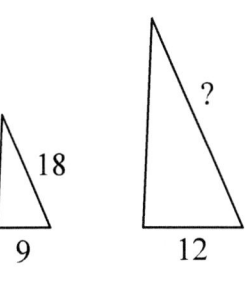

1. Small triangle: sides 3, 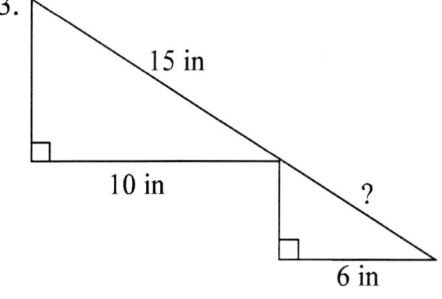 4, 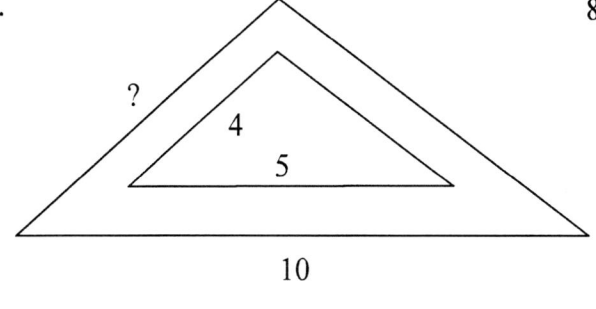 5. Larger triangle: side 6, hypotenuse ?

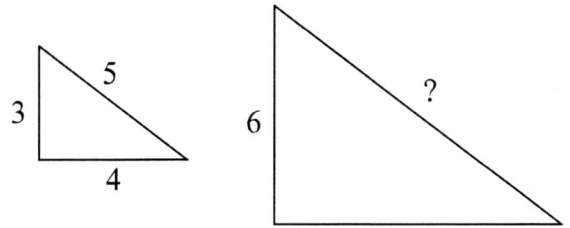

2. Small triangle: legs 9 and 18. Larger triangle: leg 12, hypotenuse ?

3. Large right triangle with hypotenuse 15 in and leg 10 in. Smaller similar right triangle with leg 6 in and hypotenuse ?

4. Outer triangle with side ? and base 10. Inner similar triangle with corresponding side 4 and base 5.

5. Isosceles triangle: two sides 12 m, 12 m, base 18 m. Similar smaller triangle with side 6 m and corresponding side ?

6. Right triangle with height 24 ft and base 25 ft. Smaller similar right triangle with base 15 ft and height ?

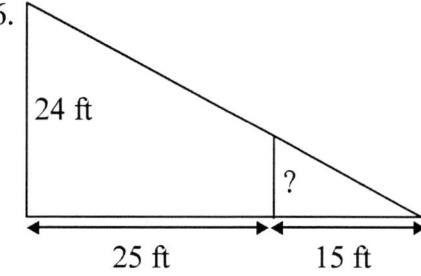

7. Triangle with segment 5 cm and 36 cm; find ? where 15 cm is the full side.

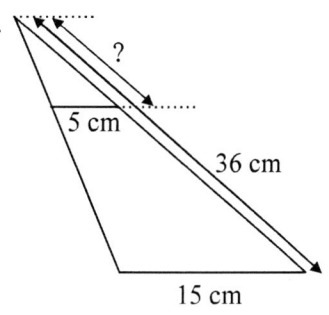

8. Two similar right triangles sharing a vertex: one with legs 12 in and 8 in, the other with leg 18 in and ?

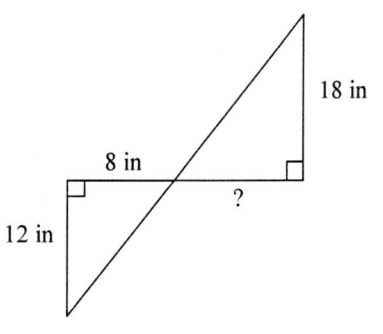

Chapter 3   Similarity

3.4   Similar Plane Figures (DOK 2)

The measures of corresponding sides of similar figures can be found by setting up a proportion.

Example 2:   The following rectangles are similar. Find the value of x.

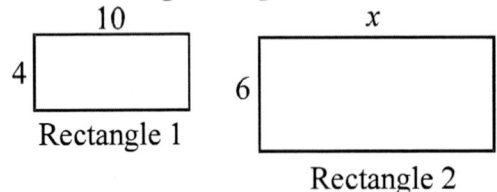

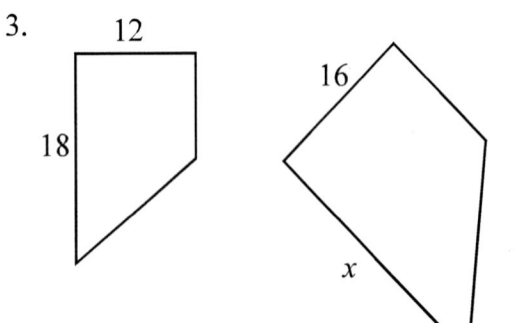

Rectangle 1: 10 by 4
Rectangle 2: x by 6

Step 1:   Set up the proportion.

$$\frac{\text{Short side rectangle 1}}{\text{Long side rectangle 1}} = \frac{\text{Short side rectangle 2}}{\text{Long side rectangle 2}}$$

Step 2: $\frac{4}{10} = \frac{6}{x}$ Cross multiply.

$4x = 60$ Divide.

$x = 15$

All of the pairs of figures below are similar. Find the missing side for each pair.

1. Trapezoid with sides 4, 12; similar trapezoid with sides 6, x.

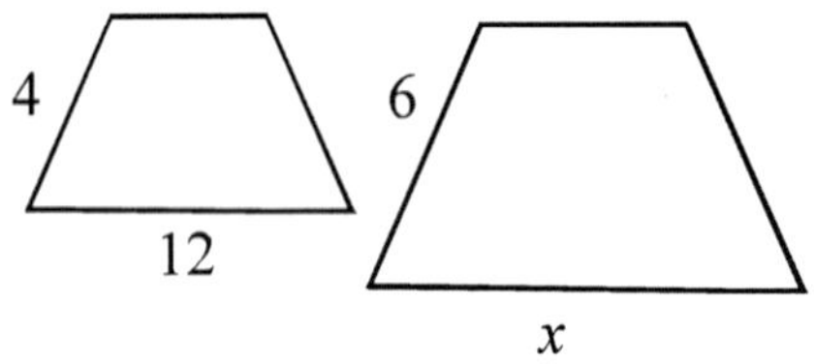

2. Parallelogram with sides 25, 15; similar parallelogram with sides x, 18.

3. Quadrilateral with sides 12, 18; similar quadrilateral with sides 16, x.

4. Pentagon with sides 3, 4; similar pentagon with sides x, 16.

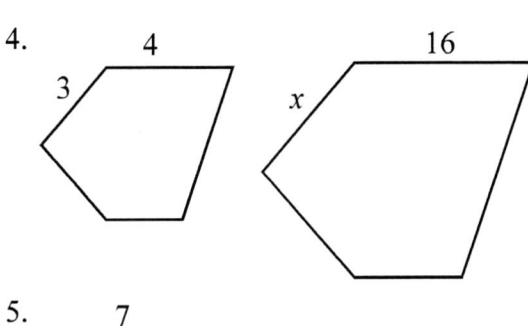

5. Rectangle with sides 3, 7; similar rectangle with sides 9, x.

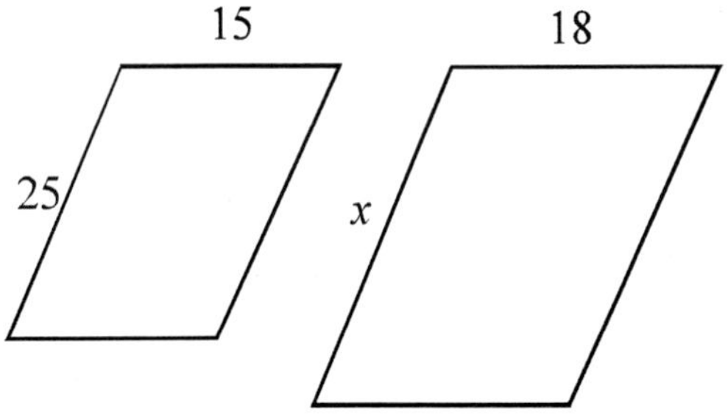

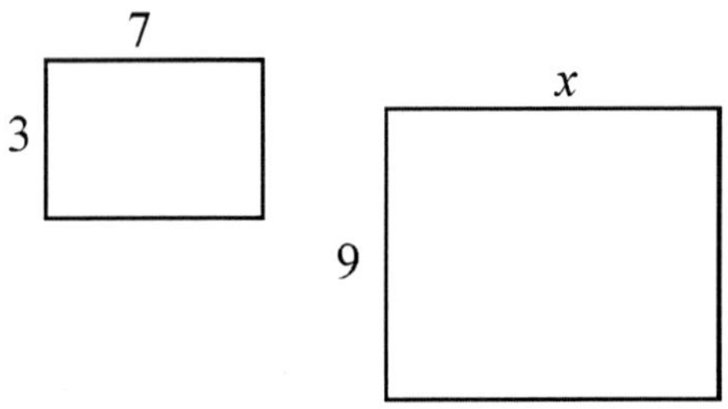

6. Figure with sides 4, x; similar figure with sides 6, 9.

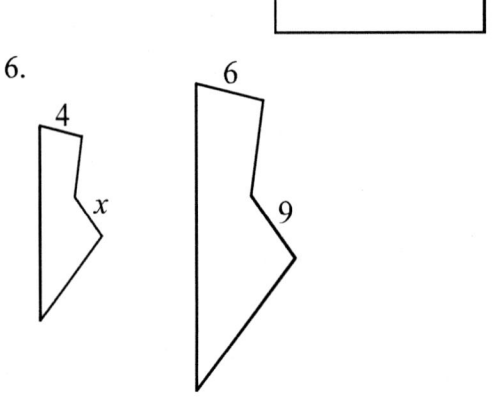

50

3.5 Geometric Relationships of Plane Figures (DOK 3)

This section illustrates what happens to the area of a figure when one or more of the dimensions is doubled or tripled.

Example 3: Sam drew a square that was 2 inches on each side for art class. His teacher said the square needed to be twice as big. When Sam doubled each side to 4 inches, what happened to the area?

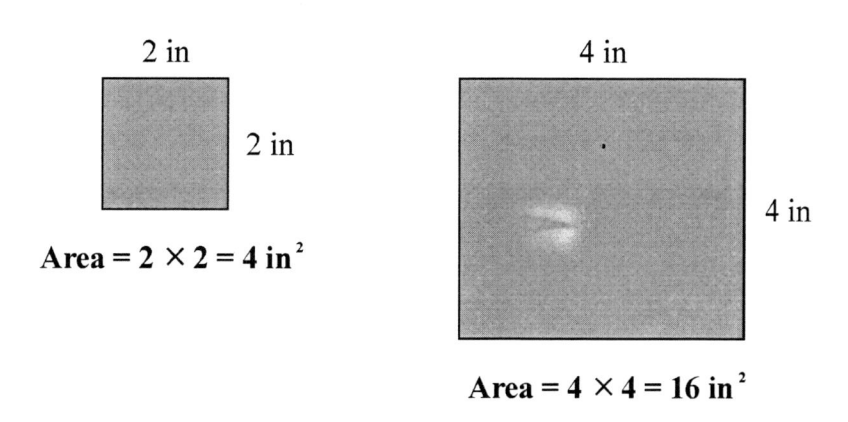

2 in

2 in

Area = 2 × 2 = 4 in^2

4 in

4 in

Area = 4 × 4 = 16 in^2

The area of the second square is 4 times larger than the first.

Example 4: Sonya drew a circle that had a radius of 3 inches for a school project. She also needed to make a larger circle that had a radius of 9 inches. When Sonya drew the bigger circle, what was the difference in area?

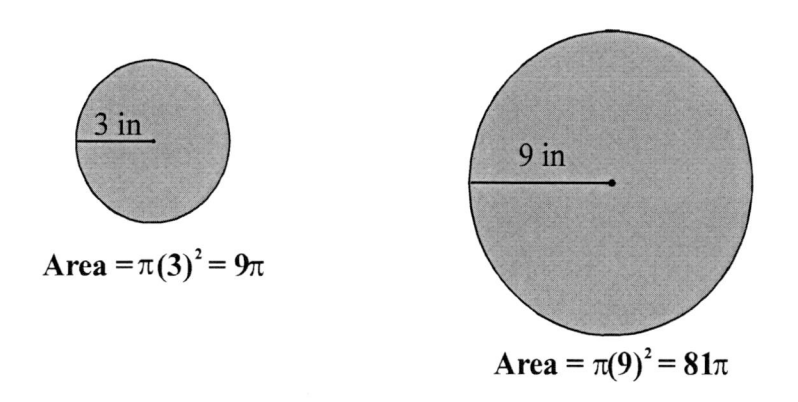

3 in

Area = $\pi(3)^2 = 9\pi$

9 in

Area = $\pi(9)^2 = 81\pi$

The area of the second circle is 9 times larger than the first.

When doubling or tripling both sides or the radius of a plane figure, the total area increases by a squared value. In other words, when both sides of the square were doubled, the area was 2^2 or 4 times larger. When the radius of the circle became 3 times larger, the area became 3^2 or 9 times larger.

Chapter 3 Similarity

Carefully read each problem below and solve.

1. Ken draws a circle with a radius of 5 cm. He then draws a circle with a radius of 10 cm. How many times larger is the area of the second circle? $A = \pi r^2$

2. Kobe draws a square with each side measuring 6 inches. He then draws a rectangle with a width of 6 inches and a length of 12 inches. How many times larger is the area of the rectangle than the area of the square? $A = l \times w$

3. Toshi draws a square 3 inches on each side. Then, he draws a bigger square that is 6 inches on each side. How many times larger is the area of the second square than the area of the first square? $A = s^2$

4. Leslie draws a triangle with a base of 5 inches and a height of 3 inches. To use her triangle pattern for a bulletin board design, it needs to be 3 times bigger. If she increases the base and the height by multiplying each by 3, how much will the area of the triangle increase? $A = \frac{1}{2}bh$

5. Heather is using 100 tiles that measure 1 foot by 1 foot to cover a 10 feet by 10 feet floor. If she used tiles that measure 2 feet by 2 feet, how many tiles would she need?

6. The area of circle B is 9 times larger than the area of circle A. If the radius of circle A is represented by x, how would you represent the radius of circle B? $A = \pi r^2$

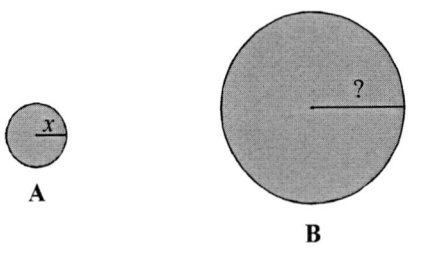

7. How many squares will it take to cover the rectangle below? $A = l \times w$, $A = s^2$

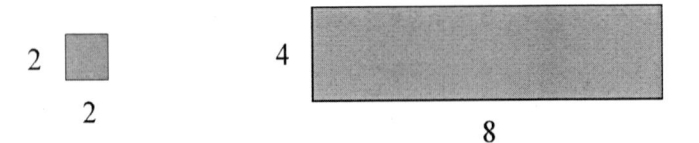

8. If the area of rhombus B is one-fourth the area of rhombus A, what are the dimensions of rhombus B? $A = s^2$

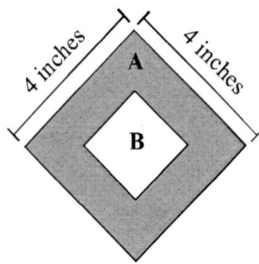

3.6 Applying Similar Plane Figures (DOK 2)

Example 5: Square $ABCD$ has an area of 25 yd^2, and square $WXYZ$ has a perimeter of 40 yards. What is the ratio of the areas?

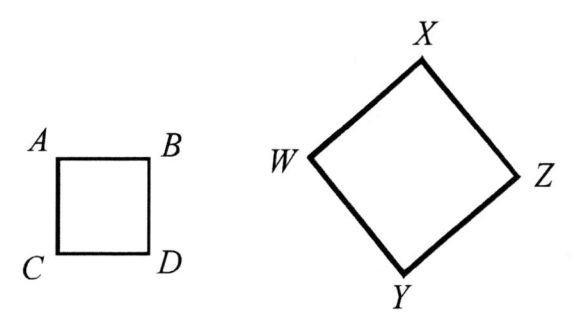

Step 1: Find the length of one side of square $WXYZ$. Divide the perimeter by 4 because there are 4 equal sides in a square.

$40 \div 4 = 10$

Step 2: Find the area of square $WXYZ$.

$s^2 = 10^2 = 100$

Step 3: Compare the areas. 25:100, then reduce the ratio to 1:4.

Answer the following questions.

1. The ratio of the diameter of two similar circles is 1 to 6. What is the ratio of their areas? $A = \pi r^2$

2. Square $ABCD$ has an area of 36 ft^2, and square $EFGH$ has a perimeter of 48 ft. What is the ratio of their areas? $A = s^2$

3. The ratio of the area of two similar right triangles is 1:9. If the area of the smaller triangle is 38.5 in^2, what is the area of the larger triangle?

Chapter 3 Similarity

Chapter 3 Review

Solve the following proportions. (DOK 1)

1. $\dfrac{8}{x} = \dfrac{1}{2}$
2. $\dfrac{2}{5} = \dfrac{x}{10}$
3. $\dfrac{x}{6} = \dfrac{3}{9}$
4. $\dfrac{4}{9} = \dfrac{8}{x}$

Solve the following similarity problems. (DOK 2 & 3)

5. What is the length of $\overline{WY}$?

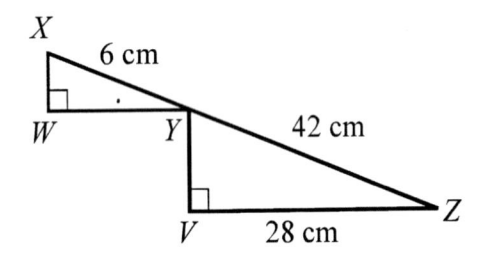

6. The following two triangles are similar. Find the length of the missing side.

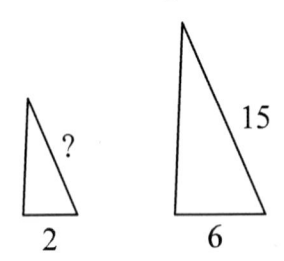

7. Find the missing side of the triangle below.

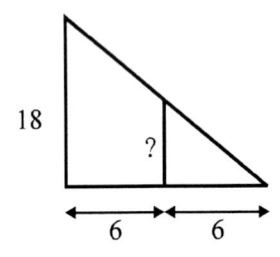

8. Are the figures below congruent, similar, or neither?

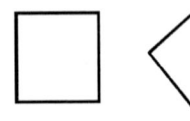

9. True or false? Explain. The triangles below are similar.

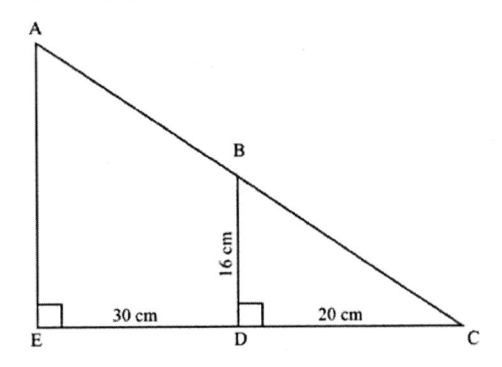

10. Find the missing side below.

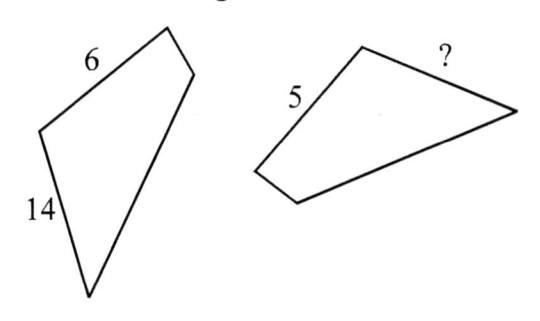

11. If you double the width of a square, how much does the area of the square increase? $A = s^2$

12. If you double each edge of a cube, how many times larger is the volume? $V = s^3$

13. It takes 8 cubic inches of water to fill a cube. If each edge of the cube is doubled, how much water is needed to fill the new cube? $V = s^3$

Chapter 3 Test

1 Solve for x. $\quad \dfrac{2}{18} = \dfrac{6}{x}$

  A 6
  B 9
  C 18
  D 54

(DOK 1)

2 Solve for x. $\quad \dfrac{x}{2} = \dfrac{91}{7}$

  A 13
  B 26
  C 52
  D 91

(DOK 1)

3 The figures below are similar. What is the measure of x?

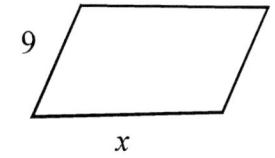

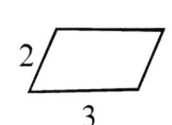

  A 10
  B 13.5
  C 26
  D Cannot be determined.

(DOK 2)

4 What is the length of $\overline{VZ}$?

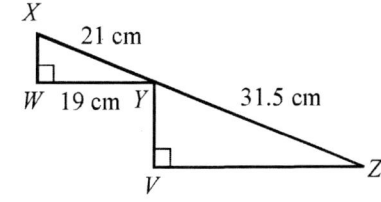

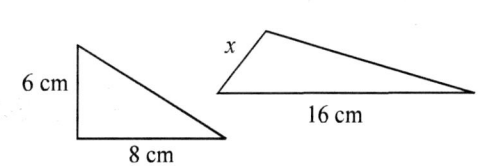

7 (triangles: 6 cm, 8 cm and 16 cm, x)

  A 5
  B 2
  C 8
  D These triangles are not similar, so it cannot be determined.

(DOK 2)

8 The ratio of the perimeters of two similar polygons is $5 : 11$.
Find the ratio of the areas.

  A $121 : 25$
  B $25 : 121$
  C $\sqrt{5} : 11$
  D $5 : \sqrt{11}$

(DOK 2)

Chapter 4
Transformations

Analytic Geometry

This chapter covers the following CCGPS standard(s):

	Content Standards
Congruence	G.CO.6, G.CO.7, G.CO.8, G.GPE.4

4.1 Reflections (DOK 2)

A **reflection** of a geometric figure is a mirror image of the object. Placing a mirror on the **line of reflection** will give you the position of the reflected image.

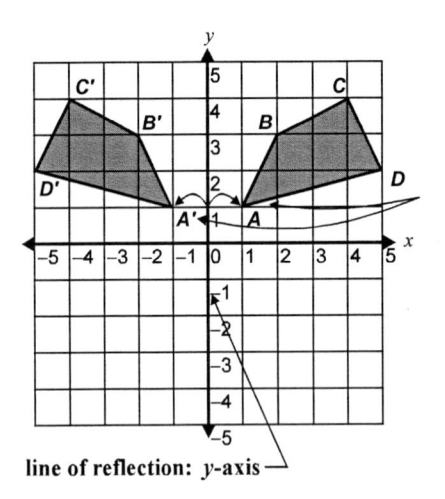

Quadrilateral $ABCD$ is reflected across the y-axis to form quadrilateral $A'B'C'D'$. The y-axis is the line of reflection. Point A' (read as A prime) is the reflection of point A, point B' corresponds to point B, C' to C, and D' to D.

Point A is $+1$ space from the y-axis. Point A's mirror image, point A', is -1 space from the y-axis.
Point B is $+2$ spaces from the y-axis. Point B' is -2 spaces from the y-axis.
Point C is $+4$ spaces from the y-axis, and point C' is -4 spaces from the y-axis.
Point D is $+5$ spaces from the y-axis, and point D' is -5 spaces from the y-axis.

4.1 Reflections (DOK 2)

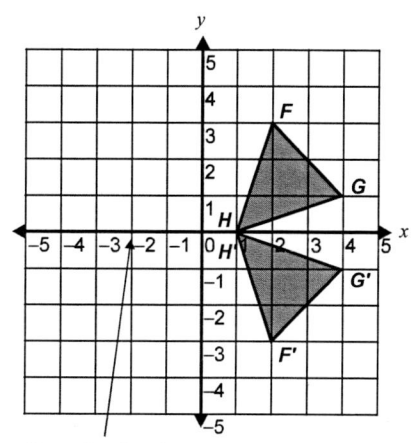

line of reflection: x-axis

Triangle FGH is reflected across the x-axis to form triangle $F'G'H'$. The x-axis is the line of reflection. Point F' is reflected from point F. Point G' corresponds to point G, and H' mirrors H.

Point F is $+3$ spaces from the x-axis. Likewise, point F' is -3 spaces from the x-axis.
Point G is $+1$ space from the x-axis, and point G' is -1 space from the x-axis.
Point H is 0 spaces from the x-axis, so point H' is also 0 spaces from the x-axis.

Reflecting Across a $45°$ Line ($y = x$)

line of reflection: line m

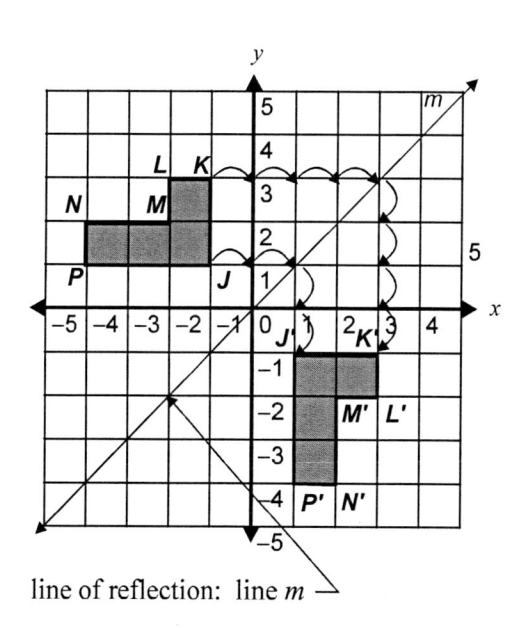

Figure $JKLMNP$ is reflected across line m to form figure $J'K'L'M'N'P'$. Line m is at a $45°$ angle. Point J corresponds to J', K to K', L to L', M to M', N to N', and P to P'. Line m is the line of reflection. **Pay close attention to how to determine the mirror image of figure $JKLMNP$ across line m described below. This method only works when the line of reflection is at a $45°$ angle.**

Point J is 2 spaces over from line m, so J' must be 2 spaces down from line m.

Point K is 4 spaces over from line m, so K' is 4 spaces down from line m, and so on.

Chapter 4 Transformations

Draw the following reflections, and record the new coordinates of the reflection. The first problem is done for you.

1. Reflect figure ABC across the x-axis. Label vertices A′, B′, and C′ so that point A′ is the reflection of point A, B′ is the reflection of B, and C′ is the reflection of C.
$A' = (-4, -2)$ $B' = (-2, -4)$ $C' = (0, -4)$

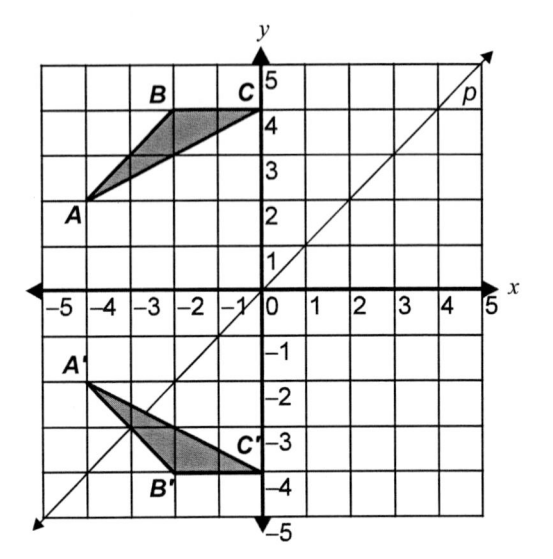

2. Reflect figure ABC across the y-axis. Label vertices A″, B″, and C″ so that point A″ is the reflection of point A, B″ is the reflection of B, and C″ is the reflection of C.
$A'' = (4, 2)$ $B'' = (2, 4)$ $C'' = (0, 4)$

3. Reflect figure ABC across line p. Label vertices A‴, B‴, and C‴ so that point A‴ is the reflection of point A, B‴ is the reflection of B, and C‴ is the reflection of C.
$A''' = (2, -4)$ $B''' = (4, -2)$ $C''' = (4, 0)$

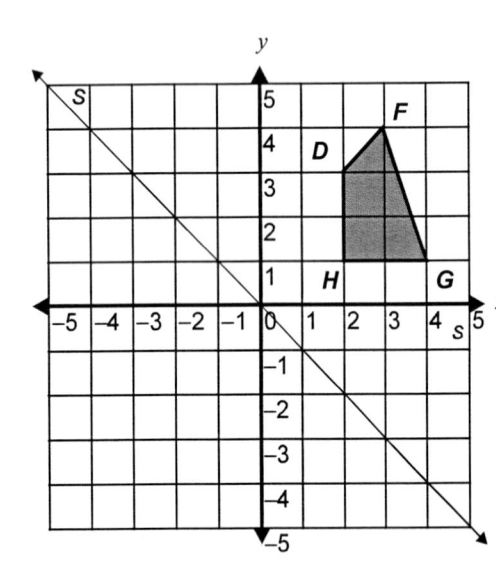

4. Reflect figure DFGH across the y-axis. Label vertices D′, F′, G′, and H′ so that point D′ is the reflection of point D, F′ is the reflection of F, G′ is the reflection of G, and H′ is the reflection of H.
$D' = (-2, 4)$ $G' = (-4, 1)$
$F' = (-3, 5)$ $H' = (-2, 1)$

5. Reflect figure DFGH across the x-axis. Label vertices D″, F″, G″, and H″ so that point D″ is the reflection of D, F″ is the reflection of F, G″ is the reflection of G, and H″ is the reflection of H.
$D'' = (2, -4)$ $G'' = (4, -1)$
$F'' = (3, -5)$ $H'' = (2, -1)$

6. Reflect figure DFGH across line s. Label vertices D‴, F‴, G‴, and H‴ so that point D‴ is the reflection of D, F‴ corresponds to F, G‴ to G, and H‴ to H.
$D''' = (-4, -2)$ $G''' = (-1, -4)$
$F''' = (-5, -3)$ $H''' = (-1, -2)$

4.1 Reflections (DOK 2)

Use the following graph for questions 7–9.

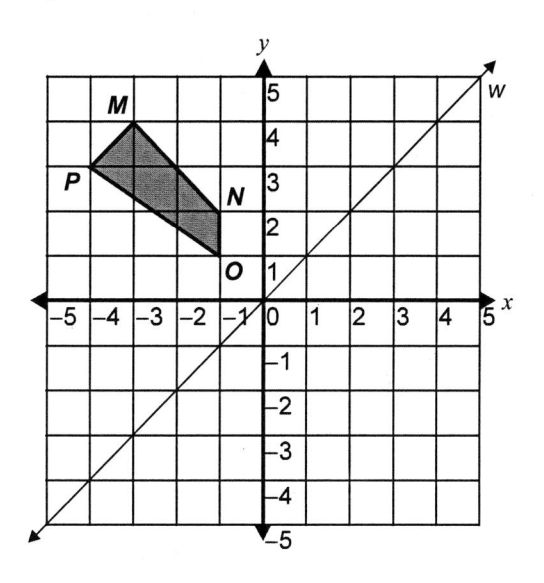

7. Reflect quadrilateral MNOP across the y-axis. Label vertices M′, N′, O′, and P′ so that point M′ is the reflection of point M, N′ is the reflection of N, O′ is the reflection of O, and P′ is the reflection of P.

   M′ = _____   O′ = _____
   N′ = _____   P′ = _____

8. Reflect figure MNOP across the x-axis. Label vertices M″, N″, O″, and P″ so that point M″ is the reflection of M, N″ is the reflection of N, O″ is the reflection of O, and P″ is the reflection of P.

   M″ = _____   O″ = _____
   N″ = _____   P″ = _____

9. Reflect figure MNOP across line w. Label vertices M‴, N‴, O‴, and P‴ so that point M‴ is the reflection of M, N‴ corresponds to N, O‴ to O, and P‴ to P.

   M‴ = _____   O‴ = _____
   N‴ = _____   P‴ = _____

Chapter 4 Transformations

4.2 Translations (DOK 2)

To make a translation of a geometric figure, first duplicate the figure and then slide it along a path.

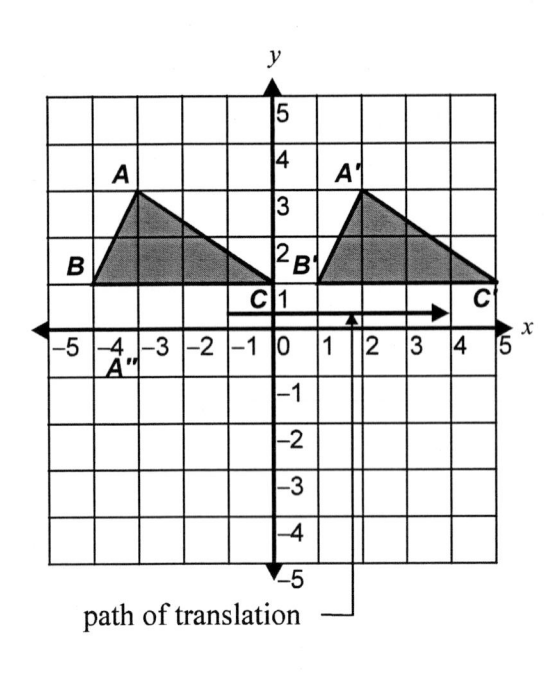

Triangle $A'B'C'$ is a translation of triangle ABC. Each point is translated 5 spaces to the right. In other words, the triangle slid 5 spaces to the right. Look at the path of translation. It gives the same information as above. Count the number of spaces across given by the path of translation, and you will see it represents a move 5 spaces to the right. Each new point is found at $(x + 5, y)$.

Point A is at $(-3, 3)$. Therefore, A' is found at $(-3 + 5, 3)$ or $(2, 3)$.

B is at $(-4, 1)$, so B' is at $(-4 + 5, 1)$ or $(1, 1)$.

C is at $(0, 1)$, so C' is at $(0 + 5, 1)$ or $(5, 1)$.

Quadrilateral $FGHI$ is translated 5 spaces to the right and 3 spaces down. The path of translation shows the same information. It points right 5 spaces and down 3 spaces. Each new point is found at $(x + 5, y - 3)$.

Point F is located at $(-4, 3)$. Point F' is located at $(-4 + 5, 3 - 3)$ or $(1, 0)$.

Point G is at $(-2, 5)$. Point G' is at $(-2 + 5, 5 - 3)$ or $(3, 2)$.

Point H is at $(-1, 4)$. Point H' is at $(-1 + 5, 4 - 3)$ or $(4, 1)$.

Point I is at $(-1, 2)$. Point I' is at $(-1 + 5, 2 - 3)$ or $(4, -1)$.

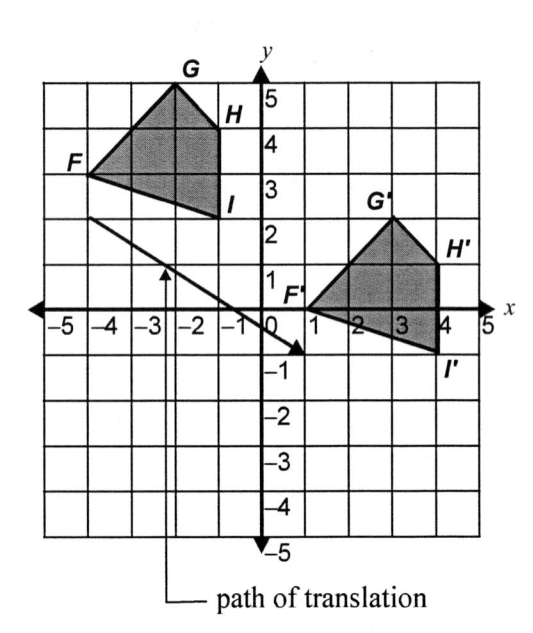

4.2 Translations (DOK 2)

Draw the following translations, and record the new coordinates of the translation. The figure for the first problem is drawn for you.

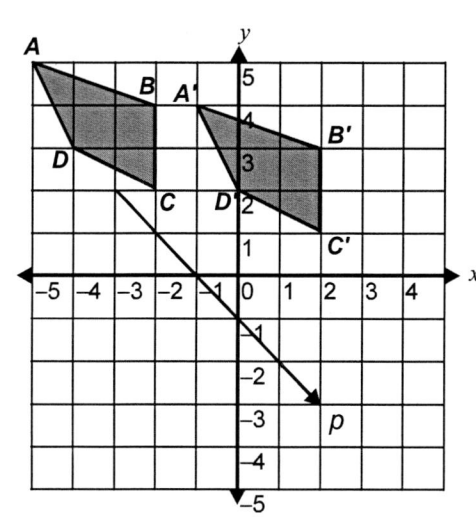

1. Translate figure ABCD 4 spaces to the right and 1 space down. Label the vertices of the translated figure A', B', C', and D' so that point A' corresponds to the translation of point A, B' corresponds to B, C' to C, and D' to D.

   A' = _____ C' = _____
   B' = _____ D' = _____

2. Translate figure ABCD 5 spaces down. Label the vertices of the translated figure A'', B'', C'', and D'' so that point A'' corresponds to the translation of point A, B'' corresponds to B, C'' to C, and D'' to D.

   A'' = _____ C'' = _____
   B'' = _____ D'' = _____

3. Translate figure ABCD along the path of translation, p. Label the vertices of the translated figure A''', B''', C''', and D''' so that point A''' corresponds to the translation of point A, B''' corresponds to B, C''' to C, and D''' to D.

   A''' = _____ C''' = _____
   B''' = _____ D''' = _____

4. Translate triangle FGH 6 spaces to the left and 3 spaces up. Label the vertices of the translated figure F', G', and H' so that point F' corresponds to the translation of point F, G' corresponds to G, and H' to H.

   F' = _____ G' = _____ H' = _____

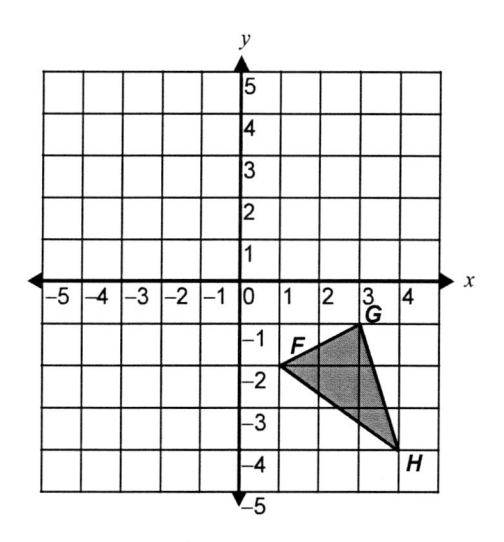

5. Translate triangle FGH 4 spaces up and 1 space to the left. Label the vertices of the translated triangle F'', G'', and H'' so that point F'' corresponds to the translation of point F, G'' corresponds to G, and H'' to H.

   F'' = _____ G'' = _____ H'' = _____

4.3 Rotations (DOK 2)

A **rotation** of a geometric figure shows motion around a point.

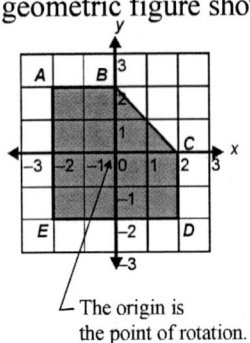

The origin is the point of rotation.

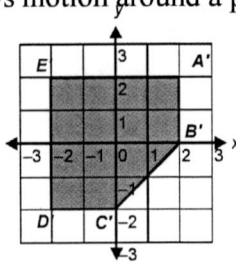

Figure ABCDE has been rotated 90° clockwise around the origin to form A'B'C'D'E'.

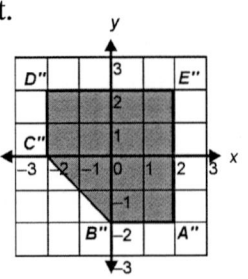

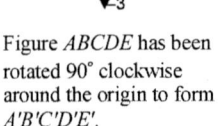

Figure ABCDE has been rotated 180° around the origin to form A''B''C''D''E''.

Draw the following rotations, and record the new coordinates of the rotation. The figure for the first problem is drawn for you.

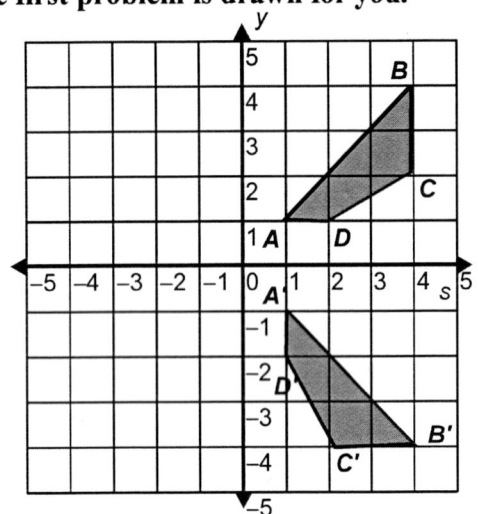

1. Rotate figure ABCD 90° clockwise around the origin. Label the vertices A', B', C', and D' so that point A' corresponds to the rotation of point A, B' corresponds to B, C' to C, and D' to D.
   $A' = $ _____ $C' = $ _____
   $B' = $ _____ $D' = $ _____

2. Rotate figure ABCD 180° clockwise around the origin. Label the vertices A'', B'', C'', and D'' so that point A'' corresponds to the rotation of point A, B'' corresponds to B, C'' to C, and D'' to D.
   $A'' = $ _____ $C'' = $ _____
   $B'' = $ _____ $D'' = $ _____

3. Rotate figure ABCD 270° clockwise around the origin. Label the vertices A''', B''', C''', and D''' so that point A''' corresponds to the rotation of point A, B''' to B, C''' to C, and D''' to D.
   $A''' = $ _____ $C''' = $ _____
   $B''' = $ _____ $D''' = $ _____

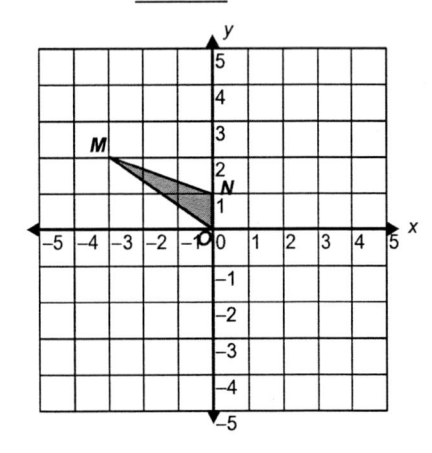

4. Rotate figure MNO 90° clockwise around point O. Label the vertices M', N', and O so that point M' corresponds to the rotation of point M and N' corresponds to N.
   $M' = $ _____ $N' = $ _____

5. Rotate figure MNO 180° clockwise around point O. Label the vertices M'', N'', and O so that point M'' corresponds to the rotation of point M and N'' corresponds to N.
   $M'' = $ _____ $N'' = $ _____

6. Rotate figure MNO 270° clockwise around point O. Label the vertices M''', N''', and O so that point M''' corresponds to the rotation of point M and N''' corresponds to N.
   $M''' = $ _____ $N''' = $ _____

4.4 Transformation Practice (DOK 2)

Answer the following questions regarding transformations.

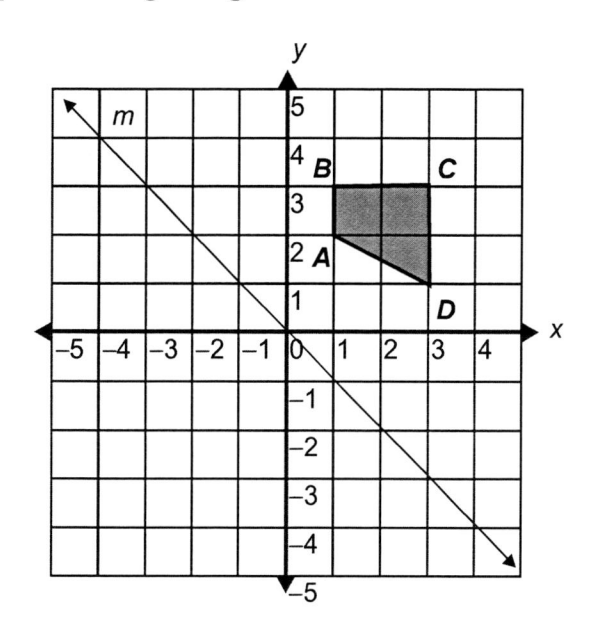

1. Translate quadrilateral *ABCD* so that point *A'*, which corresponds to point *A*, is located at coordinates $(-4, 3)$. Label the other vertices *B'* to correspond to B, *C'* to C, and *D'* to D. What are the coordinates of *A'*, *B'*, *C'*, and *D'*?

   A' = _____   C' = _____

   B' = _____   D' = _____

2. Reflect quadrilateral *ABCD* across line *m*. Label the coordinates *A''*, *B''*, *C''*, and *D''* so that point *A''* corresponds to the reflection of point *A*, *B''* corresponds to the reflection of *B*, and *C''* corresponds to the reflection of *C*. What are the coordinates of *A''*, *B''*, *C''*, and *D''*?

   A'' = _____   C'' = _____

   B'' = _____   D'' = _____

3. Rotate quadrilateral *ABCD* 90° counterclockwise around point *D*. Label the points *A'''*, *B'''*, *C'''*, and *D'''* so that *A'''* corresponds to the rotation of point *A*, *B'''* corresponds to *B*, *C'''* to *C*, and *D'''* to *D*. What are the coordinates of *A'''*, *B'''*, *C'''*, and *D'''*?

   A''' = _____   C''' = _____

   B''' = _____   D''' = _____

4.5 Algebraic Transformation Notation (DOK 2)

There are certain rules to remember for transformations when determining algebraic notation for a transformation. The point (x, y) is the notation of a point before it has gone through a transformation, the point (x', y') is the notation of a point after it has gone through a single transformation, and the point (x'', y'') is the notation of point after it has gone through two transformations.

Rules of Translations
- A point slides to the left a number of spaces - subtract a from x.
  New point: $(x', y') = (x - a, y)$
- A point slides to the right a number of spaces - add a to x.
  New point: $(x', y') = (x + a, y)$
- A point slides up b number of spaces - add b to y.
  New point: $(x', y') = (x, y + b)$
- A point slides down b number of spaces - subtract b from y.
  New point: $(x', y') = (x, y - b)$

Rules of Reflections
- If a point is reflected over the y-axis
  New point: $(x', y') = (-x, y)$
- If a point is reflected over the x-axis
  New point: $(x', y') = (x, -y)$
- If a point is reflected over the line $y = x$
  New point: $(x', y') = (y, x)$

Rules of Rotations
- If a point is rotated 90° clockwise around the origin
  New point: $(x', y') = (y, -x)$
- If a point is rotated 90° counterclockwise around the origin
  New point: $(x', y') = (-y, x)$
- If a point is rotated 180° clockwise or counterclockwise around the origin
  New point: $(x', y') = (-x, -y)$

Rules of Dilations
- If a point is dilated by a scale factor of a
  New point: $(x', y') = (ax, ay)$
- If a point is dilated by a scale factor of $\frac{1}{a}$
  New point: $(x', y') = \left(\frac{1}{a}x, \frac{1}{a}y\right)$

4.5 Algebraic Transformation Notation (DOK 2)

Example 1: What is the rule for the transformation shown below?

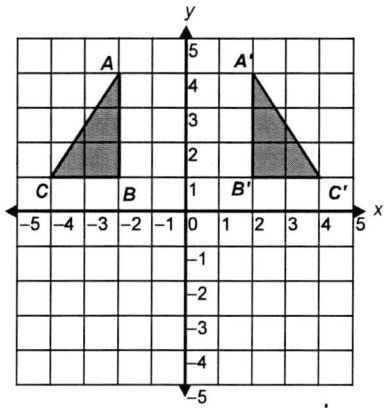

Step 1: List the original coordinates and the new coordinates.

Original Coordinates	New Coordinates
$A: (-2, 4)$	$A': (2, 4)$
$B: (-2, 1)$	$B': (2, 1)$
$C: (-4, 1)$	$C': (4, 1)$

Step 2: Determine what is different between the original and new coordinates. The signs of the values of x changed and y stayed the same, so the rule is $(x', y') = (-x, y)$.

Note: Since the transformation that occurred is a reflection over the y-axis, we could have looked at our rules on the previous page and saw that the rule for the reflection over the y-axis is $(x', y') = (-x, y)$.

Answer the following questions about algebraic transformation notation.

1. What is the rule for the transformation formed by a reflection over the x-axis?

2. What is the rule for the transformation formed by a rotation 90° clockwise around the origin?

3. What is the rule for the transformation formed by a rotation 180° counterclockwise around the origin?

4. What is the rule for the transformation formed by a translation 5 units down?

5. What is the rule for the transformation formed by a translation 6 units to the left?

6. What is the rule for the transformation formed by a rotation 90° clockwise around the origin, then a translation 5 units down?

7. What is the rule for the transformation formed by a translation 2 units right and 4 units up, then a dilation by a scale factor of 6?

8. What is the rule for the transformation formed by a rotation 90° counterclockwise around the origin, then a dilation by a scale factor of $\frac{1}{2}$?

9. What is the rule for the transformation formed by a reflection over the line $y = x$?

10. What is the rule for the transformation formed by a translation 3 units to the left, then a reflection over the y-axis?

Chapter 4 Transformations

4.6 Transforming a Shape onto Itself (DOK 2)

Shapes can be rotated or reflected. Sometimes there is no visual change when these transformations occur. The following examples show transformations on shapes where there is no visual change. Perform the transformations yourself to verify that the transformations were performed.

Example 2:   Perform the transformation on a triangle.

Reflection across the y-axis:

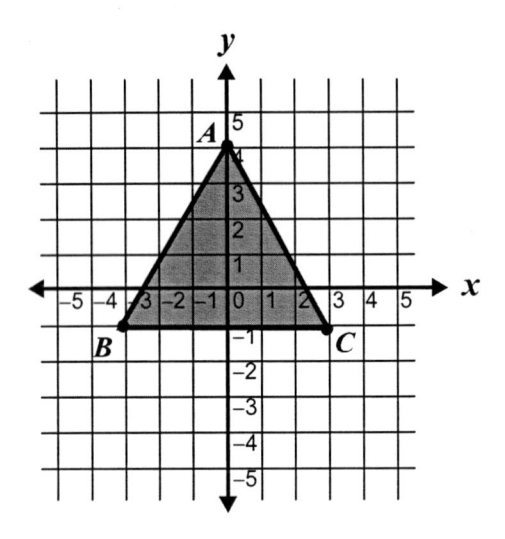

After the triangle has been reflected over the y-axis, the triangle is still the same size and in the same position. The triangle was transformed onto itself.

Example 3:   Perform the transformation on a square.

Rotate 90°:

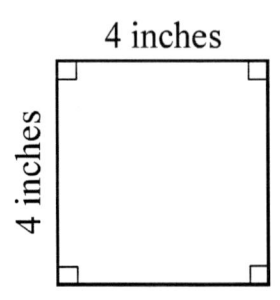

4 inches

4 inches

After the square has been rotated 90°, the square is still the same size and in the same position. The square was transformed onto itself.

66   Copyright © American Book Company

4.6 Transforming a Shape onto Itself (DOK 2)

Example 4: Perform the transformation on a regular hexagon.

Rotate 120°:

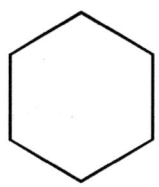

Divide 360° by the number of sides of the regular shape. In this example, the hexagon has 6 sides. $360° \div 6 = 60°$. This is the degree of rotation that will transform the hexagon onto itself. Every multiple of this degree will also rotate the shape onto itself.

After the hexagon has been rotated 120°, the hexagon is still the same size and in the same position. The hexagon was transformed onto itself.

Use the shapes on the grid below to answer questions 1–4.

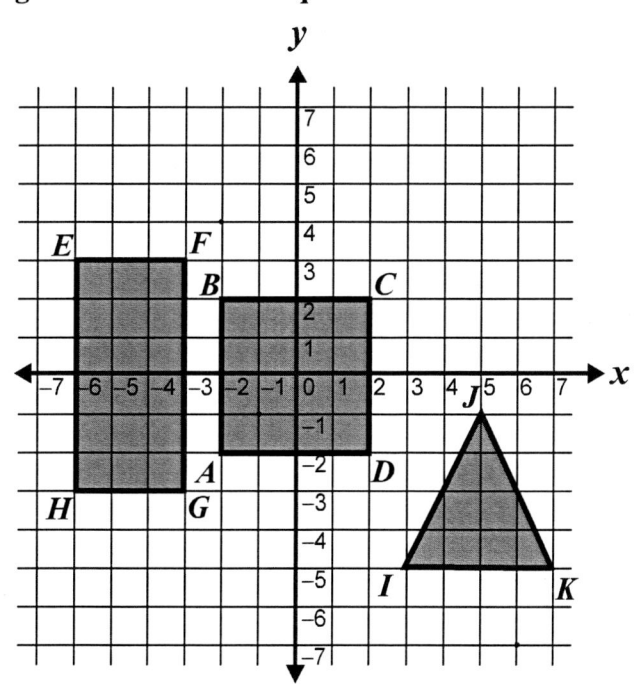

1. Which shape(s) on the grid is(are) regular?

2. Describe all transformations that will take rectangle $EFGH$ onto itself.

3. Describe all transformations that will take square $ABCD$ onto itself.

4. Describe all transformations that will take triangle IJK onto itself.

Copyright © American Book Company

Chapter 4 Transformations

4.7 Using Transformations to Prove Congruency (DOK 3)

Congruent figures have the same shape and size. Some transformations change the position of a figure through rigid motions but do not change the shape or size. When rigid motions, like translations, rotations, or reflections, are applied to any figure, the image will be congruent. The following examples show how these transformations (rigid motions) can be used to prove two figures are congruent.

Example 5: Describe the rigid motion that makes the two figures congruent.

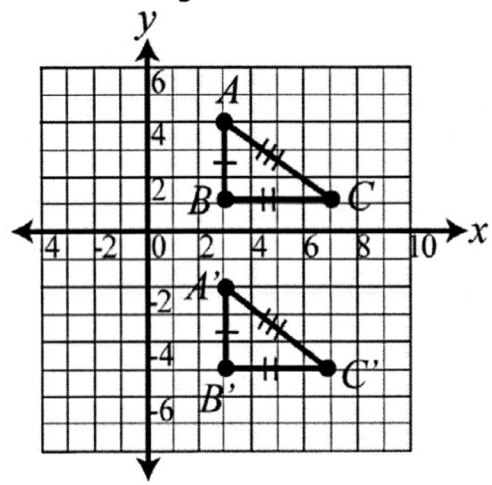

Step 1: Compare the locations of corresponding vertices in the triangles.

$A(3,4) \rightarrow (-2) - 4 = -6 \rightarrow A'(3,-2)$
$B(3,1) \rightarrow (-5) - 1 = -6 \rightarrow B'(3,-5)$
$C(7,1) \rightarrow (-5) - 1 = -6 \rightarrow C'(7,-5)$

The y-coordinates in triangle $A'B'C'$ are 6 units less than the y-coordinates in triangle ABC. Therefore, triangle ABC is translated 6 units down.

Step 2: Compare corresponding side lengths to prove the triangles are congruent.

$\overline{AB} = 3$ $\overline{A'B'} = 3$ $\overline{AB} \cong \overline{A'B'}$

$\overline{BC} = 4$ $\overline{B'C'} = 4$ $\overline{BC} \cong \overline{B'C'}$

$\overline{AC}^2 = \overline{AB}^2 + \overline{BC}^2$ $\overline{A'C'}^2 = \overline{A'B'}^2 + \overline{B'C'}^2$ Use the Pythagorean Theorem
$\overline{AC}^2 = 3^2 + 4^2$ $\overline{A'C'}^2 = 3^2 + 4^2$ to determine the side length of
$\overline{AC}^2 = 25$ $\overline{A'C'}^2 = 25$ $\overline{AC}$ and $\overline{A'C'}$.

$\overline{AC} = 5$ $\overline{A'C'} = 5$ $\overline{AC} \cong \overline{A'C'}$

Answer: Translation 6 units down, or $(x, y - 6)$

4.7 Using Transformations to Prove Congruency (DOK 3)

Example 6: Describe the rigid motion that makes the two figures congruent.

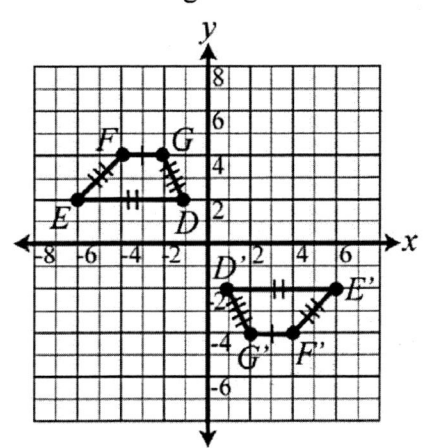

Step 1: Compare the locations of corresponding vertices in the trapezoids.
$D(-1, 2) \rightarrow (-1 \times -1, -1 \times 2) \rightarrow D'(1, -2)$
$E(-6, 2) \rightarrow (-1 \times -6, -1 \times 2) \rightarrow E'(6, -2)$
$F(-4, 4) \rightarrow (-1 \times -4, -1 \times 4) \rightarrow F'(4, -4)$

The x and y-coordinates of the transformed figure are opposite to the coordinates in the original figure. This proves trapezoid $DEFG$ is rotated 180° about the origin.

Step 2: Compare corresponding side lengths to prove the trapezoids are congruent.
$\overline{ED} = 5 \qquad \overline{E'D'} = 5$
$\overline{DG} = \sqrt{5} \qquad \overline{D'G'} = \sqrt{5}$
$\overline{GF} = 2 \qquad \overline{G'F'} = 2$
$\overline{FE} = 2\sqrt{2} \qquad \overline{F'E'} = 2\sqrt{2}$

Answer: Rotation 180° around the origin.

Example 7: Describe the rigid motion that makes the two figures congruent.

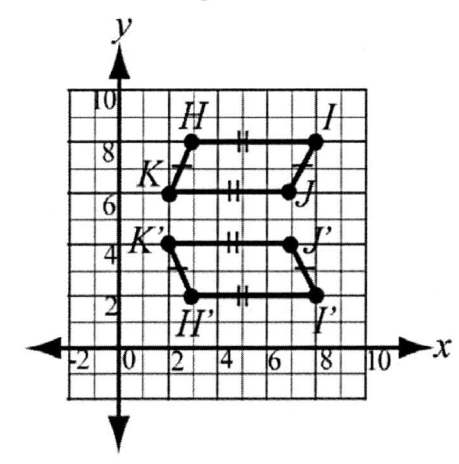

Chapter 4 Transformations

Step 1: Compare the locations of corresponding vertices in the parallelograms.
$H(3,8) \quad H'(3,2) \rightarrow \quad 8-5=3, 5-2=3$
$I(8,8) \quad I'(8,2) \rightarrow \quad 8-5=3, 5-2=3$
$J(7,6) \quad J'(7,4) \rightarrow \quad 6-5=1, 5-4=1$
$K(2,6) \quad K'(2,4) \rightarrow \quad 6-5=1, 5-4=1$

The y-coordinate of the original parallelogram are the same distance from the line $y=5$ as the y-coordinates of the transformed parallelogram. This proves parallelogram $HIJK$ is reflected across the line $y=5$.

Step 2: Compare corresponding side lengths to prove the parallelograms are congruent.
$\overline{KJ} = \overline{HI} = 5 \qquad \overline{K'J'} = \overline{H'I'} = 5$
$\overline{HK} = \overline{IJ} = \sqrt{5} \qquad \overline{H'K'} = \overline{I'J'} = \sqrt{5}$

Answer: Reflection across the line $y=5$.

Example 8: Are the two figures congruent?

Step 1: Determine the rigid motion performed on the original figure.

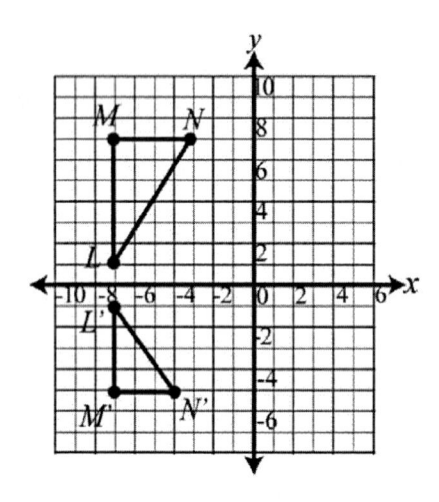

Triangle LMN is reflected across the line $x=0$.

Compare the locations of corresponding vertices in the triangles to determine if triangle $L'M'N'$ is a mirror image of triangle LMN.
$\overline{MN} = 4 \qquad \overline{M'N'} = 3 \qquad \overline{MN} \neq \overline{M'N'}$

The y-coordinates of vertices M' and N' are not mirror images of M and N. This proves the triangles are not congruent.

Answer: No, the two figures are not congruent.

4.7 Using Transformations to Prove Congruency (DOK 3)

Example 9: Are the two figures congruent?

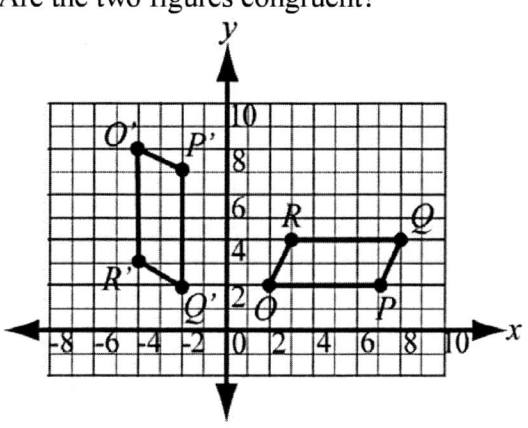

Step 1: Determine the rigid motion performed on the original figure.
Parallelogram $OPQR$ is rotated 90° about the origin.

Step 2: Compare the lengths of corresponding sides in the parallelograms to determine if figure $O'P'\ Q'R'$ is congruent to figure $OPQR$.

$\overline{OR} = \overline{PQ} = \sqrt{5}$ $\overline{O'R'} = \overline{P'Q'} = \sqrt{5}$
$\overline{RQ} = \overline{OP} = 5$ $\overline{R'Q'} = \overline{O'P'} = 5$

The side lengths of the original and transformed parallelograms are the same, therefore the parallelograms are congruent.

Answer: Yes, the two figures are congruent.

Examine the two given figures and describe the rigid motions that makes them congruent.

1.

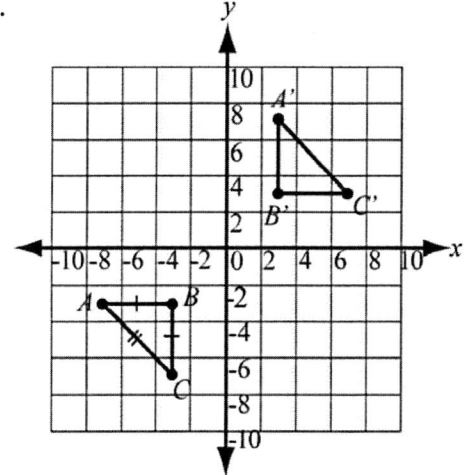

2.

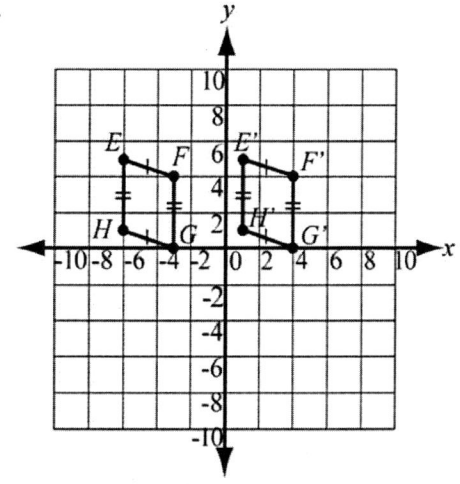

Chapter 4 Transformations

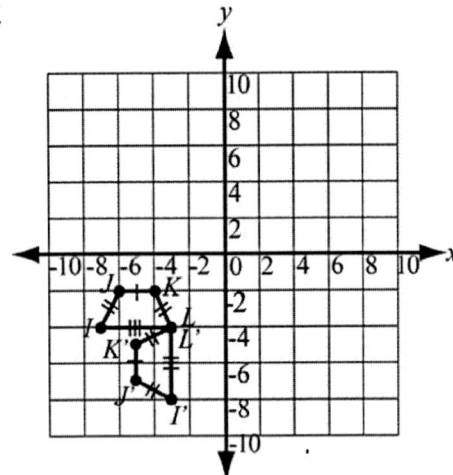

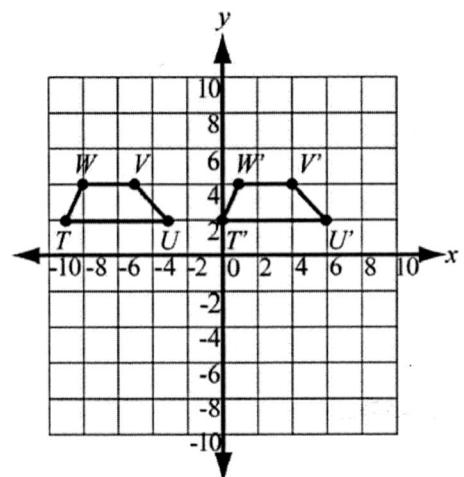

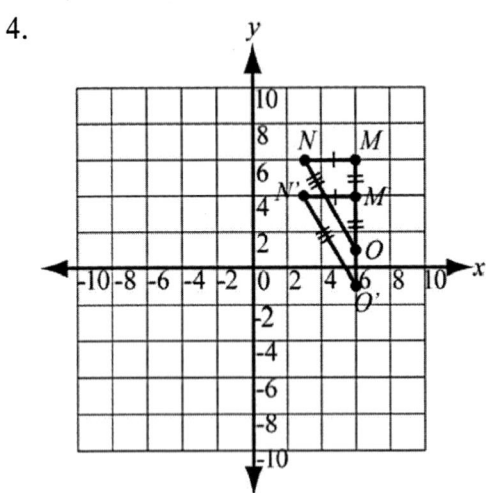

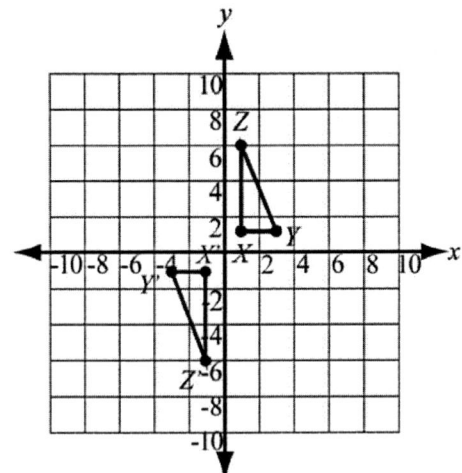

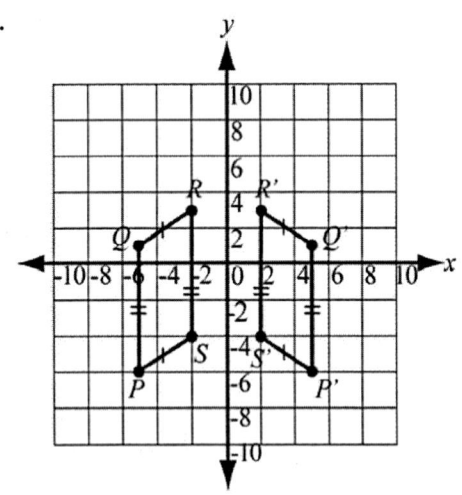

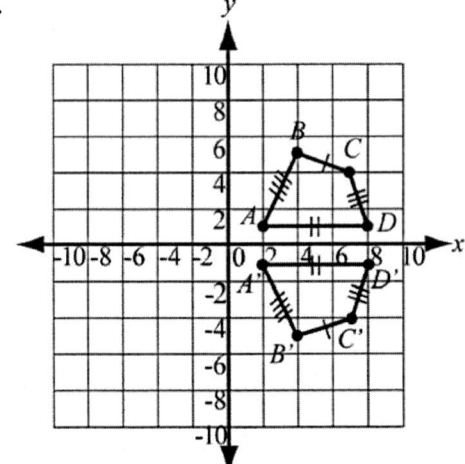

3.
4.
5.
6.
7.
8.

4.7 Using Transformations to Prove Congruency (DOK 3)

Examine the two figures and determine whether they are congruent or not congruent.

9.

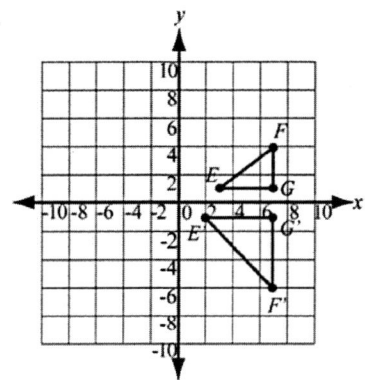

10.

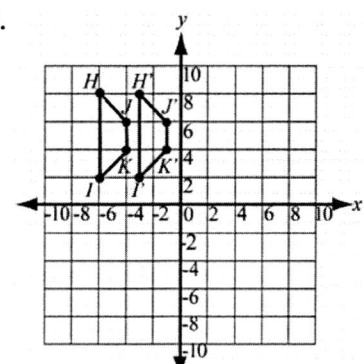

11.

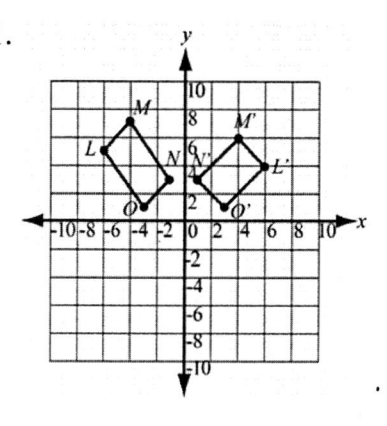

Perform the described transformation on the given figure and determine whether they are congruent or not.

12. Reflect across the y-axis, translate down 3 units

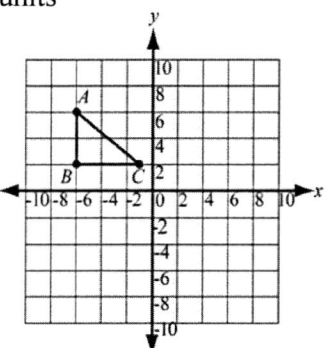

13. Rotate 90° clockwise about the origin, translate 2 units left

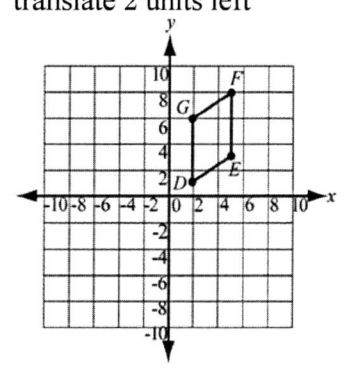

Read the description of the transformation of the figure and decide whether or not the transformed figure is congruent to the original figure.

14. Rotate 90° clockwise around the origin

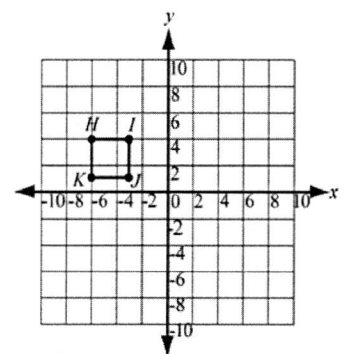

15. Reflection across the x-axis, reflection across the y-axis

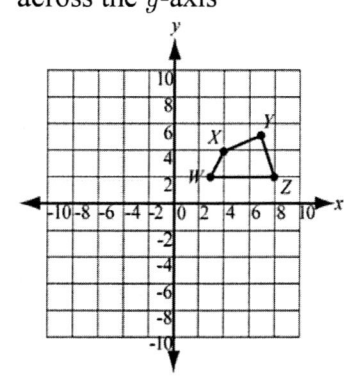

Copyright © American Book Company

4.8 Comparing Transformations (DOK 3)

Certain transformations preserve distance and angle measures, and others do not. If a change in length or angle measure happens during a transformation, the transformed figure will no longer be congruent to the original.

Transformations that result in congruency:

Reflections
Rotations
Translations

Transformations that do not result in congruency:

Horizontal/Vertical Stretch
Horizontal/Vertical Shrink
Dilations

Example 10:

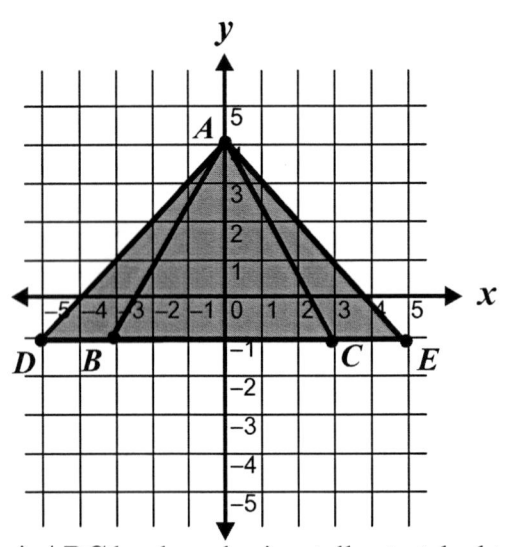

$\triangle ABC$ has been horizontally stretched to form $\triangle ADE$. Looking at the graph, we see that the lengths of sides of the triangle were not preserved during the transformation. Therefore, $\triangle ABC$ is not congruent to $\triangle ADE$.

Answer the following questions about transformations and congruency.

1. Draw a pentagon. Translate the pentagon. Was congruency preserved? Why or why not?

2. Draw a triangle. Vertically shrink the triangle. Was congruency preserved? Why or why not?

3. Draw a line segment. Rotate the line segment. Was congruency preserved? Why or why not?

4.9 Going Deeper with Transformations (DOK 3)

1. Describe the transformation needed for the pre-image to be transformed into the new image.

   Pre-Image   Post-Image

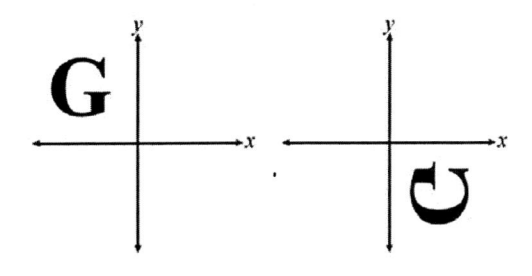

2. Describe the transformation needed for the pre-image to be transformed into the new image. Are the two images still congruent? Explain your answer.

   Pre-Image   Post-Image

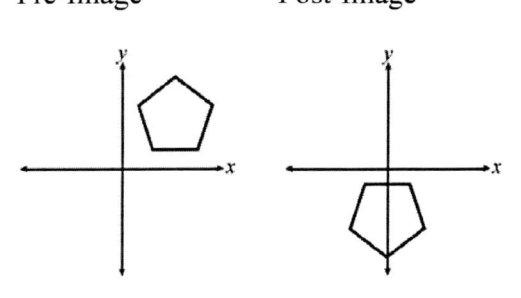

3. Using the given pre-image below construct a new image, so that it undergoes the following transformations: reflects across a 45° line, rotates 180°, and then reflects across the x-axis.

   Pre-Image

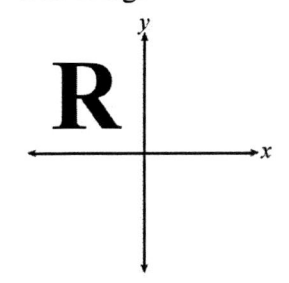

   Reflection across a 45° line

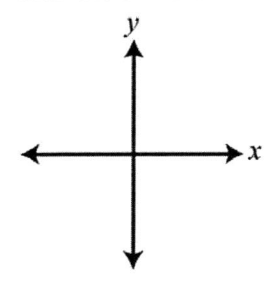

   Then rotates 180°

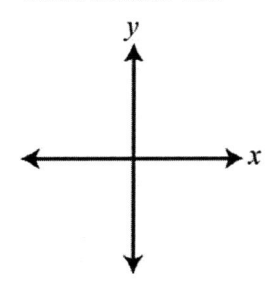

   Then reflects across the x-axis

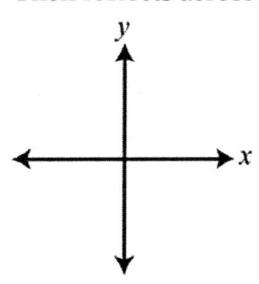

Chapter 4 Transformations

4.10   Dilations (DOK 2)

A **dilation** of a geometric figure is either an enlargement or a reduction of the figure. The point at which the figure is either reduced or enlarged is called the center of dilation. The dilation of a figure is always the product of the original and a **scale factor**. The scale factor is always a positive number that is multiplied by the coordinates of a shape's vertices, which is usually illustrated in a coordinate plane. If the scale factor is greater than one, then the resulting dilated figure will be an enlargement of the original figure. If the scale factor is less than one, then the resulting dilated figure will be a reduction of the original figure.

Example 11:   The triangle ABC has been dilated by a scale factor of $\frac{1}{4}$.

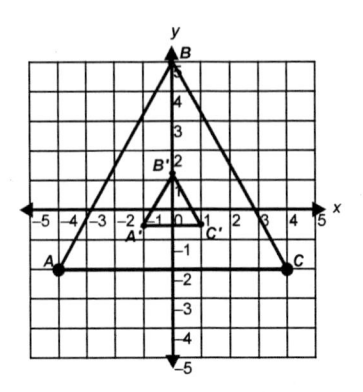

The first step in finding the dilated object is to list all the vertices of the original object, $\triangle ABC$. The next step is to multiply the coordinates of the vertices of $\triangle ABC$ by the scale factor, $\frac{1}{4}$, to find the coordinates of the dilated figure. Lastly, draw the dilated object on the coordinate plane as shown above.

$A : (-4, -2)$ $A' : \left(-1, -\frac{1}{2}\right)$
$B : (0, 5)$ $B' : \left(0, \frac{5}{4}\right)$
$C : (4, -2)$ $C' : \left(1, -\frac{1}{2}\right)$

Note: Since the scale factor is less than one, the dilated figure $A'B'C'$ is a reduction of original triangle, ABC.

Circle the coordinate plane that contains the shape that has been dilated.

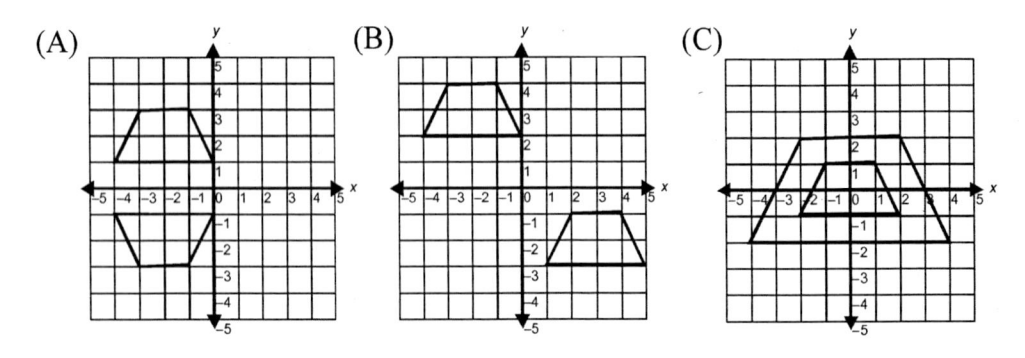

(A)   (B)   (C)

76   Copyright © American Book Company

4.10   Dilations (DOK 2)

On your own graph paper sketch the dilated and original figures.

For the questions 1–6, find the coordinates of the vertices of the dilated figure.

1. $A: (-3, 1)$
   $B: (-1, 4)$
   $C: (1, 4)$
   $D: (3, 1)$
   Scale factor: 4

2. $A: (-6, 5)$
   $B: (3, 5)$
   $C: (3, -4)$
   $D: (-6, -4)$
   Scale factor: $\frac{1}{3}$

3. $A: (-10, 0)$
   $B: (0, 10)$
   $C: (8, 5)$
   Scale factor: $\frac{4}{5}$

4. $A: (-1, 7)$
   $B: (1, 7)$
   $C: (5, 5)$
   $D: (5, \frac{1}{2})$
   $E: (1, -3)$
   $F: (-1, -3)$
   $G: (-5, \frac{1}{2})$
   $H: (-5, 5)$
   Scale factor: 2

5. $A: (-8, 7)$
   $B: (-4, 7)$
   $C: (-2, 3)$
   $D: (-6, 3)$
   Scale factor: $\frac{3}{2}$

6. $A: (-4, 12)$
   $B: (6, -2)$
   $C: (-14, -2)$
   Scale factor: $\frac{1}{2}$

For questions 7–10, find the scale factor.

7. $A: (-3, 2)$ $A': (-10.5, 7)$
   $B: (1, 2)$ $B': (3.5, 7)$
   $C: (1, -3)$ $C': (3.5, -10.5)$
   $D: (-3, -3)$ $D': (-10.5, -10.5)$

8. $A: (-6, 9)$ $A': (-2, 3)$
   $B: (3, 12)$ $B': (1, 4)$
   $C: (6, 3)$ $C': (2, 1)$
   $D: (-9, 0)$ $D': (-3, 0)$

9. $A: (0, -3)$ $A': (0, -2)$
   $B: (6, 0)$ $B': (4, 0)$
   $C: (0, 3)$ $C': (0, 2)$

10. $A: (-2, 6)$ $A': (-10, 30)$
    $B: (2, 6)$ $B': (10, 30)$
    $C: (3, 3)$ $C': (15, 15)$
    $D: (2, 0)$ $D': (10, 0)$
    $E: (-2, 0)$ $E': (-10, 0)$
    $F: (-3, 3)$ $F': (-15, 15)$

For questions 11 and 12, determine whether or not $A'B'C'D'$ is a dilation of $ABCD$.

11. $A: (-2, 5)$ $A': (-1, 2)$
    $B: (8, 8)$ $B': (4, 4)$
    $C: (12, 0)$ $C': (6, 0)$
    $D: (2, -6)$ $D': (1, -3)$

12. $A: (0, 8)$ $A': (-2, 6)$
    $B: (5, 8)$ $B': (3, 6)$
    $C: (5, -3)$ $C': (3, -1)$
    $D: (0, -3)$ $D': (-2, -1)$

4.11 Using Dilations to Prove Similarity (DOK 2)

A **dilation** is a transformation whose pre-image and image are similar. A dilation is therefore a similarity transformation.

The following is a dilation.   The following is **not** a dilation.

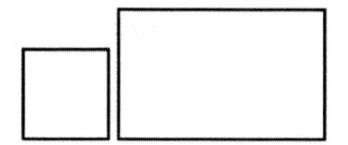

Another way to verify similarity, particularly among triangles, is to measure their angles. If their angles are equal and their sides are proportional, then the shapes are dilated and similar.

Example 12: A dilation takes a line not passing through the center of the dilation to a parallel line, and leaves a line passing through the center unchanged. Look at $\triangle ABC$ to the right. $\triangle ABC$ has been dilated by a factor of $\frac{1}{2}$ about the origin to form $\triangle ADE$. After the dilation, $\overline{BC}$ is parallel to $\overline{DE}$, but $\overline{AD}$ and $\overline{AB}$ are not parallel because they are collinear. $\overline{AE}$ and $\overline{AC}$ are also not parallel because they are collinear.

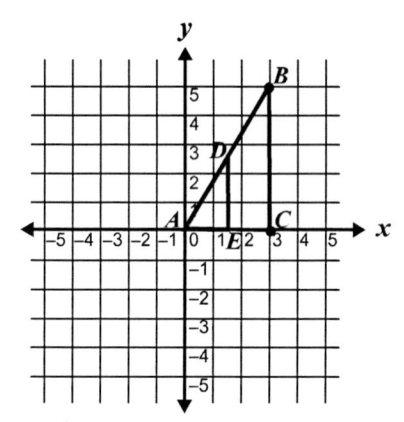

Solve the following problems about dilations.

1. Measure the angles to verify whether the following images are similar.

2. Find the missing side length.
   15
   20
   x
   12

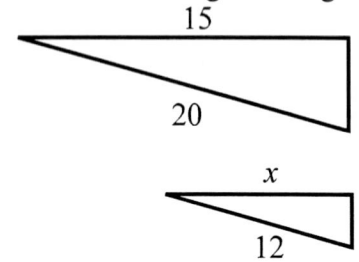

3. Determine whether the following images are similar.

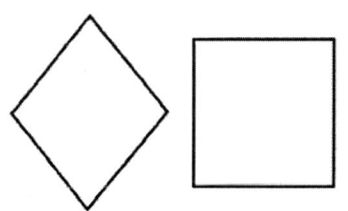

4. Find the missing side length.
   12 cm 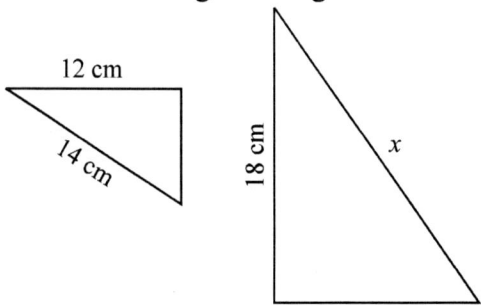
   14 cm
   18 cm
   x

Chapter 4 Review

(DOK 2)

1. Draw the reflection of image *ABCD* over the *y*-axis. Label the points *A′*, *B′*, *C′*, and *D′*. List the coordinates of these points below.

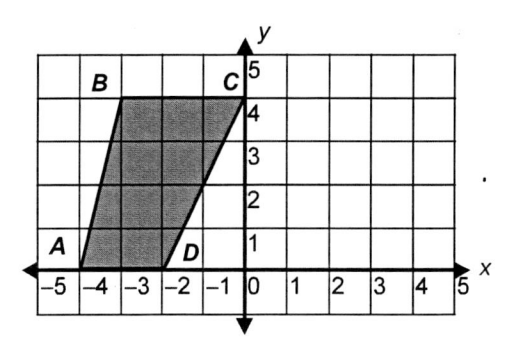

2. *A′* = _____
3. *B′* = _____
4. *C′* = _____
5. *D′* = _____

6. Rotate the figure below 180° about the origin. Label the points *A′*, *B′*, *C′*, *D′*, *E′*, and *F′*. List the coordinates of these points below.

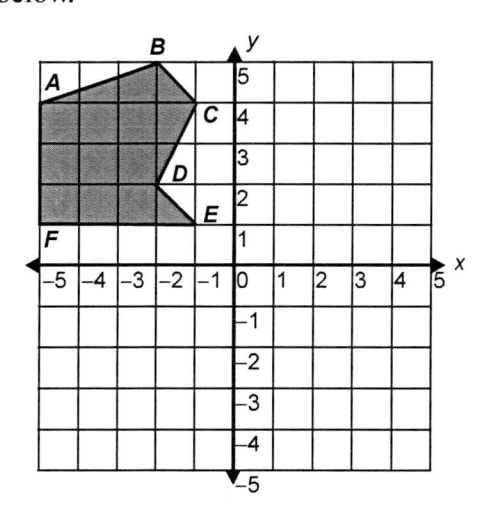

7. *A′* = _____
8. *B′* = _____
9. *C′* = _____
10. *D′* = _____
11. *E′* = _____
12. *F′* = _____

13. Use the translation described by the arrow to translate the polygon below. Label the points *P′*, *Q′*, *R′*, *S′*, *T′*, and *U′*. List the coordinate of each.

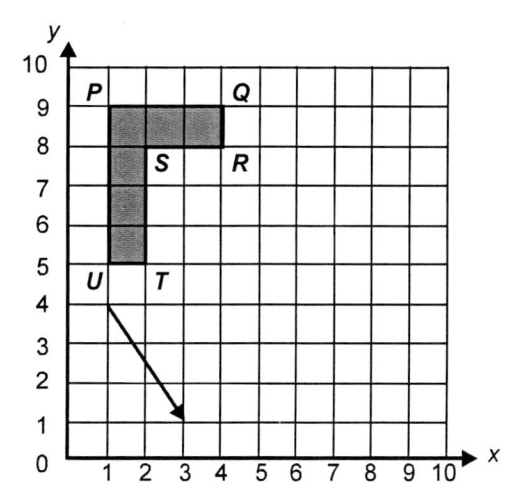

14. *P′* = _____
15. *Q′* = _____
16. *R′* = _____
17. *S′* = _____
18. *T′* = _____
19. *U′* = _____

20. A point at $(-3, 2)$ is moved to $(0, 0)$. If a point at $(1, 1)$ is moved in the same way, what will its new coordinates be?

21. What is the rule for the transformation formed by a reflection over the *y*-axis?

22. What is the rule for the transformation formed by a dilation by a scale factor of 3, then a rotation 90° clockwise around the origin?

23. What is the rule for the transformation formed by a translation 2 units up and 7 units to the left, then a reflection over the line $y = x$?

Chapter 4   Transformations

Chapter 4 Test

1 If the figure below were reflected across the y-axis, what would be the coordinates of point C?

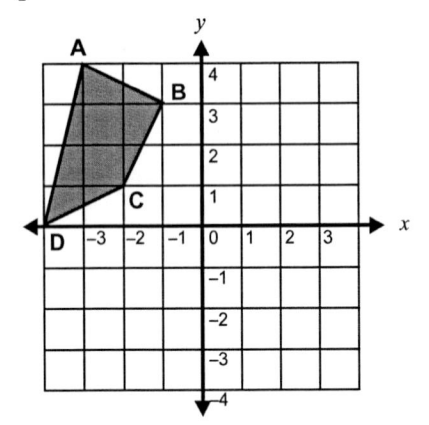

A $(2,-1)$

B $(2,1)$

C $(-2,-1)$

D $(-2,1)$

(DOK 2)

2 If the figure below is translated in the direction described by the arrow, what will be the new coordinates of point D after the transformation?

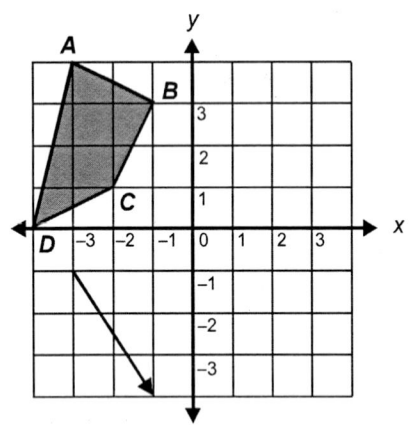

A $(-2,-4)$

B $(-1,-3)$

C $(-1,-4)$

D $(-2,-3)$

(DOK 2)

3 Figure 1 goes through a transformation to form Figure 2. Which of the following descriptions fits the transformation(s) shown below?

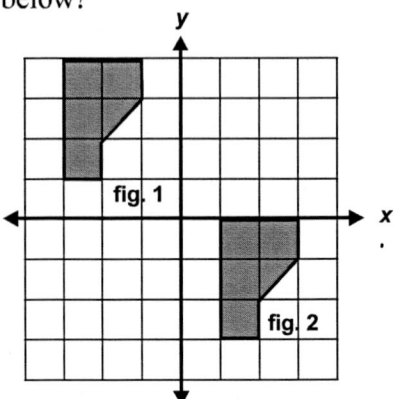

A reflection across the x-axis

B reflection across the origin

C 180° clockwise rotation around the origin

D translation right 4 units and down 4 units

(DOK 2)

4 Sammy plots the point $(-4,3)$ on a coordinate grid. He reflects this point over the y-axis, then over the x-axis. What are the coordinates of the new reflected point?

A $(-4,-3)$
B $(-4,3)$
C $(4,-3)$
D $(4,3)$

(DOK 2)

5 What is the rule for the transformation formed by a translation 4 units down, then a rotation 90° clockwise around the origin?

A $(x',y')=(x,-y+4)$

B $(x',y')=(x,y-4)$

C $(x',y')=(y-4,-x)$

D $(x',y')=(-y+4,-x)$

(DOK 2)

6 Figure 1 goes through a transformation to form figure 2. Which of the following descriptions fits the transformation shown?

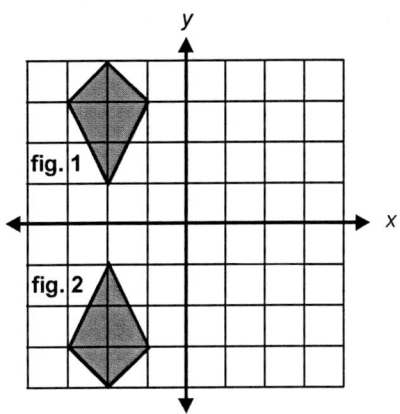

A reflection across the x-axis

B reflection across the y-axis

C 270° clockwise rotation around the origin

D translation down 2 units (DOK 2)

7 Figure 1 goes through a transformation to form figure 2. Which of the following descriptions fits the transformation shown?

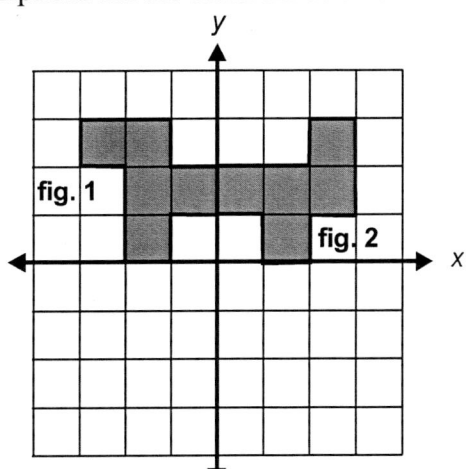

A reflection across the y-axis

B 270° clockwise rotation around the origin

C translation right 3 units

D 90° clockwise rotation around the origin

(DOK 2)

8 What is the rule for the transformation formed by a dilation by a scale factor of 5?

A $(x', y') = \left(\frac{1}{5}x, \frac{1}{5}y\right)$

B $(x', y') = (5x, 5y)$

C $(x', y') = (5x, y)$

D $(x', y') = (x, 5y)$ (DOK 2)

9 Describe the transformation needed for the pre-image to be transformed into the post-image, and determine if the two figures are still congruent or not.

Pre-Image   Post-Image

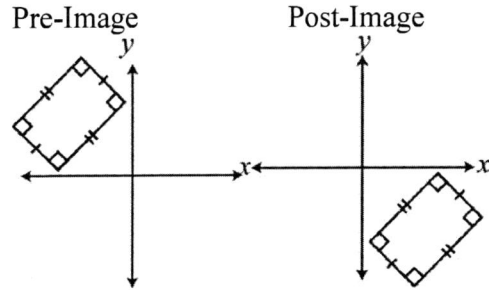

A Yes, the images are still congruent because all the side lengths and angle measurements are the same. The image has been rotated 180° and reflected across the y-axis.

B Yes, the images are still congruent because all the side lengths and angle measurements are the same. The image has been reflected across the x-axis and then reflected across the y-axis.

C Yes, the images are still congruent because all the side lengths and angle measurements are the same. The image has been rotated 270° and then reflected across a 45° line.

D No, the images are not congruent because the side lengths and angle measurements have changed due to transformations. The image has been reflected across the x-axis and then reflected across the y-axis.

(DOK 3)

Chapter 5
Triangle Trigonometry

This chapter covers the following CCGPS standard(s):

	Content Standards
Similarity, Right Triangles, and Trigonometry	G.SRT.6, G.SRT.7, G.SRT.8

5.1 Pythagorean Theorem (DOK 2 & 3)

Pythagoras was a Greek mathematician and philosopher who lived around 600 B.C. He started a math club among Greek aristocrats called the Pythagoreans. Pythagoras formulated the **Pythagorean theorem,** which states that in a **right triangle**, the sum of the squares of the legs of the triangle is equal to the square of the hypotenuse. Most often, this formula is written as $a^2 + b^2 = c^2$, where c is the length of the hypotenuse and a and b are the lengths of the legs. **This relationship is only true for right triangles.**

Example 1: Find the length of side c, the hypotenuse.

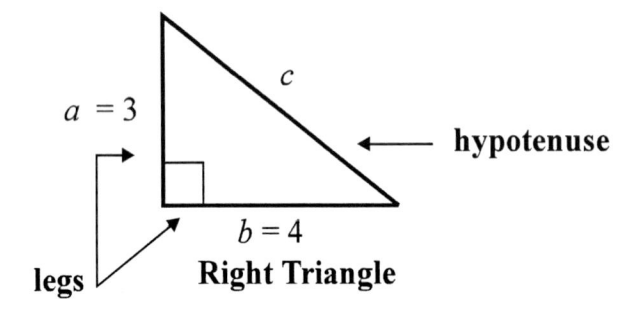

Formula:
$$a^2 + b^2 = c^2$$
$$3^2 + 4^2 = c^2$$
$$9 + 16 = c^2$$
$$25 = c^2$$
$$\sqrt{25} = \sqrt{c^2}$$
$$5 = c$$

5.1 Pythagorean Theorem (DOK 2 & 3)

Proof of the Pythagorean Theorem Using Similar Triangles

Use the figure below to prove the Pythagorean theorem. It is given that the three triangles shown (ABC, ACD, and CBD) are similar triangles.

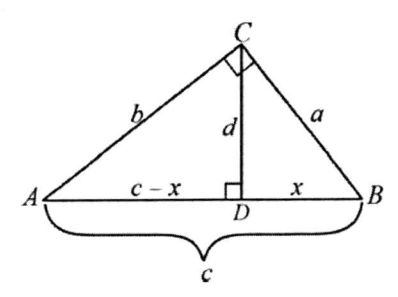

Using the fact that corresponding parts of similar triangles are proportional, we can prove that $a^2 + b^2 = c^2$.

$$\frac{\text{hypotenuse of } \triangle CBD}{\text{base of } \triangle CBD} = \frac{\text{hypotenuse of } \triangle ABC}{\text{base of } \triangle ABC} \qquad \frac{a}{x} = \frac{c}{a}$$

Cross multiply to eliminate the denominators. $\qquad a^2 = cx$

$$\frac{\text{hypotenuse of } \triangle ACD}{\text{height of } \triangle ACD} = \frac{\text{hypotenuse of } \triangle ABC}{\text{height of } \triangle ABC} \qquad \frac{b}{c-x} = \frac{c}{b}$$

Cross multiply to eliminate the denominators. $\qquad b^2 = c^2 - cx$

Substitute a^2 for cx, since $a^2 = cx$. $\qquad b^2 = c^2 - a^2$

Add a^2 to both sides of the equation. $\qquad b^2 + a^2 = c^2 - a^2 + a^2$

Therefore, we have proven that $\qquad a^2 + b^2 = c^2$.

Find the hypotenuse of the following triangles. Use the Pythagorean theorem, $a^2 + b^2 = c^2$, and round the answers to two decimal places.

1. 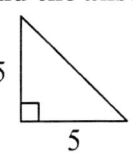 (legs 5 and 5) $c = \underline{\qquad}$

2. 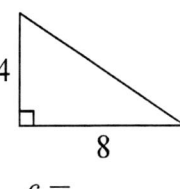 (legs 4 and 8) $c = \underline{\qquad}$

3. 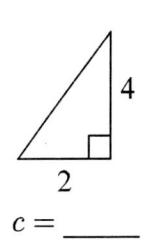 (legs 4 and 2) $c = \underline{\qquad}$

4. 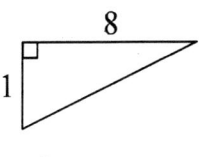 (legs 8 and 1) $c = \underline{\qquad}$

5. 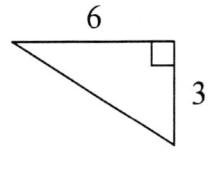 (legs 6 and 3) $c = \underline{\qquad}$

6. 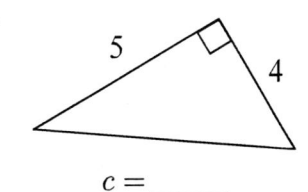 (legs 5 and 4) $c = \underline{\qquad}$

7. 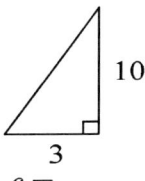 (legs 10 and 3) $c = \underline{\qquad}$

8. 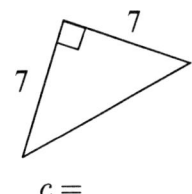 (legs 7 and 7) $c = \underline{\qquad}$

9. 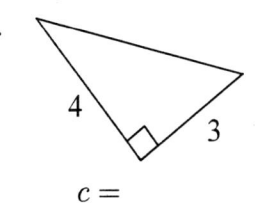 (legs 4 and 3) $c = \underline{\qquad}$

Chapter 5   Triangle Trigonometry

5.2   Finding the Missing Leg of a Right Triangle (DOK 2)

In a right triangle, if the length of the hypotenuse and one of the legs is known, use the Pythagorean theorem to solve for the unknown side.

Example 2:   Find the measure of b.

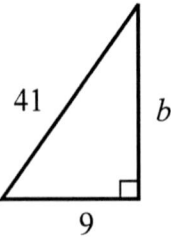

In the Pythagorean theorem, $a^2 + b^2 = c^2$, a and b are the legs and c is always the hypotenuse.

$9^2 + b^2 = 41^2$
$81 + b^2 = 1681$
$b^2 = 1681 - 81$
$b^2 = 1600$
$\sqrt{b^2} = \sqrt{1600}$
$b = 40$

Practice finding the measure of the missing leg in each right triangle below. Simplify square roots.

1.

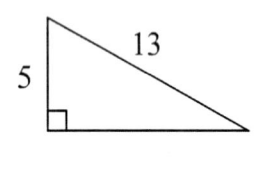

2.

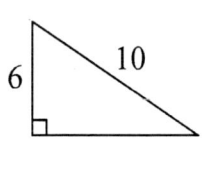

3.

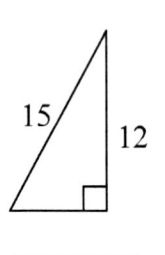

4.

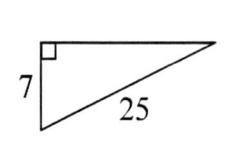

5.

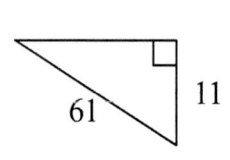

6.

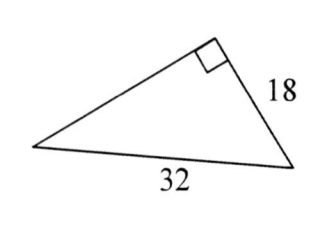

7.

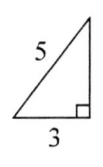

8.

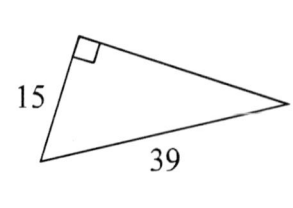

9.

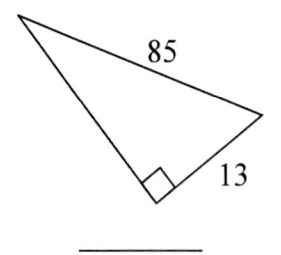

5.3 Applications of the Pythagorean Theorem (DOK 3)

The Pythagorean theorem can be used to determine the distance between two points. Recall that the formula is $a^2 + b^2 = c^2$.

Example 3: Find the distance between point B and point A given that the measurement of each square is 1 inch long and 1 inch wide.

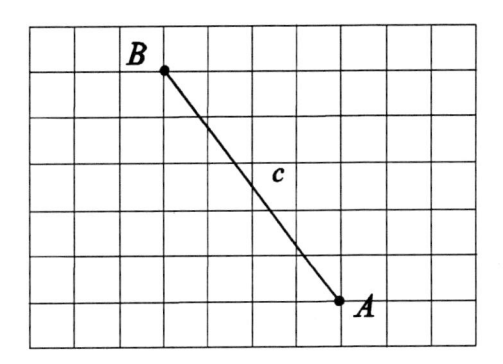

Step 1: Draw a straight line between the two points. Call this side c.

Step 2: Draw two more lines, one from point B and one from point A. These lines will make a 90° angle. The two new lines will be labeled a and b. Now use the Pythagorean theorem to find the distance from Point B to Point A.

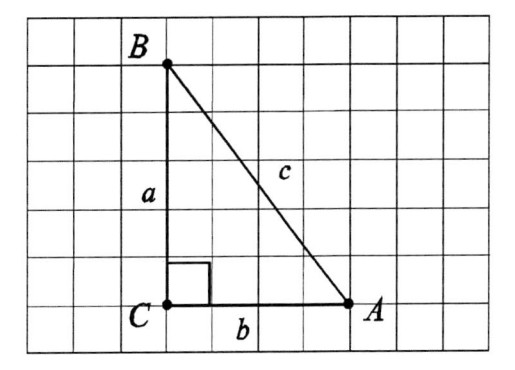

Step 3: Find the length of a and c by counting the number of squares. The length of $a = 5$ units and $b = 4$ units. Now, substitute these values found into the Pythagorean Theorem.

$$\begin{aligned} a^2 + b^2 &= c^2 \\ 5^2 + 4^2 &= c^2 \\ 25 + 16 &= c^2 \\ 41 &= c^2 \\ \sqrt{41} &= \sqrt{c^2} \\ \sqrt{41} &= c \text{ or } c \approx 6.4 \end{aligned}$$

Copyright © American Book Company

85

Chapter 5   Triangle Trigonometry

Use the Pythagorean theorem to find the distances. Round your answers to two decimal points.

Below is a diagram of the mall. Use the grid to help answer questions 1 and 2. Each square is 25 feet × 25 feet.

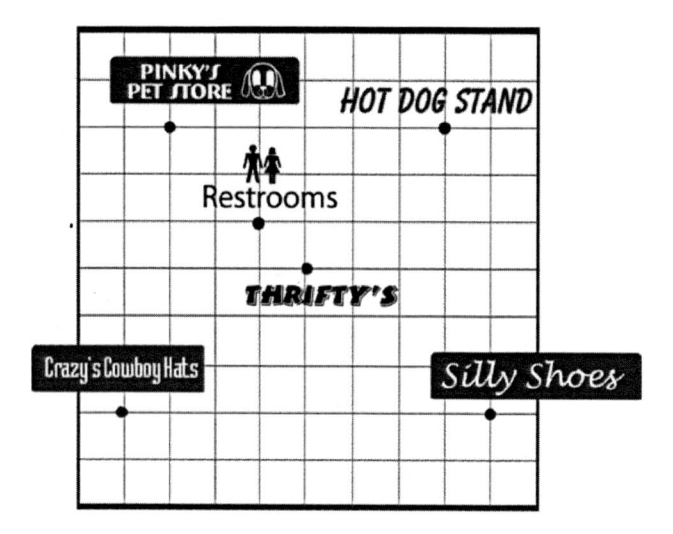

1. Marty walks from Pinky's Pet Store to the restroom to wash his hands. How far did he walk?

2. Betty needs to meet her friend at Silly Shoes, but she wants to get a hot dog first. If Betty is at Thrifty's, how far will she walk to meet her friend?

Below is a diagram of a football field. Use the grid on the football field to help find the answers to questions 3 and 4. Each square is 10 yards × 10 yards.

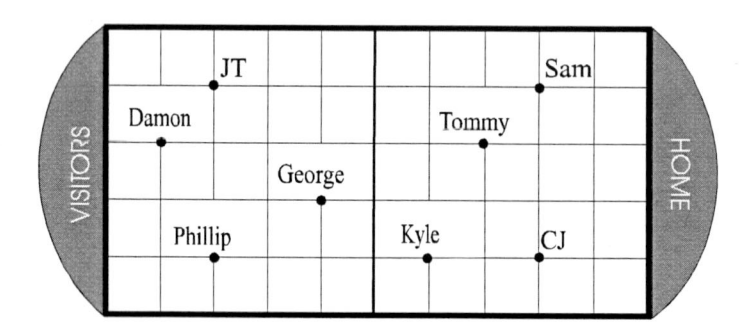

3. George must throw the football to a teammate before he is tackled. If CJ is the only person open, how far must George be able to throw the ball?

4. Phillip wants to throw to Damon for a touchdown. How far must he throw to reach Damon?

5.4 Special Right Triangles (DOK 2)

Two types of right triangles are **special right triangles** if they have fixed ratios among their sides.

45-45-90 Triangles

In a 45-45-90 triangle, the two sides opposite the 45° angles will always be equal. The length of the hypotenuse is $\sqrt{2}$ times the length of one of the sides opposite a 45° angle.

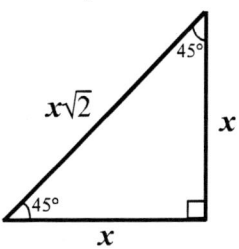

Example 4: What are the lengths of sides a and b?

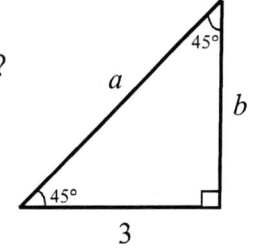

Step 1: The two sides opposite the 45° angles are equal. Therefore, side $b = 3$.

Step 2: The hypotenuse is $\sqrt{2}$ times the length of a side opposite a 45° angle. Therefore, $a = 3 \times \sqrt{2}$.
Simplify: $a = 3\sqrt{2}$.

30-60-90 Triangles

In a 30-60-90 triangle, the side opposite the 30° angle is the shortest leg. The side opposite the 60° angle is $\sqrt{3}$ times as long as the shortest leg, and the hypotenuse is twice as long as the shortest leg.

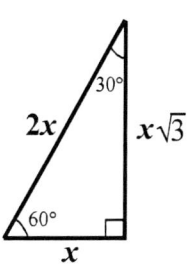

Example 5: What are the lengths of sides a and b?

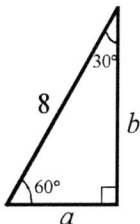

Step 1: The hypotenuse is 2 times the side opposite the 30° angle.
Write the above statement using algebra and then solve.
$8 = 2a$
$\dfrac{8}{2} = \dfrac{2a}{2}$
$4 = a$

Step 2: Now that it is known that the shortest leg has a length of 4, the side opposite the 60° angle can be calculated.
$b = a \times \sqrt{3} = 4 \times \sqrt{3}$
$b = 4\sqrt{3}$

Chapter 5 Triangle Trigonometry

Find the missing side of each of the special right triangles. Simplify your answers.

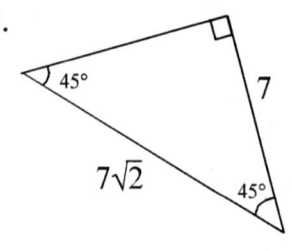

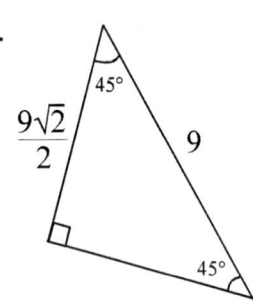

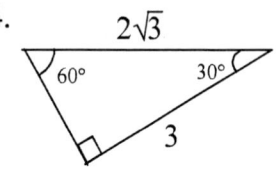

1. sides: 4, $4\sqrt{3}$, angles 60°, 30°

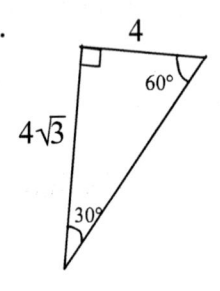

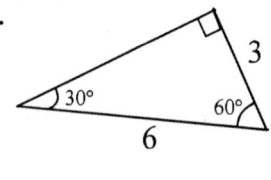

2. sides: $\frac{9\sqrt{2}}{2}$, 9, angles 45°, 45°

3. sides: 7, $7\sqrt{2}$, angles 45°, 45°

4. sides: $2\sqrt{3}$, 3, angles 60°, 30°

5. sides: 3, 6, angles 30°, 60°

6. sides: $5\sqrt{2}$, $5\sqrt{2}$, angles 45°, 45°

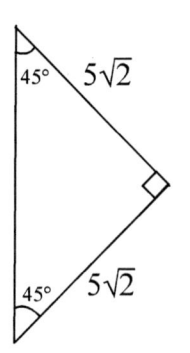

Find the lengths of sides *a* and *b* in each of the special right triangles.

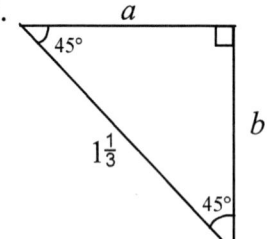

7. *a*, *b*, $\frac{2}{3}$, angles 60°, 30°

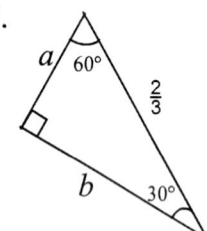

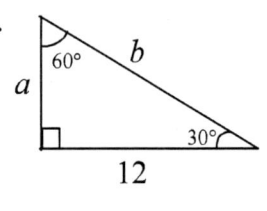

8. *a*, *b*, $\frac{4}{5}$, angles 45°, 45°

9. *a*, *b*, $1\frac{1}{3}$, angles 45°, 45°

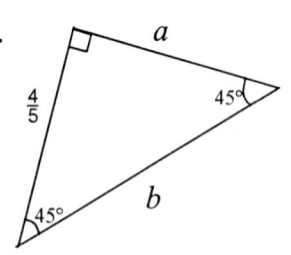

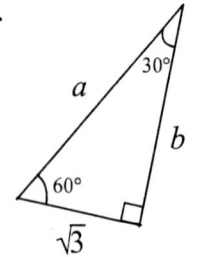

10. *a*, *b*, $\sqrt{3}$, angles 30°, 60°

11. *a*, *b*, 12, angles 60°, 30°

12. *a*, *b*, $8\sqrt{2}$, angles 45°, 45°

88   Copyright © American Book Company

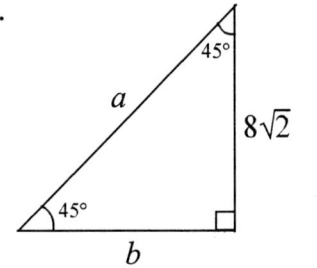

5.5 Introduction to Trigonometric Ratios (DOK 2 & 3)

Trigonometry is a mathematical topic that applies the relationships between sides and angles in right triangles. Recall that a right triangle has one 90° angle and two acute angles. Consider the right triangle shown below. Note that the angles are labeled with capital letters. The sides are labeled with lowercase letters that correspond to the angles opposite them.

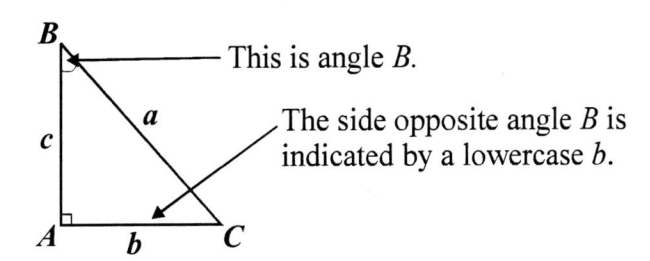

- This is angle B.
- The side opposite angle B is indicated by a lowercase b.

Trigonometric ratios are ratios of the measures of two sides of a right triangle and are related to the acute angles of a right triangle, not the right angle. The value of a trigonometric ratio is dependent on the size of the acute angle and the ratio of the lengths of the sides of the triangle.

We will consider the three basic trigonometric ratios in this section: **sine, cosine, and tangent**. Definitions and descriptions of the sine, cosine, and tangent functions are presented below.

Trigonometric Functions

$$\sin A = \frac{\text{length of side opposite } A}{\text{length of hypotenuse}} = \frac{\text{opp.}}{\text{hyp.}} = \frac{a}{c}$$

$$\cos A = \frac{\text{length of side adjacent to } A}{\text{length of hypotenuse}} = \frac{\text{adj.}}{\text{hyp.}} = \frac{b}{c}$$

$$\tan A = \frac{\text{length of side opposite } A}{\text{length of side adjacent to } A} = \frac{\text{opp.}}{\text{adj.}} = \frac{a}{b}$$

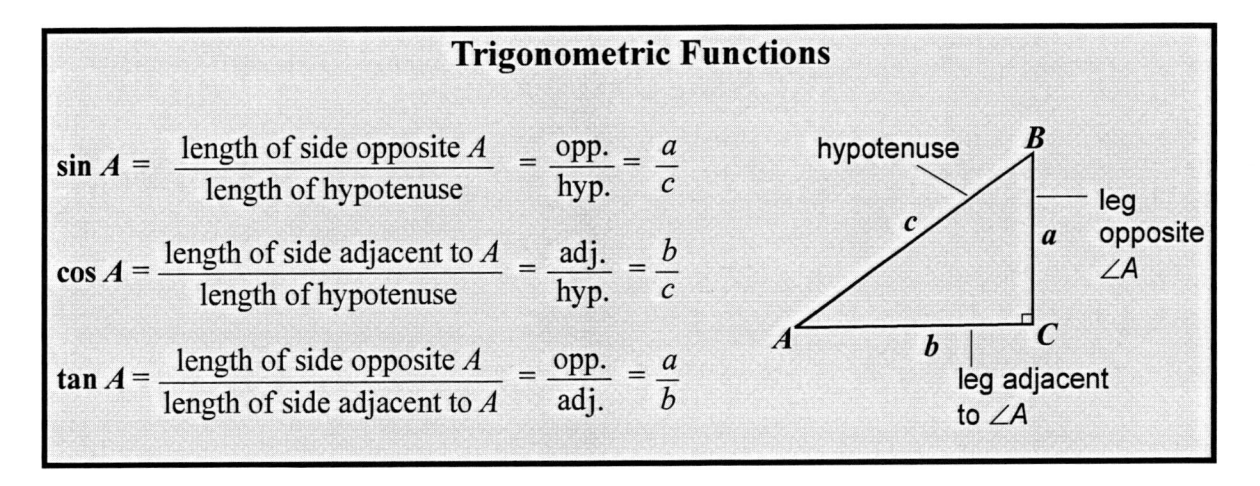

Example 6: For right triangle ABC, find $\sin A, \cos A, \tan A, \sin C, \cos C,$ and $\tan C$.

$$\sin A = \frac{\text{opp.}}{\text{hyp.}} = \frac{3}{5} = 0.6 \qquad \sin C = \frac{\text{opp.}}{\text{hyp.}} = \frac{4}{5} = 0.8$$

$$\cos A = \frac{\text{adj.}}{\text{hyp.}} = \frac{4}{5} = 0.8 \qquad \cos C = \frac{\text{adj.}}{\text{hyp.}} = \frac{3}{5} = 0.6$$

$$\tan A = \frac{\text{opp.}}{\text{adj.}} = \frac{3}{4} = 0.75 \qquad \tan C = \frac{\text{opp.}}{\text{adj.}} = \frac{4}{3} = 1.\overline{3}$$

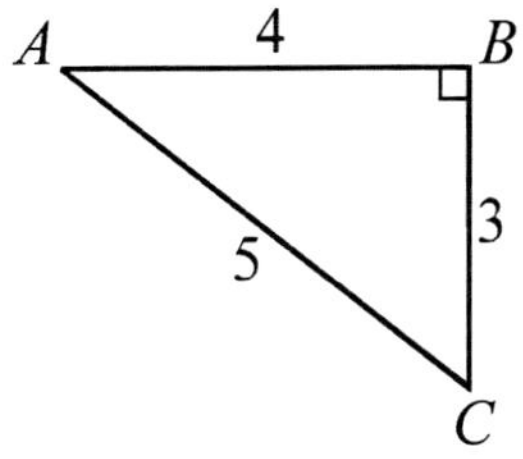

Chapter 5   Triangle Trigonometry

Find $\sin A$, $\cos A$, $\tan A$, $\sin B$, $\cos B$, **and** $\tan B$ **in each of the following right triangles. Express answers as fractions and as decimals rounded to three decimal places.**

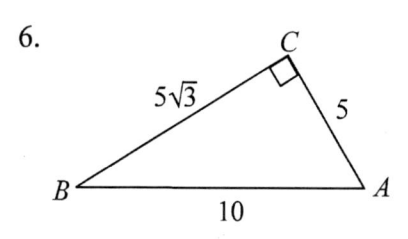

1. Triangle with B and C at top (right angle at C), A at bottom. $BC = 12$, $CA = 9$, $BA = 15$.

2. Triangle with C and A at top (right angle at C), B at bottom. $CA = 21$, $CB = 20$, $AB = 29$.

3. Triangle with B at top, C at right (right angle at C), A at bottom left. $BC = 9$, $BA = 41$, $CA = 40$.

4. Triangle with B at top right, C at bottom right (right angle at C), A at bottom left. $BA = 17$, $BC = 8$, $AC = 15$.

5. Triangle with A at top, C at left (right angle at C), B at bottom right. $AC = 1$, $AB = \sqrt{10}$, $CB = 3$.

6. Triangle with C at top (right angle at C), B at bottom left, A at bottom right. $BC = 5\sqrt{3}$, $CA = 5$, $BA = 10$.

Once the values of the trigonometric ratios are found, then the measures of the angles within the triangle can be found using the arcsine and arccosine. The arcsine and arccosine can also be written as $\sin^{-1}$ and $\cos^{-1}$. The arc functions' identities can be defined as:

$$\arcsin(\sin(A)) = A$$
$$\arccos(\cos(A)) = A$$
$$\arctan(\tan(A)) = A$$

Example 7:   For right triangle ABC, where $\sin A = 0.6$, find the measures of angles A and C. Round to the nearest whole number.

Step 1:   Using your calculator, arcsine is $\sin^{-1}$.**
$\sin A = 0.6$
$\sin^{-1}(\sin A) = \sin^{-1}(0.6)$ Take the arcsine of both sides.
$A = 37°$

Step 2:   Since all the angles in a triangle add up to $180°$, and $A = 37°$ and $B = 90°$, then
$A + B + C = 180°$.
$37° + 90° + C = 180°$
$C = 180° - 37° - 90° = 53°$
Therefore, $A = 37°$, $B = 90°$, and $C = 53°$.

**   To find the arcsine, arccosine, or arctangent using a scientific calculator, first enter the trig ratio without rounding, such as $\frac{3}{5}$ from the example above. Then, type 2nd SIN, 2nd COS, or 2nd TAN. If there isn't a 2nd button, you will have to press the inverse button and then the SIN, COS, or TAN button. The inverse button is usually abbreviated INV. When finding an angle using a scientific calculator, you must always remember to be in degree mode. To check this, look at the top of the screen, and it will show DEG in small print. If DEG is not shown, then press DRG or the degree button until DEG shows on the top of the screen. In other calculators (like graphing calculators), you must hit the function keys first before entering the ratio to find the angle measure. For example, you may need to type 2nd SIN in the calculator, then $\frac{3}{5}$ to find the measure of A in $\sin\ A = 0.6$.

5.5 Introduction to Trigonometric Ratios (DOK 2 & 3)

Find the measures of the angles given the trigonometric function. Round your answers to the nearest degree.

1. $\sin A = 0.4$
2. $\tan x = 1$
3. $\sin b = 0.7$
4. $\cos C = \frac{\sqrt{2}}{2}$
5. $\tan A = -1.5$
6. $\cos y = -1$
7. $\sin B = -0.6$
8. $\cos A = 0$
9. $\tan z = 2.6$
10. $\tan c = 50$
11. $\sin x = \frac{\sqrt{2}}{2}$
12. $\cos x = 0.1$
13. $\tan y = 0$
14. $\cos a = -0.4$
15. $\sin C = 1$

Example 8: Find the values of the sine, cosine, and tangent functions of both acute angles in the right triangle ABC shown below.

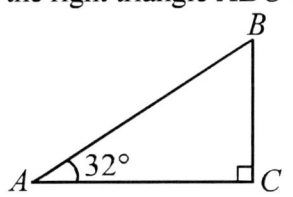

Step 1: Find the third angle.
$A + B + C = 180°$
$32° + B + 90° = 180°$
$B + 90° = 180° - 32° - 90° = 58°$

Step 2: Plug the angle values into $\sin A$, $\cos A$, $\tan A$, $\sin B$, $\cos B$, and $\tan B$.
$\sin A = \sin 32° = 0.5299$ $\sin B = \sin 58° = 0.8480$
$\cos A = \cos 32° = 0.8480$ $\cos B = \cos 58° = 0.5299$
$\tan A = \tan 32° = 0.6249$ $\tan B = \tan 58° = 1.600$

Find $\sin A$, $\cos A$, $\tan A$, $\sin B$, $\cos B$, and $\tan B$ in each of the following right triangles. Express answers as decimals rounded to three decimal places.

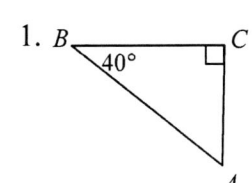

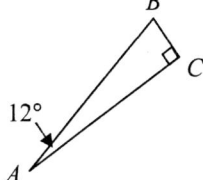

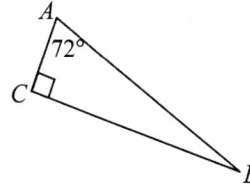

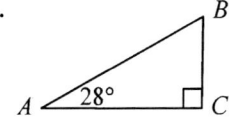

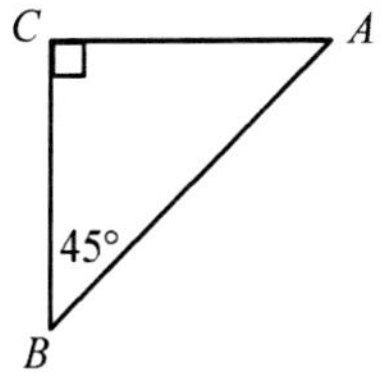

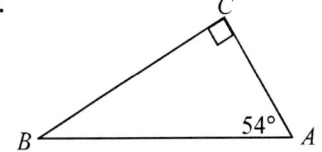

Chapter 5   Triangle Trigonometry

When given one acute angle and one side of a right triangle, the other angle and two sides can be found using trigonometric functions.

Example 9:   Find the third angle and the other two sides of the triangle.

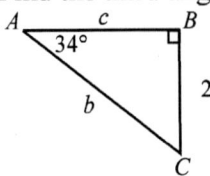

Step 1:   Find the third angle. Since all the angles in a triangle add up to 180°, and $A = 34°$ and $B = 90°$, then
$A + B + C = 180°$
$34° + 90° + C = 180°$
$C = 180° - 34° - 90° = 56°$

Step 2:   Find the missing sides. This can be done several different ways using sine, cosine, or tangent. We are going to use sine to find b and tangent to find c.

$$\sin A = \frac{\text{opp.}}{\text{hyp.}} \qquad \tan A = \frac{\text{opp.}}{\text{adj.}}$$

$$\sin 34° = \tfrac{2}{b} \qquad \tan 34° = \tfrac{2}{c}$$

$$0.5592 = \tfrac{2}{b} \qquad 0.6745 = \tfrac{2}{c}$$

$$0.5592b = 2 \qquad 0.6745c = 2$$

$$\frac{0.5592b}{0.5592} = \frac{2}{0.5592} \qquad \frac{0.6745c}{0.6745} = \frac{2}{0.6745}$$

$$b = 3.58 \qquad c = 2.97$$

$C = 56°, b = 3.58$, and $c = 2.97$

Note: After you have calculated the second side using one of the trigonometric ratios, you can use the Pythagorean theorem to find the third side.
$3.58^2 = 2^2 + c^2 \longrightarrow c^2 = 3.58^2 - 2^2 \longrightarrow c^2 = 8.8164 \longrightarrow c = 2.97$

Find the missing sides and angles using the information given.

1.

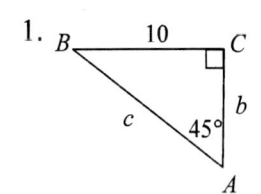

2.

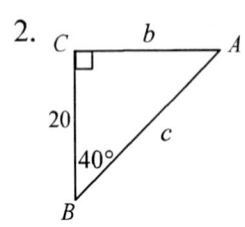

3.

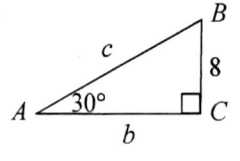

4.

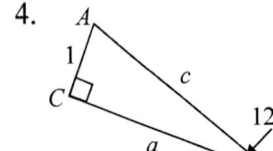

5.

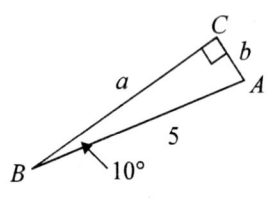

6.

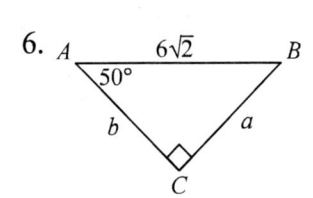

92   Copyright © American Book Company

5.5 Introduction to Trigonometric Ratios (DOK 2 & 3)

Use the pictures to help solve the problems.

1. An F-22 is flying over two control towers. There is a point above the two towers where the fighter pilot can get a clean signal to both the towers. If he is 1,120 feet from tower one and is making a 59° angle with the two towers, find the distance, x, the F-22 is from the second tower and find the distance, y, between the two towers.

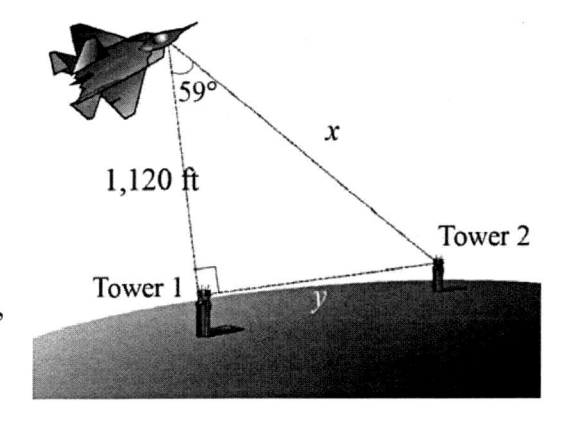

2. Sandra is trying to use her old cell phone to call her best friend. She can be no more than 3 miles from a tower in order to get a signal for her phone. If the telephone tower is 252 feet tall, find the angle of elevation, m, when Sandra's phone is the maximum 3 miles from the tower.
HINT: Convert 3 miles to feet first.
1 mile = 5,280 feet

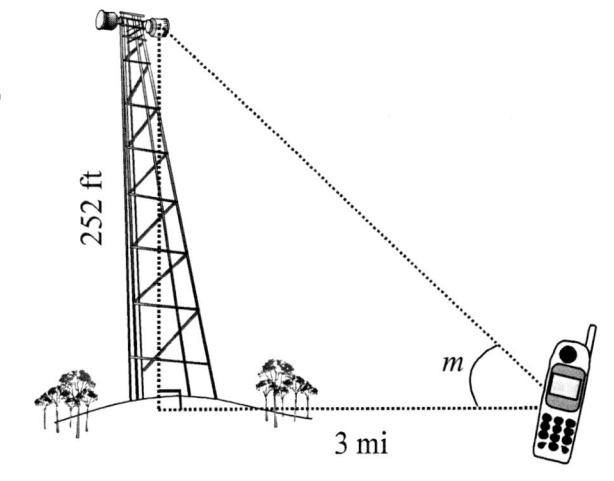

3. Sir Stephen is returning from fighting a war. His first concern on his homecoming journey is to see if his family's banner still flies above their castle. If the flag rises 95 feet above the ground, and he emerges from the forest 185 feet from the tower, at what angle, m, is his line of sight to the banner?

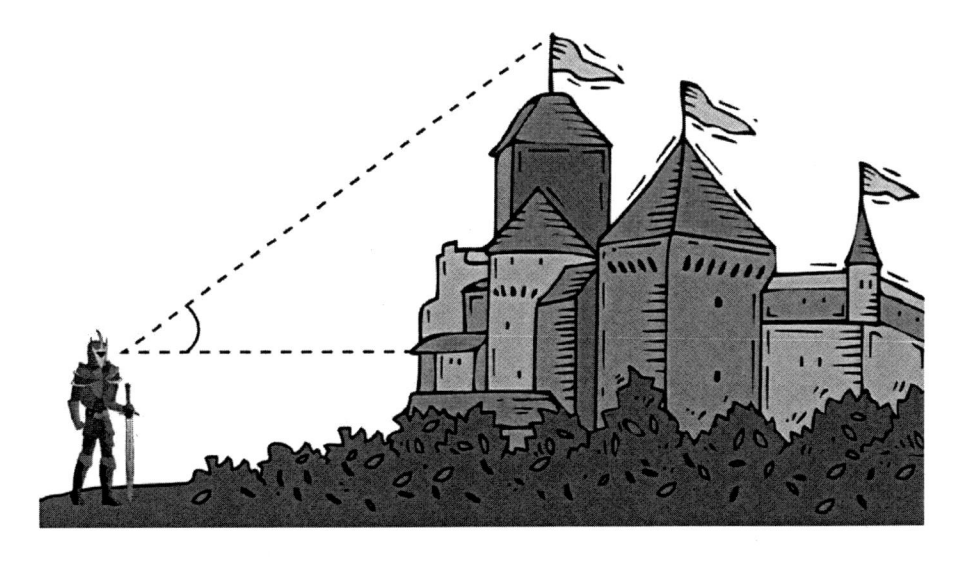

Chapter 5   Triangle Trigonometry

5.6   More Similar Triangles (DOK 3)

Trigonometric ratios can be useful when finding missing information in similar triangles.

Example 10: $\triangle ABC$ is similar to $\triangle DEF$. Use trig ratios to find the measure of $\angle F$.

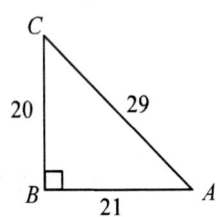

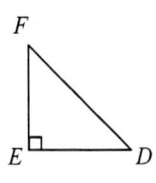

Step 1:   Since $\triangle ABC \sim \triangle DEF$, we know that the measures of the angles in $\triangle ABC$ are equal to the measures of the angles in $\triangle DEF$. Therefore, $\angle A \cong \angle D$, $\angle B \cong \angle E$, and $\angle C \cong \angle F$. In order to find the measure of $\angle F$, we can just find the measure of $\angle C$.

Step 2:   To find the measure of $\angle C$, use a trig ratio. We will find sine of $\angle C$.
$$\sin C = \frac{\text{opp.}}{\text{hyp.}} = \frac{21}{29}$$

Step 3:   Take the arcsine of both sides to find C. (Make sure your calculator is in DEGREE mode.)
$$\sin^{-1}(\sin C) = \sin^{-1}\left(\frac{21}{29}\right)$$
$C \approx 46.4°$ Since $\angle C \cong \angle F$, the measure of $\angle F \approx 46.4°$.

Use trigonometric ratios to solve the similar triangle problems.

1. $\triangle ABC \sim \triangle DEF$. Find the measures of $\angle E$ and $\angle D$.

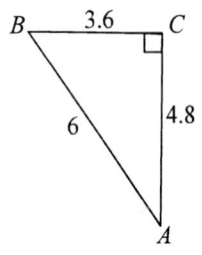

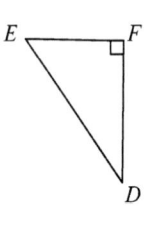

2. $\triangle ABC \sim \triangle ECD$. Find the measure of $\overline{CD}$.

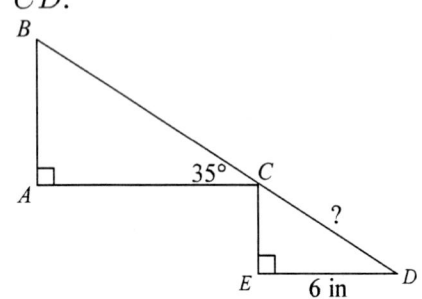

3. $\triangle ABC \sim \triangle DEF$. Find the measures of $\overline{FD}$ and $\overline{EF}$.

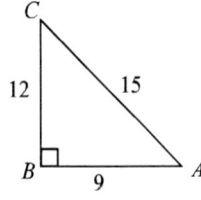

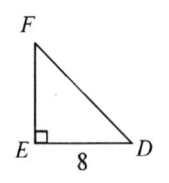

4. $\triangle ABC \sim \triangle ADE$. Find the measure of $\angle E$.

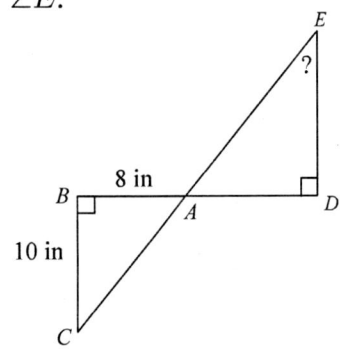

5.7 More Trigonometric Ratios (DOK 2)

Two angles are **complementary** if the sum of the measures of the angles is 90°.

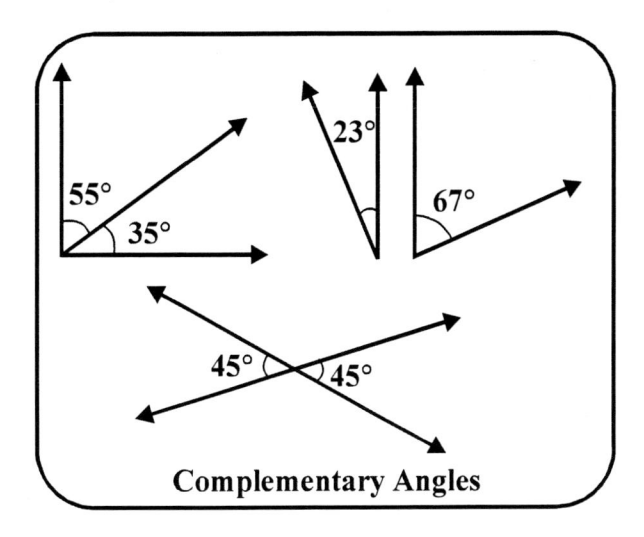

Complementary Angles

In a right triangle, one angle equals 90°. The other two angles add together to equal 90°. These two angles in the right triangle are complementary angles. In the triangle below, angle B and angle C are complementary angles. $\angle B + \angle C = 90°$. This can also be applied to trigonometric ratios.

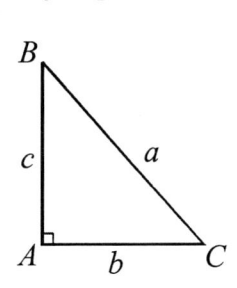

$$\sin B = \frac{b}{a} \qquad \sin C = \frac{c}{a} = \cos B$$

$$\cos B = \frac{c}{a} \qquad \cos C = \frac{b}{a} = \sin B$$

$$\tan B = \frac{b}{c} \qquad \tan C = \frac{c}{b} = \frac{1}{\tan B}$$

Example 11: Find the measure of $\angle C$.

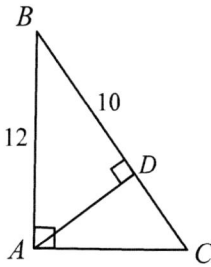

Step 1: $\triangle DAB$ is a right triangle. Using a trig ratio, we can find the measure of $\angle B$.

$$\cos B = \frac{\text{adjacent}}{\text{hypotenuse}} = \frac{10}{12}$$

$$\cos^{-1}(\cos B) = \cos^{-1}\left(\frac{10}{12}\right)$$

$$B = 33.6°$$

Chapter 5 Triangle Trigonometry

Step 2: △ABC is a also right triangle. Since ∠A = 90°, ∠B + ∠C = 90°.

∠B + ∠C = 90°   Substitute the measure of ∠B into the equation.
33.6° + ∠C = 90°   Solve for ∠C.
∠C = 90° − 33.6°
∠C = 56.4°

The measure of ∠C is 56.4°.

Use trigonometric ratios to solve the triangle problems.

1. Find the measure of ∠B.

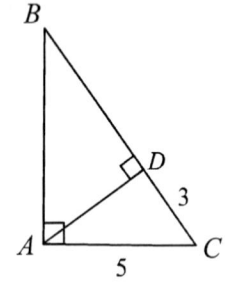

2. Find the measure of ∠B.

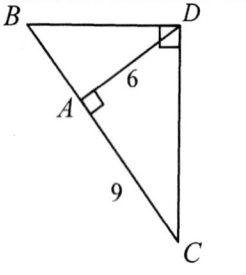

3. If $\sin C = \frac{4}{5}$, what is $\cos B$?

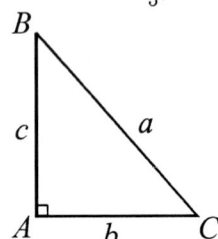

4. Find the measure of ∠A.

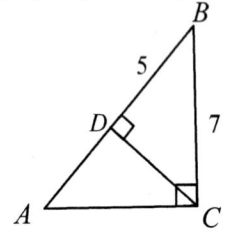

5. Find the measure of ∠ADB.

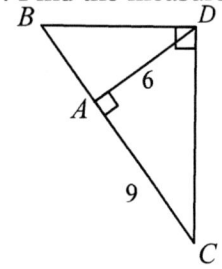

6. Find the measure of ∠DAB.

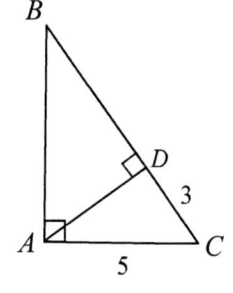

7. If $\cos C = \frac{10}{17}$, what is $\sin B$?

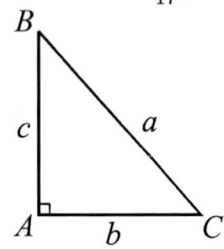

8. Find the measure of ∠ACD.

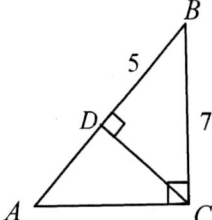

5.8 Trigonometric Ratios Applied to Other Shapes (DOK 2)

Example 12: Find the area of the trapezoid.

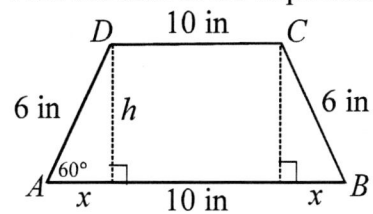

[Trapezoid ABCD with DC = 10 in (top), AD = BC = 6 in (legs), AB = 10 in between the two feet of the altitudes, with x on each side of AB, angle A = 60°, height h from D perpendicular to AB.]

Step 1: Define the area of a trapezoid. $A = \frac{1}{2}h(b_1 + b_2)$

We need to find the h and the value of x to find b_2 to solve for the area.

Step 2: Find the height. We are given the hypotenuse and measurement $\angle A$.

$$\sin(x) = \frac{\text{opp}}{\text{hyp}} \qquad \cos(x) = \frac{\text{adj}}{\text{hyp}} \qquad \tan(x) = \frac{\text{opp}}{\text{adj}}$$

We have the measures of the opposite leg and the hypotenuse, so we'll use the sine ratio to find h.

$$\sin(60) = \frac{h}{6} \rightarrow 0.866 = \frac{h}{6} \rightarrow 5.196 = h$$

Step 3: Find the value of x to determine b_2.

$$\cos(60) = \frac{x}{6} \rightarrow 0.5 = \frac{x}{6} \rightarrow 3 = x$$

$b_2 = x + 10 + x = 3 + 10 + 3 = 16$

Step 4: Now that we have solved for all of the unknowns, plug the known variables into the equation and solve.

$A = \frac{1}{2}h(b_1 + b_2) = \frac{1}{2} \times 5.196\,(10 + 16) = 67.55 \text{ in}^2$

Find the area. (Figures are not drawn to scale.)

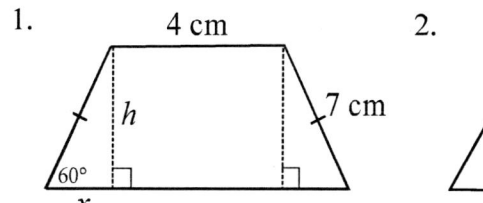

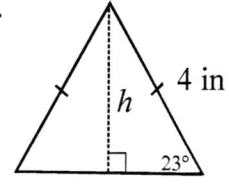

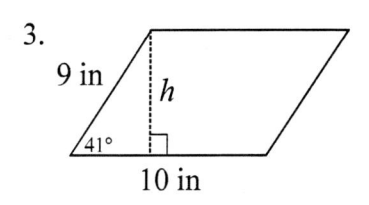

1. [Trapezoid with top 4 cm, slant side 7 cm, angle 60°, height h, base x]

2. [Isoceles triangle with slant side 4 in, height h, base angle 23°]

3. [Parallelogram with slant side 9 in, base 10 in, angle 41°, height h]

Chapter 5 Triangle Trigonometry

Chapter 5 Review

Find the length of a and b in each of the special right triangles. Simplify your answers. (DOK 2)

1.

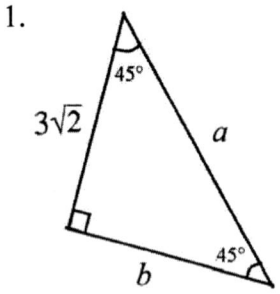

   45°, $3\sqrt{2}$, a, b, 45°

2.

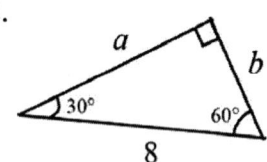

   a, b, 30°, 60°, 8

3.

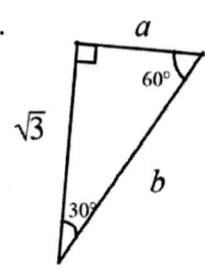

   a, 60°, $\sqrt{3}$, b, 30°

4.

   a, 45°, b, $4\sqrt{2}$, 45°

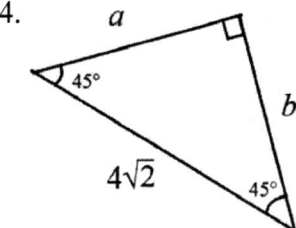

For both of the right triangles shown in questions 5 and 6, find the three trigonometric ratios for $\angle A$. (Hint: $\angle C = 90°$) (DOK 2)

5. C — 27 — B, 36, 45, A

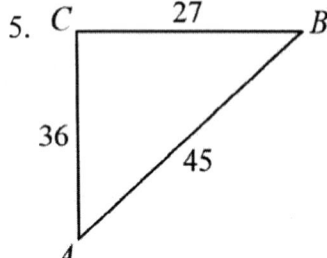

6. C, a, 23, B, 32, A

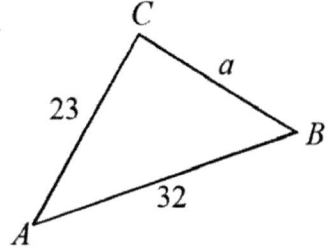

For questions 7 and 8, find the value of x. (DOK 2)

7. $\sin x = 0.5$

8. $\tan x = -1$

For questions 9 and 10, find the missing angle and sides. (DOK 2 & 3)

9. A, 39°, b, C, 3, c, B

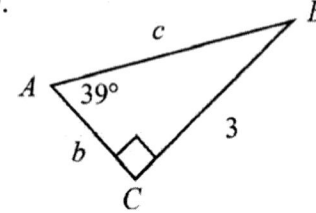

10. A, 45°, 6, c, C, a, B

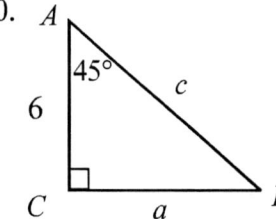

98

Chapter 5 Review

11. If the ratio of the lengths of the legs of a right triangle is 7 : 13, what is the sine of the angle formed by the shorter leg and the hypotenuse?

12. The length of one leg of a right triangle is 9 times the length of the other leg. What is the measurement to the nearest degree of the angle formed by the longer leg and the hypotenuse?

13. If the length of a leg of a right triangle is 55 percent of the length of the hypotenuse, what is the measurement to the nearest degree of the angle formed by the leg and the hypotenuse?

Solve the following problems. (DOK 2 & 3)

14. Find the height of the parallelogram.

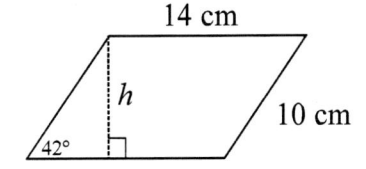

15. $\triangle ABC \sim \triangle DEF$. Find the measures of $\overline{FD}$ and $\overline{EF}$.

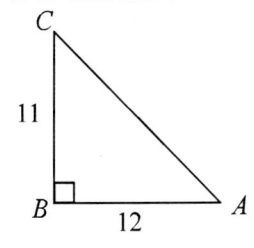

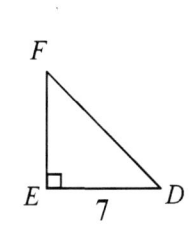

16. Find the measure of $\angle A$.

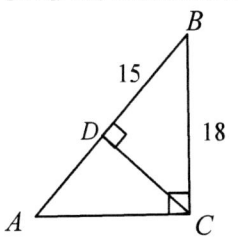

17. If $\tan C = \frac{9}{11}$, what is $\tan B$?

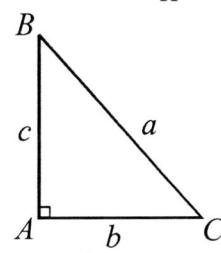

18. $\triangle ABC \sim \triangle DEF$. Find the measures of $\angle F$ and $\angle D$.

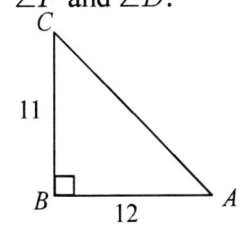

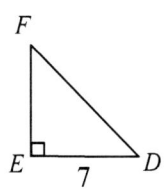

19. Find the measure of $\angle ACD$.

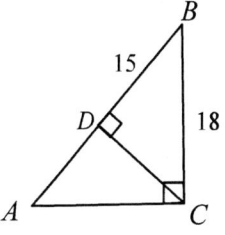

20. Find the missing side. Use $a^2 + b^2 = c^2$.

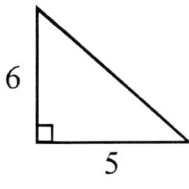

21. Find the measure of the missing leg of the right triangle below. Use $a^2 + b^2 = c^2$.

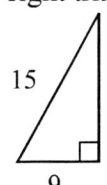

Chapter 5 Triangle Trigonometry

Chapter 5 Test

1 What is the measurement of $\angle Z$ to the nearest degree?

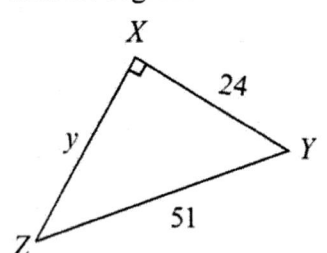

A 25°
B 28°
C 30°
D 62°
(DOK 2)

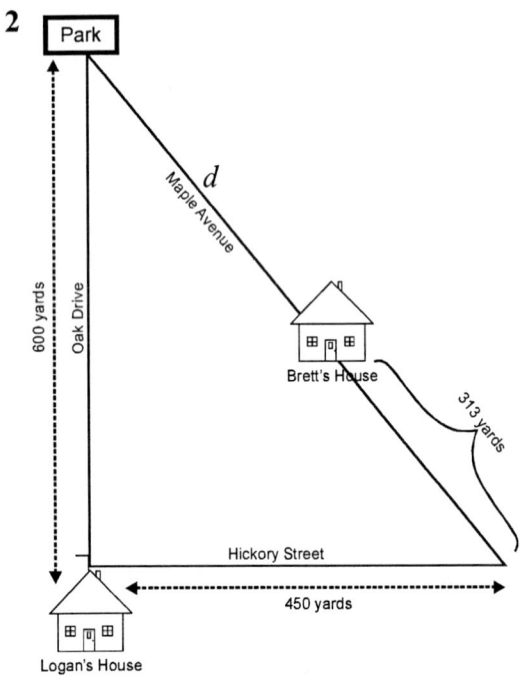

2 Logan enjoys taking his dog to the park. He walks down Hickory St., turns onto Maple Ave. to meet his friend, Brett, and then continues on Maple Ave. to the park. What is the approximate distance (d) from Brett's house to the park?

A 350 yards
B 437 yards
C 687 yards
D 532 yards
(DOK 3)

3 What is the measure of $\overline{AB}$?

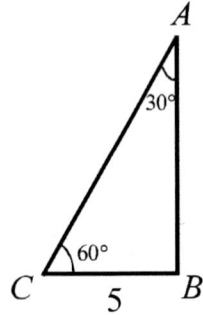

A $5\sqrt{3}$
B 10
C 5
D 2.5
(DOK 2)

4 Solve $\sin x = 0.5$ for x. Round your answer to the nearest degree.

A 27°
B 30°
C 60°
D 0°
(DOK 2)

5 Solve $\tan x = 0.5$ for x. Round your answer to the nearest degree.

A 27°
B 30°
C 60°
D 0°
(DOK 2)

6 What is the measure of $\angle Z$?

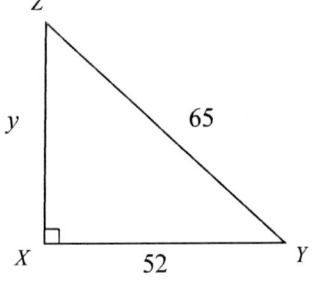

A 0.014°
B 36.87°
C 51.34°
D 53.13°
(DOK 2)

100 Copyright © American Book Company

Chapter 5 Test

7 Harrison steps outside his house to see the hot air balloon pass by. He raises his eyes at a 35° angle to view the balloon. If the balloon is 5,000 feet above the ground, *about* how far is it from Harrison?

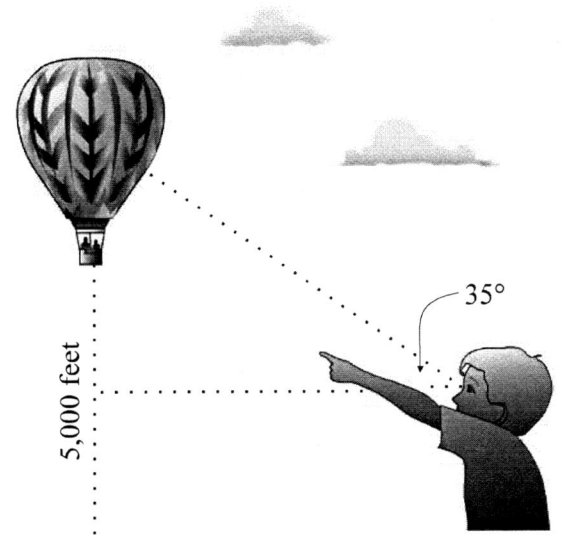

HINT: Harrison's eye level is 5.2 feet from the ground. $\sin(35°) \approx 0.57$ and $\cos(35°) \approx 0.82$

A 6,100 feet
B 8,700 feet
C 7,100 feet
D 2,900 feet   (DOK 3)

8 If $\cos C = \frac{4}{7}$, what is $\sin B$?

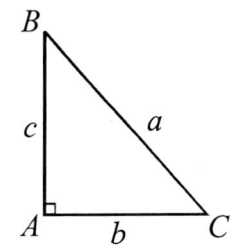

A $\frac{4}{7}$
B $\frac{7}{4}$
C $\frac{6}{7}$
D Cannot be determined with the information given.   (DOK 2)

9 What is the measure of $\angle F$ if $\triangle ABC \sim \triangle DEF$?

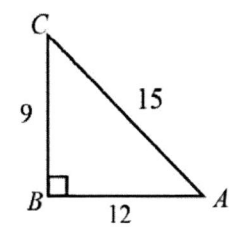

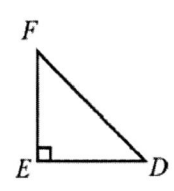

A 53.1°
B 36.9°
C 45°
D 29.4°   (DOK 3)

10 Approximately what is the measure of the hypotenuse of the triangle if its legs measure 12 inches and 13 inches? Use $a^2 + b^2 = c^2$.

A 14 in
B 157 in
C 18 in
D 313 in   (DOK 2)

11 What is the measure of the missing side in the triangle? Use $a^2 + b^2 = c^2$.

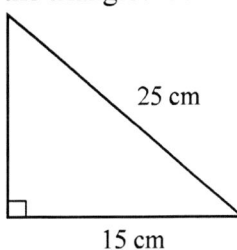

A 5 cm
B 29 cm
C 10 cm
D 20 cm   (DOK 2)

Copyright © American Book Company

101

Chapter 6
Circles

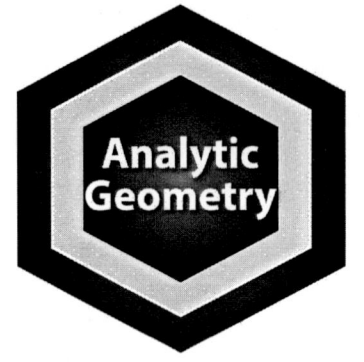

This chapter covers the following CCGPS standard(s):

	Content Standards
Circles	G.C.1, G.C.2, G.C.3, G.C.4, G.C.5, G.GPE.1

6.1 Parts of a Circle (DOK 2)

A **circle** is defined as all points in a plane that are an equal distance from a point called the **center**. The circle is named by the center point.

A **chord** is a segment that has its endpoints on the circle. The longest chord going through the center is the **diameter**. The distance from the center to the circle is the **radius**. When the radius is perpendicular to a chord, it bisects the chord. The distance around the circle is the **circumference**.

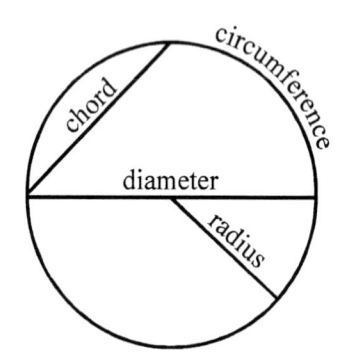

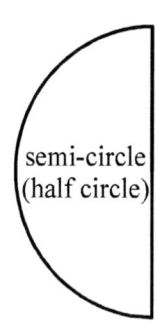

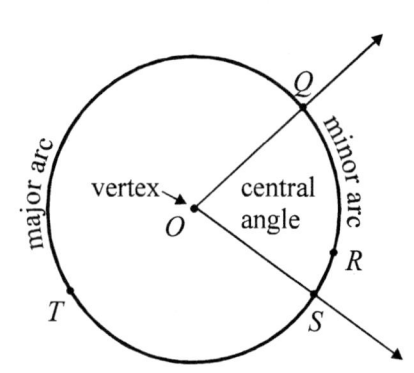

The points Q and S separate the circle into **arcs**. The arc lies on the circle itself. It does not include any points inside or outside the circle. $\overset{\frown}{QRS}$ or $\overset{\frown}{QS}$ is a **minor arc** because it is less than a semicircle. A minor arc can be named by 2 or 3 points. $\overset{\frown}{QTS}$ is a **major arc** because it is more than a semicircle. A major arc must be named by 3 points.

A **central angle** of a circle has the center of the circle as its vertex. The rays of a central angle each contain a radius of the circle. $\angle QOS$ is a central angle.

An **inscribed angle** is an angle whose vertex lies on the circle and whose sides contain **chords** of the circle. $\angle ABC$ in Figure 1 is an inscribed angle. Inscribed angles with rays that have the endpoints of the diameter are always right (90°) angles. A line is **tangent** to a circle if it only

6.1 Parts of a Circle (DOK 2)

touches the circle at one point, which is called the point of tangency, as in Figure 2. A tangent will always be perpendicular to the radius at the point it touches the circle. A **circumscribed angle** is an angle whose vertex lies outside the circle and whose sides contain tangents of the circle. $\angle ABC$ in Figure 3 is a circumscribed angle. A **secant**, shown in Figure 4, is a line that intersects with a circle at two points. Every secant forms a chord. In Figure 4, secant $\overleftrightarrow{AB}$ forms chord $\overline{AB}$.

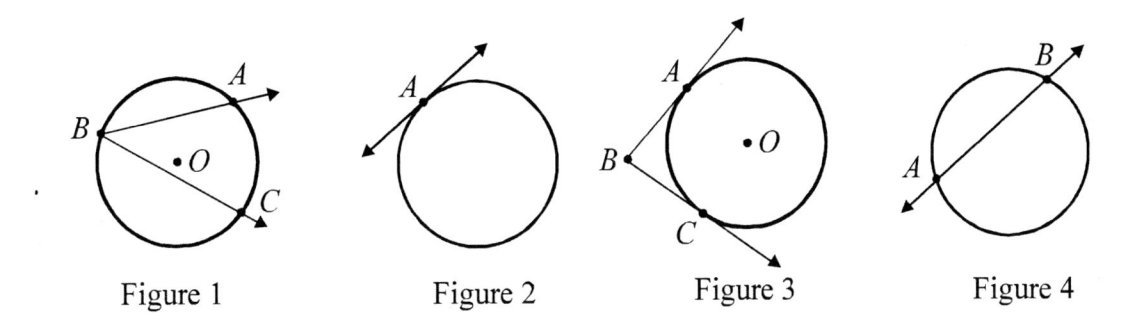

Figure 1   Figure 2   Figure 3   Figure 4

Refer to Circle P below to answer questions 1–5.

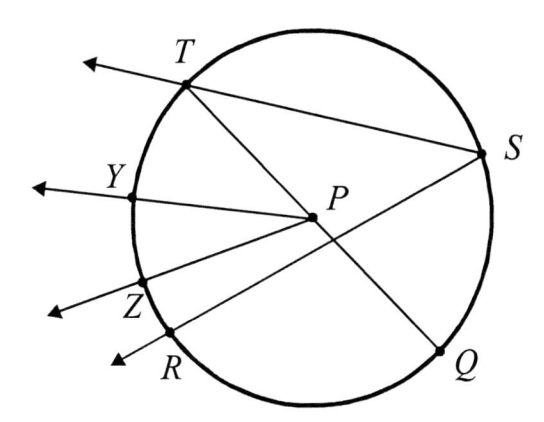

1. Identify the 2 line segments that are chords of the circle but not diameters.

2. Identify the largest major arc of the circle that contains point S.

3. Identify the center of the circle.

4. Identify the 2 inscribed angles.

5. Identify the central angles.

Copyright © American Book Company

103

Chapter 6 Circles

Look at the diagram and identify each of the following.
The center of the circle is marked by the the letter I.

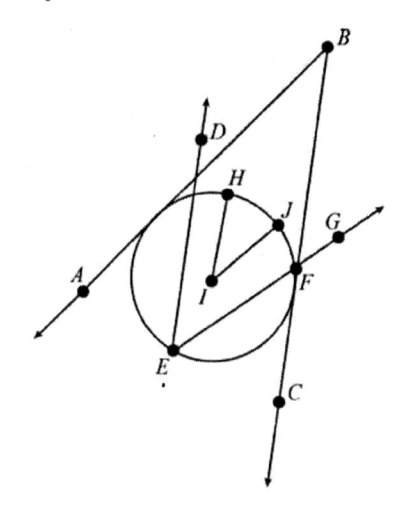

6. $\overrightarrow{BA}$

7. $\angle ABC$

8. $\angle HIJ$

9. $\overline{IJ}$

10. $\angle DEG$

11. $\overline{EF}$

6.2 Circumference (DOK 2)

Circumference, C - the distance around the outside of a circle

Diameter, d - twice as long as the radius, a line segment passing through the center of a circle from one side of the circle to the other

Radius, r - half as long as the diameter, a line segment from the center of a circle to the edge of a circle

Pi, π - the ratio of a circumference of a circle to its diameter. Pi is irrational, but is often approximated by 3.14 or $\frac{22}{7}$.

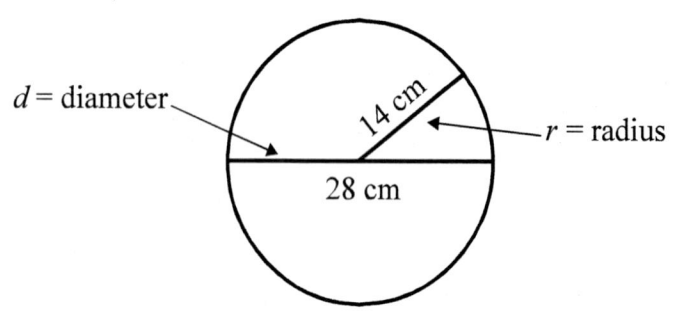

The formula for the circumference of a circle is $C = 2\pi r$ or $C = \pi d$. (The formulas are equal because the diameter is equal to twice the radius, $d = 2r$.)

6.3 Area of a Circle (DOK 2)

Example 1: Find the approximate circumference of the circle above.

$C = \pi d$ Use $\pi \approx 3.14$.
$C \approx 3.14 \times 28$
$C \approx 87.92 \text{ cm}$

$C = 2\pi r$
$C \approx 2 \times 3.14 \times 14$
$C \approx 87.92 \text{ cm}$

Use the formulas given above to find the approximate circumferences of the following circles. Use $\pi \approx 3.14$.

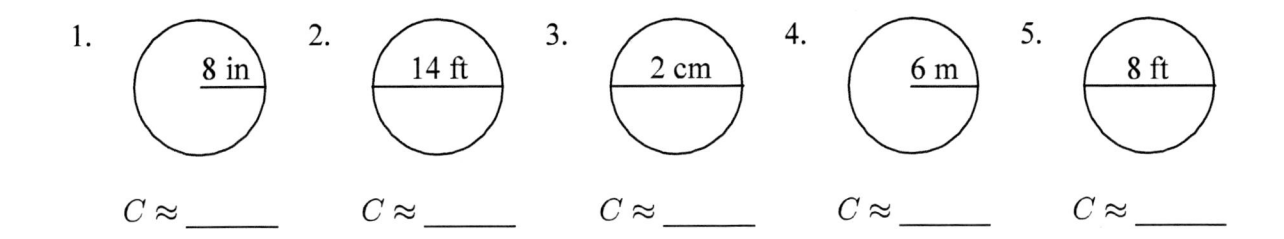

1. 8 in $C \approx$ _____
2. 14 ft $C \approx$ _____
3. 2 cm $C \approx$ _____
4. 6 m $C \approx$ _____
5. 8 ft $C \approx$ _____

Use the formulas given above to find the approximate circumferences of the following circles. Use $\pi \approx \frac{22}{7}$.

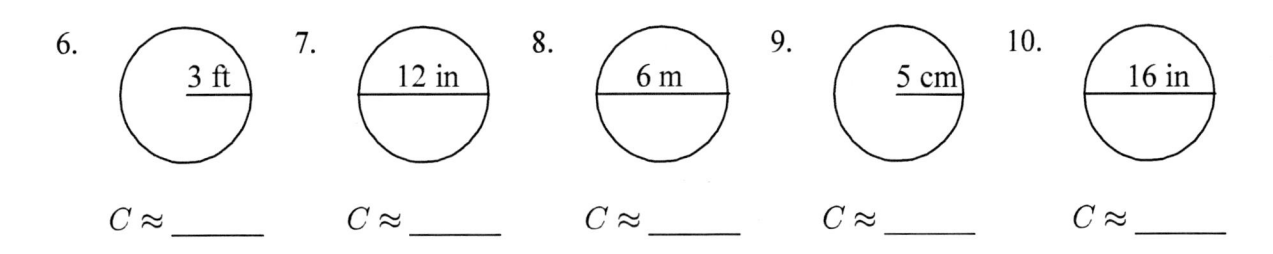

6. 3 ft $C \approx$ _____
7. 12 in $C \approx$ _____
8. 6 m $C \approx$ _____
9. 5 cm $C \approx$ _____
10. 16 in $C \approx$ _____

6.3 Area of a Circle (DOK 2)

The formula for the area of a circle is $A = \pi r^2$. The area is how many square units of measure would fit inside a circle.

For the proof of the area of a circle, we will use a dissection argument. First, we need to split the circle into sections as shown and then place them side by side. The sum of the two lengths of the new shape is equal to the circumference (C) of the circle, $2\pi r$. The shape has a width r which is equal to the radius of the circle.

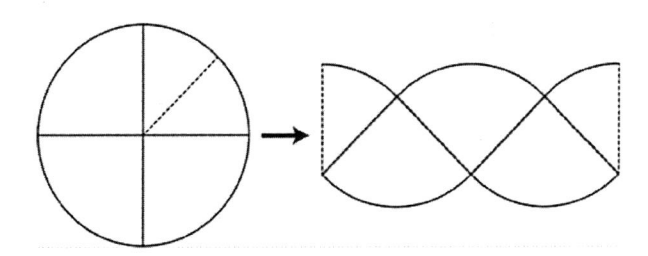

When the sections are placed side by side, they form a shape that is similar to a parallelogram.

Copyright © American Book Company

Chapter 6 Circles

If these sections were dissected smaller and smaller, the shape they would form when placed side by side would look more and more like a perfect rectangle.

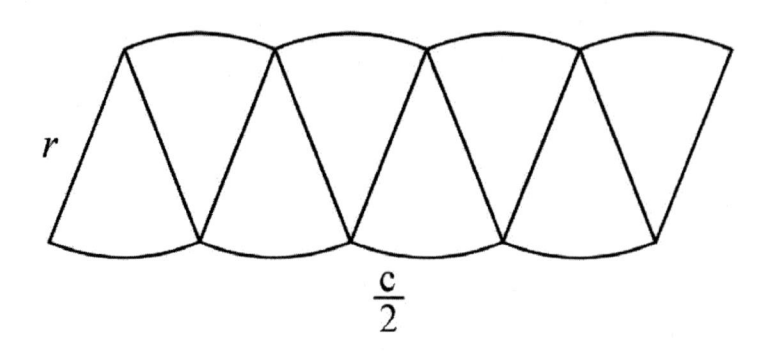

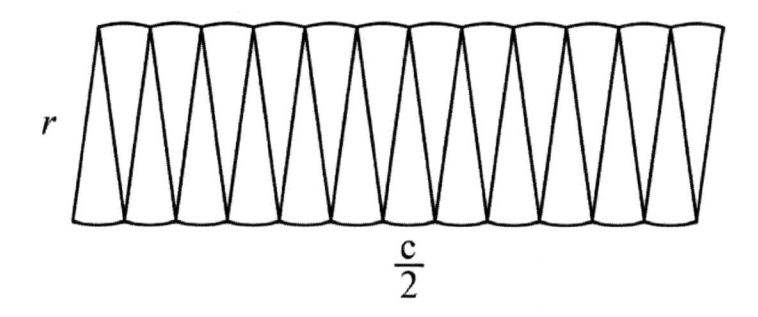

The width of this shape is r, the radius of the circle, and the length is πr. The area of the new shape is equal to that of the original circle and since the area of a rectangle is $l \times w$, therefore, area of the shape and the circle is $(\pi r)r = \pi r^2$

Example 2: Find the area of the circle, using both approximate values for π.

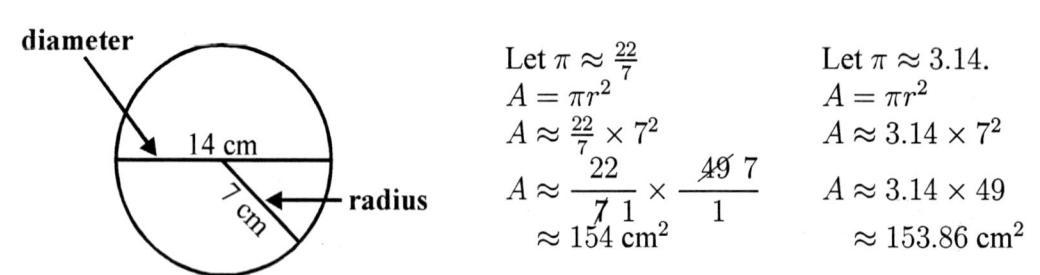

Let $\pi \approx \frac{22}{7}$
$A = \pi r^2$
$A \approx \frac{22}{7} \times 7^2$
$A \approx \frac{22}{\underset{1}{7}} \times \frac{\overset{49}{\cancel{49}}\;7}{1}$
$\approx 154 \text{ cm}^2$

Let $\pi \approx 3.14$.
$A = \pi r^2$
$A \approx 3.14 \times 7^2$
$A \approx 3.14 \times 49$
$\approx 153.86 \text{ cm}^2$

6.3   Area of a Circle (DOK 2)

Find the area of the following circles. Remember to include units.

	$\pi \approx 3.14$	$\pi \approx \frac{22}{7}$
1. 5 in	$A \approx$ _____	$A \approx$ _____
2. 16 ft	$A \approx$ _____	$A \approx$ _____
3. 8 cm	$A \approx$ _____	$A \approx$ _____
4. 3 m	$A \approx$ _____	$A \approx$ _____

Fill in the chart below. Include appropriate units.

	Radius	Diameter	Area $\pi \approx 3.14$	Area $\pi \approx \frac{22}{7}$
5.	9 ft			
6.		4 in		
7.	8 cm			
8.		20 ft		
9.	14 m			
10.		18 cm		
11.	12 ft			
12.		6 in		

Copyright © American Book Company

6.4 Proving Similarity of Circles (DOK 2)

A circle is a line that forms a closed loop, with all points on it having equal distance from the center of the circle. All circles are similar. In order to prove this, we must show that the ratio between any circle's circumference and its diameter is the constant, pi (π).

Example 3: Show that a circle of radius 7 and a circle of radius 5 are similar.

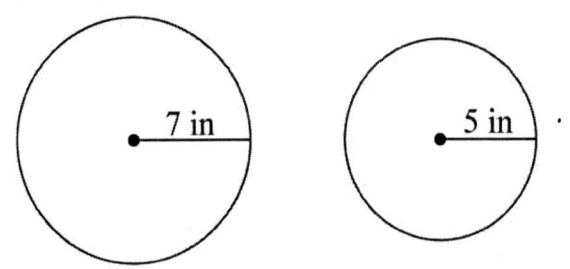

Proof: The circumference of a circle is equal to the diameter times pi: $C = d\pi$.

Step 1: $C_1 = 44, C_2 = 31.4$

Step 2: $C_1 \div 14 = 3.14, C_2 \div 10 = 3.14; 3.14 = \pi$

Step 3: In every circle, the ratio of its circumference to its diameter is always equal to π.

Therefore, all circles are similar. In addition, every circle has an angle measure of 360°.

6.5 Constructing a Tangent Line To The Circle (DOK 2)

A tangent to a circle intersects the circle at exactly one point, and it is perpendicular to the radius of the circle that intersects the circle at that exact point.

In the diagram below, $\overline{AC}$ is the tangent to circle with center O. $\overline{OB}$ is the radius of this circle. The circle, the tangent, and the radius all intersect at only one point, B.

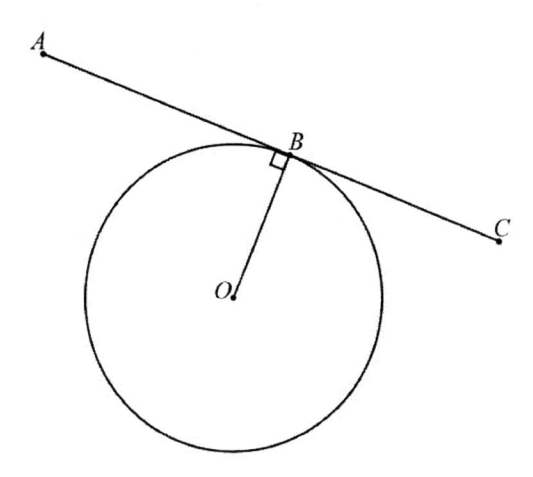

Example 4: Given the circle and an external point, find the tangent line.

Step 1: Draw a straight line between the center of the circle and the given point.

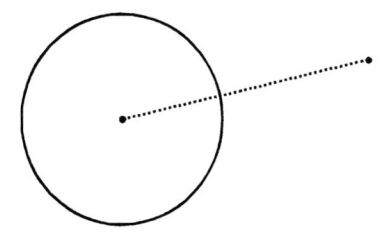

Step 2: Find the midpoint of this line by constructing the line's perpendicular bisector.

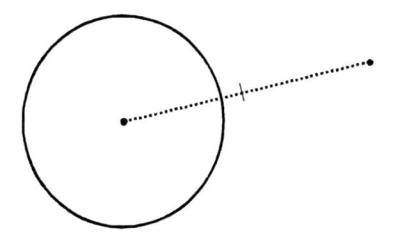

Copyright © American Book Company

Chapter 6   Circles

Step 3:   Using a compass, place the compass on the midpoint, and the pencil on the center of the circle. Without changing the width, draw an arc across the circumference at the two possible tangent points.

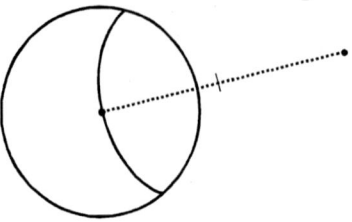

Step 4:   Identify the tangent points and draw a line from these points to the external point.

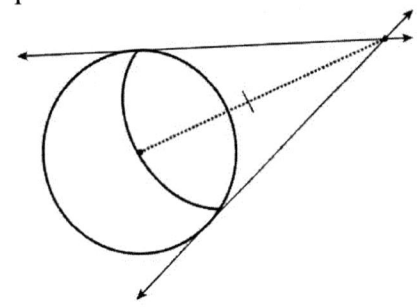

Given the circle and a point outside the circle, find the tangent line.

1.

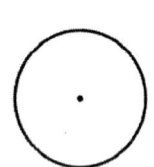

2.

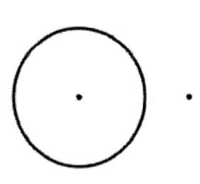

3.

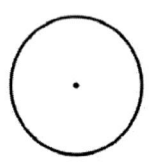

4.

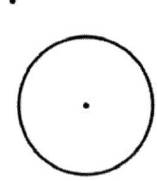

5.

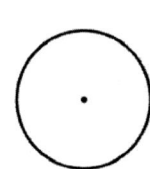

6.6 Constructing an Inscribed Circle of a Triangle (DOK 2)

Example 5: Construct an inscribed circle in a triangle.

Step 1: Bisect two angles of the triangle. They intersect at the center of the inscribed circle.

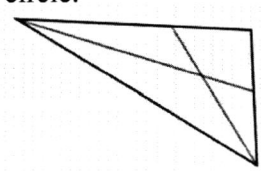

Step 2: Construct a perpendicular segment from the center point to a point on any side of the triangle.

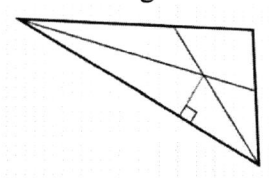

Step 3: Place the compass on the center point. Adjust the length so it lies on the intersection of the perpendicular segment and the triangle. Draw the circle.

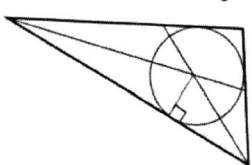

Construct an inscribed circle in the following triangles.

1.

2.

3.

4.

5.

Chapter 6   Circles

6.7   Constructing a Circumscribed Circle of a Triangle (DOK 2)

Example 6:   Construct a circumscribed circle on a triangle.

Step 1:   Construct perpendicular bisectors for any two sides of the triangle. They intersect at the center of the circumscribed circle.

Step 2:   Place the compass on the center point. Adjust the length so it lies on any vertex of the triangle. Draw the circle.

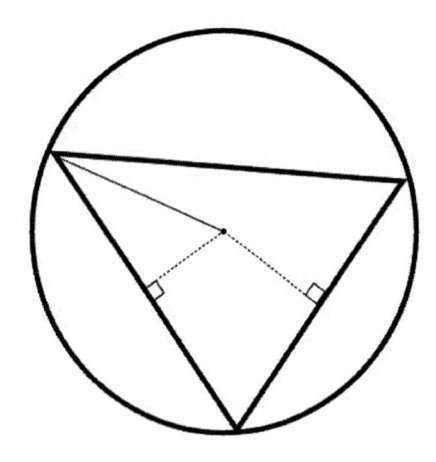

Construct a circumscribed circle on the following triangles.

1.

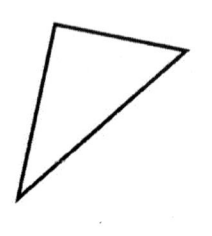

2.

3.

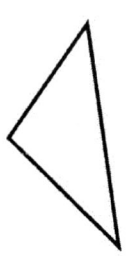

4.

5.

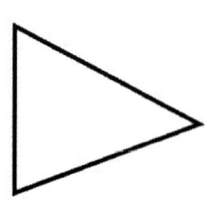

6.8 Arc Measures (DOK 2)

The measure of a minor arc of a central angle is the same as the measure of its central angle.
In the circle at right, $\angle AOC$ measures $80°$.
Therefore, $m\overset{\frown}{AC} = 80$.

A complete rotation about the center point of a circle is $360°$.

The measure of a major arc of a central angle is 360 minus the measure of its central angle.
In the circle at right, $m\overset{\frown}{ADC} = 360 - 80 = 280$.

The measure of a semicircle is 180°.
In the circle at right, $\overline{AD}$ is a diameter of the circle.
Therefore, $\overset{\frown}{ACD}$ is a semicircle and $m\overset{\frown}{ACD} = 180$.

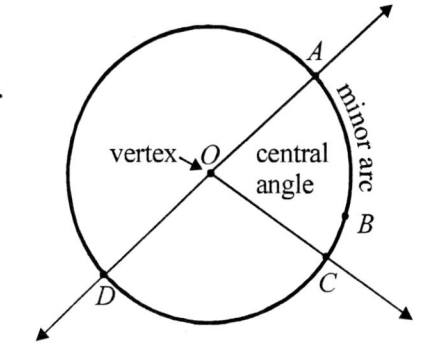

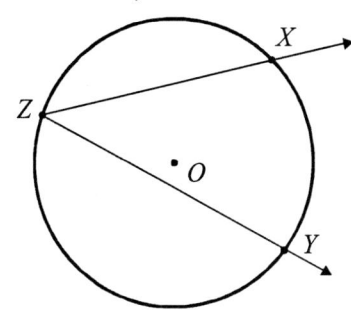

The measure of the minor arc of an inscribed angle is two times the measure of the inscribed angle.
In the circle at left, $\angle XZY$ measures $45°$.
Therefore, $m\overset{\frown}{XY} = 2 \times 45$.
Simplified, $m\overset{\frown}{XY} = 90$.

In the circle below, $m\angle KOJ = 26°$, $m\angle MON = 37°$, and $\overline{KM}$ and $\overline{JL}$ are diameters. Find each measure.

1. $m\overset{\frown}{NM} = $ _____
2. $m\overset{\frown}{KJ} = $ _____
3. $m\overset{\frown}{LM} = $ _____
4. $m\overset{\frown}{JN} = $ _____
5. $m\overset{\frown}{LKJ} = $ _____
6. $m\overset{\frown}{LKN} = $ _____
7. $m\overset{\frown}{MJK} = $ _____
8. $m\overset{\frown}{MNJ} = $ _____

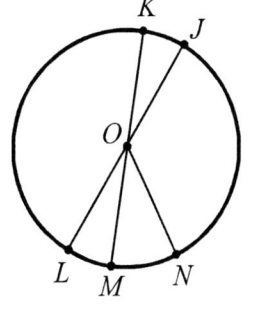

In the circle below, $\overline{AC}$ is a diameter. $m\angle DAC = 50°$, $m\angle BCD = 105°$, $m\overset{\frown}{AD} = 80$, and $m\overset{\frown}{BC} = 50$. Find each measure.

9. $m\angle ACD = $ _____
10. $m\angle BAC = $ _____
11. $m\angle BCA = $ _____
12. $m\angle ABC = $ _____
13. $m\overset{\frown}{ABC} = $ _____
14. $m\angle CDA = $ _____
15. $m\overset{\frown}{CD} = $ _____
16. $m\overset{\frown}{BCD} = $ _____

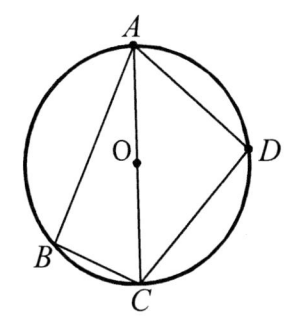

6.9 Properties of Angles of a Quadrilateral (DOK 2)

Example 7: Opposite angles in a quadrilateral are supplementary. The proof is shown below.

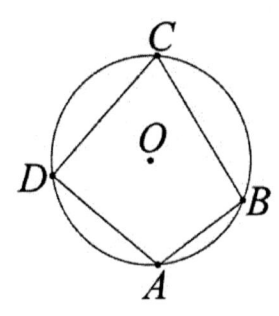

Proof: An inscribed angle is half the measure of the intercepted arc.

$$m\angle DAB = \tfrac{1}{2} m\widehat{DCB}$$
$$m\angle DCB = \tfrac{1}{2} m\widehat{DAB}$$

We know that a circle is 360°. So $m\widehat{DCB} + m\widehat{DAB} = 360°$.

$$m\angle DAB + m\angle DCB = \tfrac{1}{2} m\widehat{DCB} + \tfrac{1}{2} m\widehat{DAB} = \tfrac{1}{2}\left(m\widehat{DCB} + m\widehat{DAB}\right) = \tfrac{1}{2}(360°) = 180°$$

Therefore, opposite angles in a quadrilateral inscribed in a circle equal 180°, which means they are supplementary.

Use the figure below to answer the following questions.

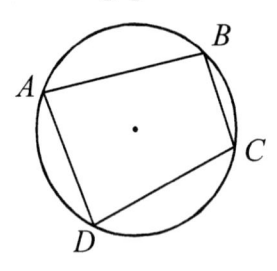

1. If $m\angle DAB = 75°$, what does $m\angle BCD$ equal?

2. If $m\angle BCD = 100°$, what does $m\angle DAB$ equal?

3. If $m\angle ABC = 112°$, what does $m\angle ADC$ equal?

4. If $m\angle BCD = 92°$, what does $m\angle BAD$ equal?

5. If $m\angle DAB = 83°$, what does $m\angle BCD$ equal?

6. If $m\angle ADC = 88°$, what does $m\angle CBA$ equal?

6.10 More Arc Lengths (DOK 3)

Arcs possess two properties:

1. Arc length is a portion of the circle's circumference.

2. They have measurable curvature, as long as the corresponding central angle is known.

Here is a set of proportions worth remembering.

$$\frac{\text{portion of circle}}{\text{whole of circle}} = \frac{\text{measure of central angle}}{360°} = \frac{\text{arc length}}{\text{circumference}}$$

The diagram shows two circles that share the same center, otherwise referred to as concentric. Their radii are r_1 and r_2, and the arc lengths corresponding to the central angle θ are s_1 and s_2. The central angle can be measured in degrees or radians. The radian measure of the angle is defined by the constant of proportionality and is equal to the corresponding arc length.

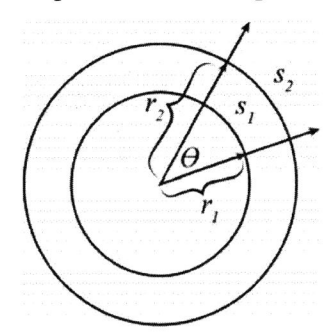

We know that all circles are similar. Since the two segments of these two circles share a common angle, they too must be similar.

Now consider the arc lengths, s_1 and s_2, as a portion of the circumferences of their respective circles.

Smaller (inner) circle: $C_1 = 2\pi r_1$, arc length: $s_1 = \dfrac{\theta}{360°} 2\pi r_1$

Larger (outer) circle: $C_2 = 2\pi r_2$, arc length: $s_2 = \dfrac{\theta}{360°} 2\pi r_2$

When we compare the arc lengths, we get:

$$\frac{s_1}{s_2} = \frac{\dfrac{\theta}{360°} 2\pi r_1}{\dfrac{\theta}{360°} 2\pi r_2}$$

Therefore, $\dfrac{s_1}{s_2} = \dfrac{r_1}{r_2}$.

This proves that the length of the arc intercepted by an angle is proportional to the radius.

6.11 More Circle Properties (DOK 3)

An angle created by two secants, a secant and a tangent, or by two tangents has a measure equal to half the difference of the corresponding arc measures.

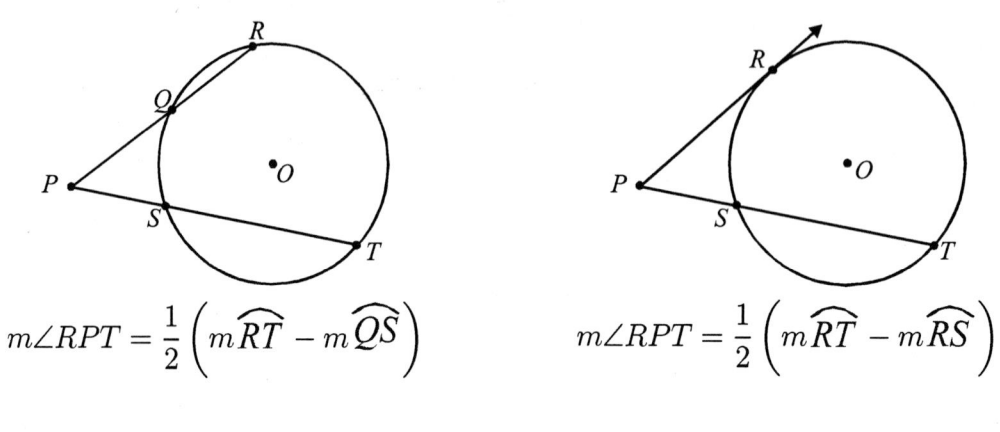

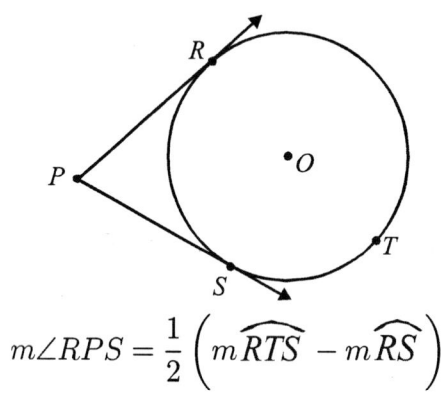

$$m\angle RPT = \frac{1}{2}\left(m\widehat{RT} - m\widehat{QS}\right)$$

$$m\angle RPT = \frac{1}{2}\left(m\widehat{RT} - m\widehat{RS}\right)$$

$$m\angle RPS = \frac{1}{2}\left(m\widehat{RTS} - m\widehat{RS}\right)$$

If two chords intersect, the angles created have a measure equal to the average of the corresponding arc measures.

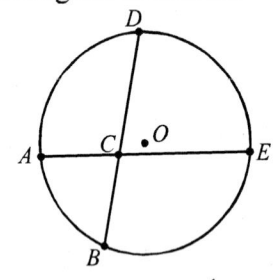

$$m\angle ACB = m\angle DCE = \frac{1}{2}\left(m\widehat{AB} + m\widehat{DE}\right)$$
$$m\angle ACD = m\angle BCE = \frac{1}{2}\left(m\widehat{AD} + m\widehat{BE}\right)$$

Two angles that correspond to the same arc are equal.

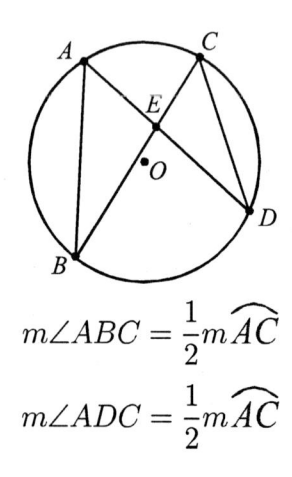

$$m\angle ABC = \frac{1}{2}m\widehat{AC}$$
$$m\angle ADC = \frac{1}{2}m\widehat{AC}$$

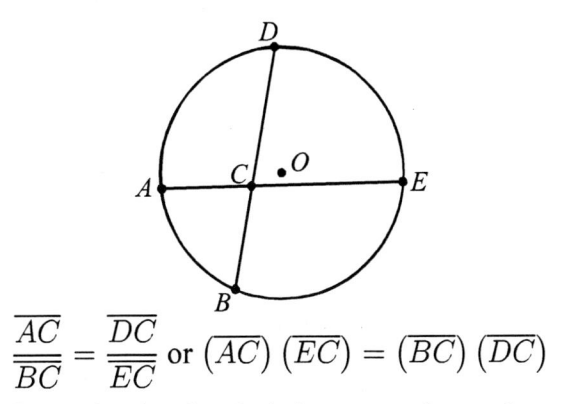

6.11 More Circle Properties (DOK 3)

$\dfrac{\overline{AC}}{\overline{BC}} = \dfrac{\overline{DC}}{\overline{EC}}$ or $(\overline{AC})(\overline{EC}) = (\overline{BC})(\overline{DC})$

If two chords of a circle intersect, the products of each line segment in the chords are equal.

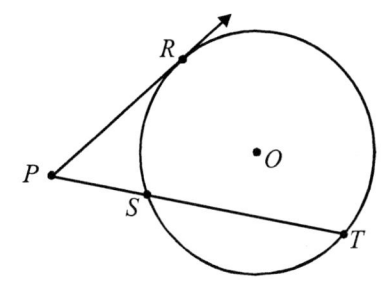

$\dfrac{\overline{PT}}{\overline{PR}} = \dfrac{\overline{PR}}{\overline{PS}}$ or $(\overline{PR})^2 = \overline{PT} \times \overline{PS}$

If a secant and a tangent intersect outside of a circle, the square of the tangent line segment is equal to the product of the outside secant segment and the secant.

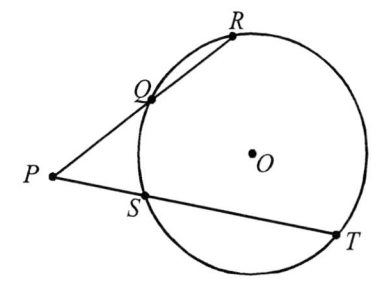

$\dfrac{\overline{PT}}{\overline{PQ}} = \dfrac{\overline{PR}}{\overline{PS}}$ or $(\overline{PT})(\overline{PS}) = (\overline{PQ})(\overline{PR})$

If two secants intersect outside of a circle, the product of the secant and its outside secant segment is equal to the product of the other secant and its outside secant segment.

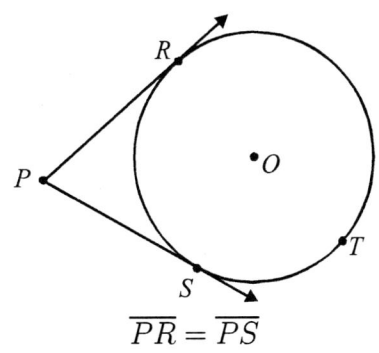

$\overline{PR} = \overline{PS}$

If two tangent lines intersect outside a circle, the line segments of each tangent are equal.

Copyright © American Book Company

117

Chapter 6   Circles

Use the circles to find the measures of the angles and arcs.

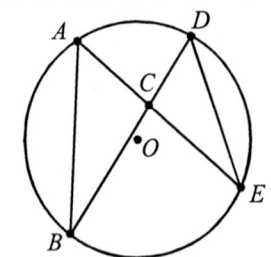

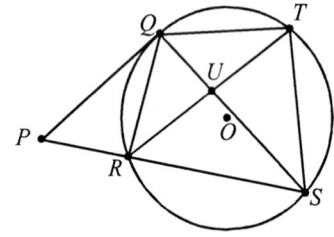

$m\angle BDE = 50°$
$m\angle ABD = 25°$
$m\widehat{AB} = 120$

1. $m\widehat{AD}$
2. $m\angle ACD$
3. $m\angle AED$
4. $m\angle ACB$
5. $m\widehat{DE}$

$\overline{PQ}$ is tangent to circle O.
$m\angle QTR = 20°$
$m\angle TSQ = 25°$
$m\angle TRS = 60°$

6. $m\widehat{QR}$
7. $m\angle TUS$
8. $m\angle QPS$
9. $m\widehat{QT}$
10. $m\angle RUS$

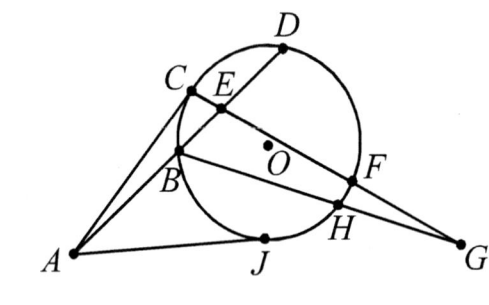

$\overline{AC}$ and $\overline{AJ}$ are tangent to circle O.
$\overline{AJ} = 5.477$
$\overline{CF} = 8$
$\overline{DE} = 3$
$\overline{EF} = 6$
$\overline{BH} = 7$
$\overline{BG} = 10$

11. $\overline{BE}$
12. $\overline{AB}$
13. $\overline{AC}$
14. $\overline{GF}$

6.12 Area of Sectors (DOK 2)

A **sector** of a circle is a region bounded by a central angle and its intercepted arc. In the circle below, $\angle AOB$ is a central angle measuring $80°$. Therefore, $\overset{\frown}{AB}$ is $80°$. $\angle AOB$ and $\overset{\frown}{AB}$ form a sector of the circle.

Example 8: Find the area of the sector formed by $\angle AOB$ and $\overset{\frown}{AB}$.

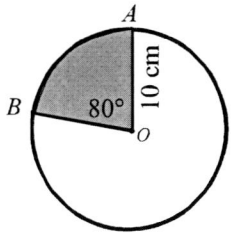

Step 1: The area of a sector is a fraction of the area of the circle. First, find the area of the circle.
$A = \pi r^2$
$A \approx 3.14 \times 10^2 = 3.14 \times 100 = 314 \text{ cm}^2$

Step 2: Next, find the fraction of the circle that the sector occupies. Remember that the sum of the measures of the central angles of a circle is $360°$. The fraction that the sector occupies is the measure of the central angle, denoted by the letter N, divided by 360.

Fraction that sector occupies $= \dfrac{N}{360} = \dfrac{80}{360} = \dfrac{2}{9}$

Step 3: Now, calculate the area of the sector.
$A = \dfrac{N}{360}\pi r^2 \approx \dfrac{2}{9} \times 3.14 \text{ cm}^2 \times 10^2 = \dfrac{628}{9} \text{ cm}^2$
Simplified, $A \approx 69\frac{7}{9} \text{ cm}^2$.

Each of the following is a measurement for a central angle. Calculate the fraction of a circle that the central angle occupies. Simplify your answers.

1. 30°
2. 2°
3. 16°
4. 52°
5. 45°
6. 72°
7. 120°
8. 270°
9. 15°
10. 108°
11. 60°
12. 90°

Find the approximate area of the sector bounded by $\angle XYZ$ and $\overset{\frown}{XZ}$ in each of the following circles. Use the formula $A = \dfrac{N}{360}\pi r^2$. Use $\pi \approx 3.14$.

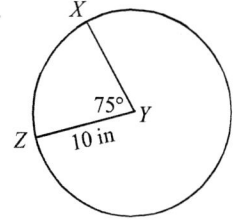

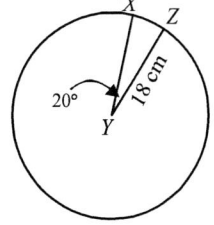

13. (36°, 5 cm)
14. (75°, 10 in)
15. (20°, 18 cm)
16. (100°, 2 ft)

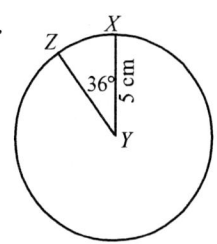

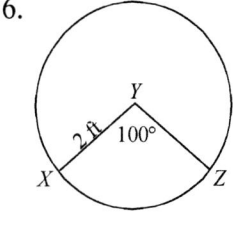

Chapter 6 Circles

6.13 Finding Equations of Circles Using the Pythagorean Theorem (DOK 2)

A **circle** with a radius, r, and a center, (a, b), that is graphed on a coordinate grid can be defined by the equation $(x - a)^2 + (y - b)^2 = r^2$. If the center of the circle is at the origin, $(0, 0)$, then the equation can be simplified to $x^2 + y^2 = r^2$.

Example 9: Find the equation of the circle with the center: at $(-1, 4)$ and a radius of 15.

Step 1: The center is $(-1, 4)$, so $a = -1$ and $b = 4$. The first part of the equation is $(x + 1)^2 + (y - 4)^2$.

Step 2: The radius is 15, so $r^2 = 15^2$ or 225.
The equation of the circle is $(x + 1)^2 + (y - 4)^2 = 225$.

*Note: The radius, r, of a circle can be found when given the circumference, C, or area, A, by using the formulas $C = 2\pi r$ and $A = \pi r^2$.

Find the equations of the following circles.

1. Center: $(-2, 10)$; Radius: 13
2. Center: $(12, -1)$; Radius: 21
3. Center: $(-8, -9)$; Radius: 7
4. Center: $(17, 3)$; Radius: 11
5. Center: $(2, -4)$; Radius: 3
6. Center: $(20, -5)$; Radius: 15

7. Find the equation of the following circle where C is the center and A is on the circle.

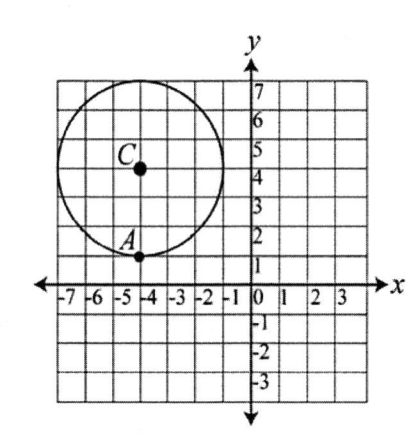

8. A circle has a center at the point $(2, -2)$ and a circumference of 8π units. Find the equation of the circle.

9. A circle has a center at the point $(-5, 3)$ and an area of 31π units^2. Find the equation of the circle.

10. The origin of a coordinate grid lies on a circle with a center at the point $(3, 4)$. Find the equation of the circle.

Chapter 6 Review

(DOK 2)

1. In the circle below, $\overline{AE}$ is a diameter, $\angle DAE$ measures 30°, and $m\widehat{BC} = 45$. What is the measure of $\widehat{DE}$ and $\angle BOC$?

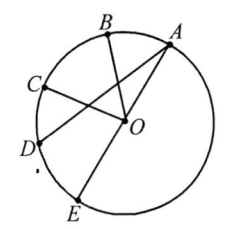

2. Which line is tangent to the circle?

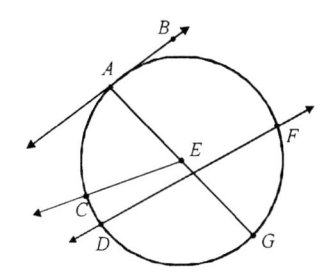

3. In the diagram below, $\overline{PQ}$ and $\overline{PR}$ are tangent to circle O. Find $m\overline{PR}$.

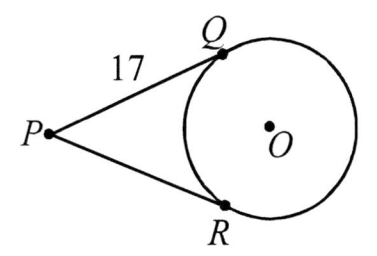

4. Find $m\overline{AC}$.

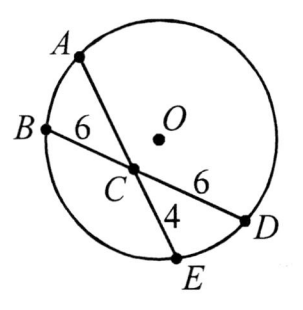

For questions 5 and 6, find the equation of the circle given the center and radius. (DOK 3)

5. Center: $(12, -17)$; Radius: 24

6. Center: $(-3, 22)$; Radius: $\frac{1}{9}$

7. True or false? All circles are similar. Explain.

8. Draw a circumscribed angle.

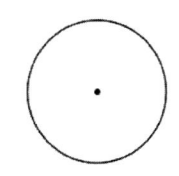

9. Draw an inscribed angle.

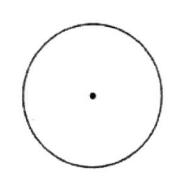

10. Draw a central angle.

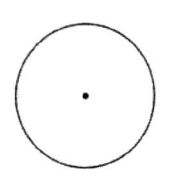

11. Draw a chord.

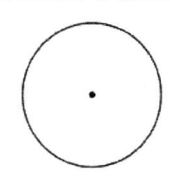

12. True or false? The tangent line is perpendicular to the radius that intersects the circle at the same point.

Chapter 6 Circles

(DOK 3)

13. Given the circle and the external point, find the tangent line.

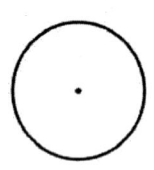

14. Draw an inscribed circle.

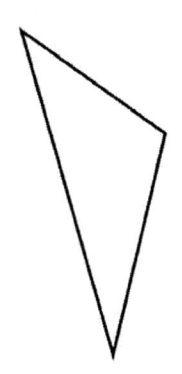

15. Draw an inscribed circle.

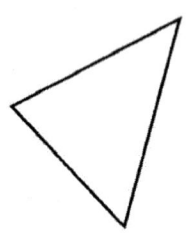

16. Construct a circumscribed circle on the triangle.

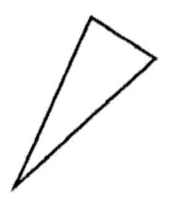

17. True or False? Explain. The perpendicular bisectors intersect at the center of the triangle. .

(DOK 3)

Chapter 6 Test

1 What is the first step to constructing the tangent line of a circle from an external point?

A Draw the tangent line.
B Draw a straight line between the center of the circle and the given external point.
C Identify the tangent points.
D Find the midpoint of the line between the center of the circle and the given external point.
(DOK 2)

2 Which of the following is a drawing of an inscribed angle?

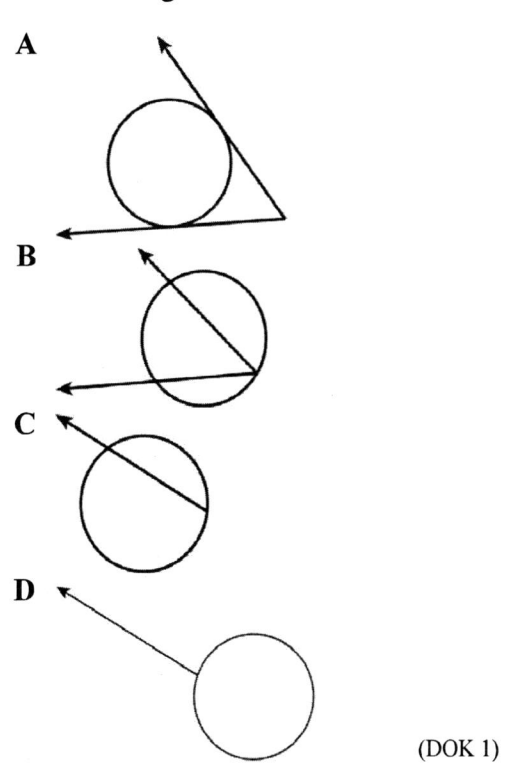

A
B
C
D
(DOK 1)

3 Which is an angle whose vertex lies on a circle and whose sides contain chords of the circle?

A central angle
B tangent
C inscribed angle
D secant
(DOK 1)

4 What is the measure of the central angle of a sector that is 15% of a circle?

A 43.2°
B 90°
C 54°
D 112.5°
(DOK 1)

5 A line that touches a circle only at one point is called a

A tangent.
B chord.
C secant.
D inscribed line.
(DOK 1)

6 What does the measure of $\widehat{ABC}$ equal?

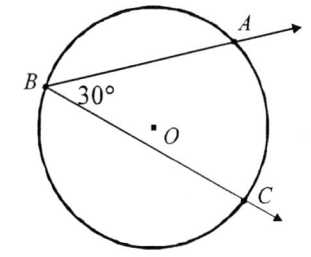

A 60°
B 330°
C 300°
D 210°
(DOK 2)

Use the circle below to answer questions 7 and 8.

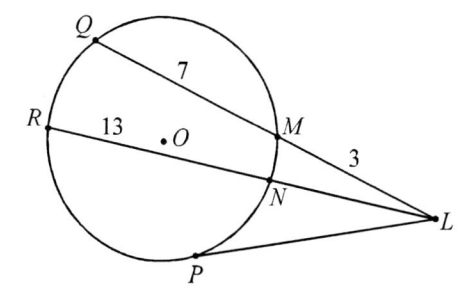

Chapter 6 Circles

7 Find $\overline{LN}$.

   A -15
   B 2
   C -2
   D 15 (DOK 2)

8 If $\overline{LP}$ is tangent to circle O, find the length of $\overline{LP}$.

   A $\sqrt{30}$
   B $\sqrt{10}$
   C $\sqrt{15}$
   D $\sqrt{21}$ (DOK 2)

9 The origin of a coordinate grid lies on a circle with a center at the point $(-5, 12)$. What is the equation of the circle?

   A $(x-5)^2 + (y+12)^2 = 13$
   B $(x+5)^2 + (y-12)^2 = 13$
   C $(x-5)^2 + (y+12)^2 = 169$
   D $(x+5)^2 + (y-12)^2 = 169$ (DOK 3)

10 If you decide to divide a pie that has a diameter of 8 inches into 6 equal slices, what is the area of each slice?
$A = \dfrac{N}{360}\pi r^2 \quad \pi \approx 3.14$

   A 8.37 in^2
   B 4.19 in^2
   C 33.49 in^2
   D 50.24 in^2 (DOK 3)

11 A circle has a center at the point $(11, -16)$ and a circumference of 78π units. What is the equation of the circle?

   A $(x-11)^2 + (y+16)^2 = 1521$
   B $(x+11)^2 + (y-16)^2 = 1521$
   C $(x-11)^2 + (y+16)^2 = 6084$
   D $(x+11)^2 + (y-16)^2 = 6084$ (DOK 3)

12 Which of the following represents an inscribed circle of a triangle?

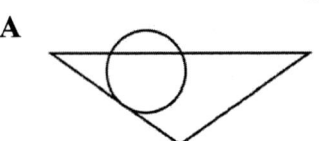

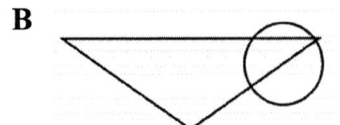

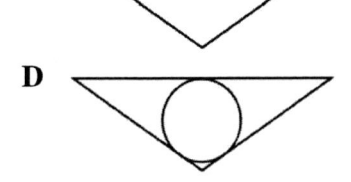

   A
   B
   C
   D (DOK 1)

13 What is the first step in constructing an inscribed circle of a triangle?

   A Trisect three angles of the triangle.
   B Draw the circle.
   C Bisect two angles of the triangle.
   D Construct a perpendicular from the center point to a point on any side of the triangle. (DOK 2)

14 What is the first step to constructing a circumscribed circle on a triangle?

   A Find two perpendicular bisectors to two sides of the triangle.
   B Bisect to angles of the triangle.
   C Draw the circle around the triangle.
   D Find the parallel bisectors of two sides of the triangle. (DOK 2)

Chapter 7
Solid Geometry

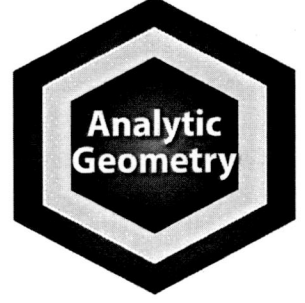

This chapter covers the following CCGPS standard(s):

	Content Standards
Geometric Measurement and Dimension	G.GMD.1, G.GMD.2, G.GMD.3

7.1 Types of 3-Dimensional Figures (DOK 1)

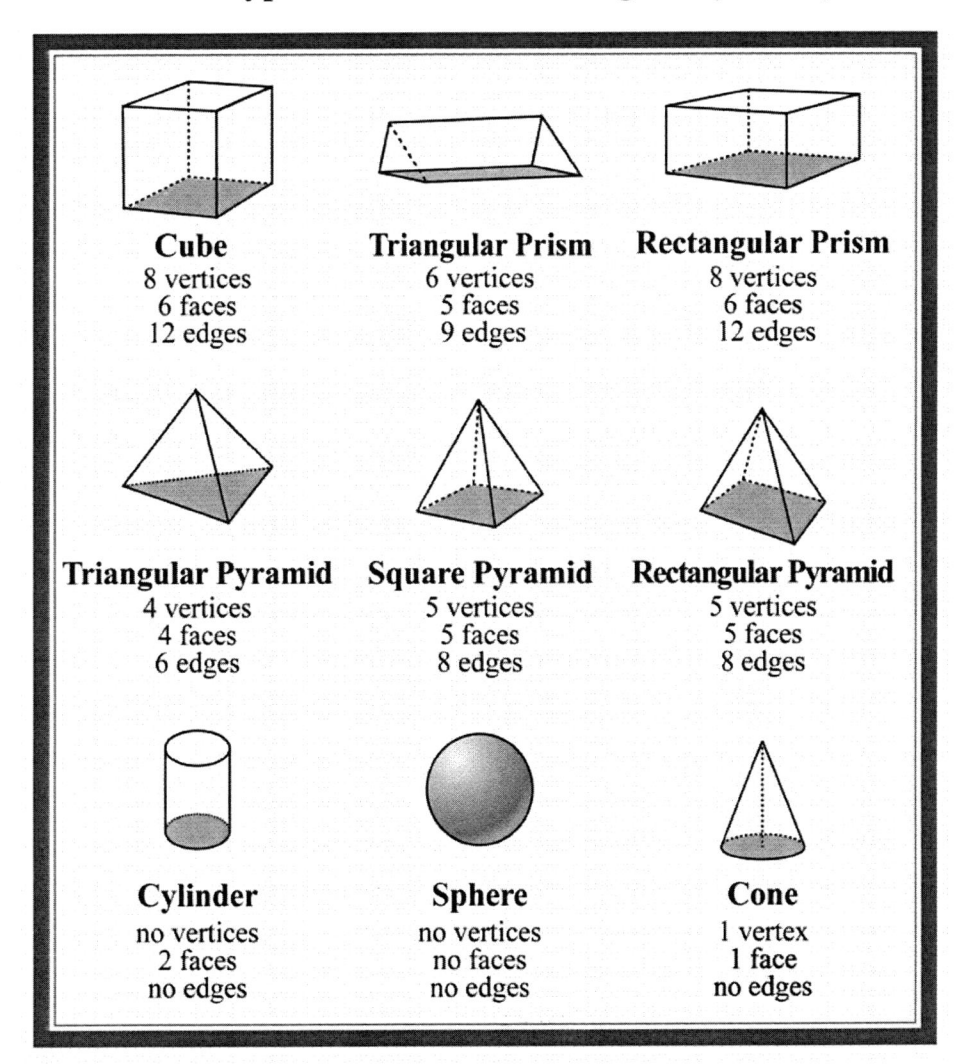

Cube
8 vertices
6 faces
12 edges

Triangular Prism
6 vertices
5 faces
9 edges

Rectangular Prism
8 vertices
6 faces
12 edges

Triangular Pyramid
4 vertices
4 faces
6 edges

Square Pyramid
5 vertices
5 faces
8 edges

Rectangular Pyramid
5 vertices
5 faces
8 edges

Cylinder
no vertices
2 faces
no edges

Sphere
no vertices
no faces
no edges

Cone
1 vertex
1 face
no edges

Chapter 7   Solid Geometry

Pyramids: The faces (except the base sometimes) of all pyramids are triangles. A pyramid is named by the shape of its base.

Prism: A prism has two parallel and congruent bases. A prism is named by the shape of its base.

Name the 3-dimensional figures below.

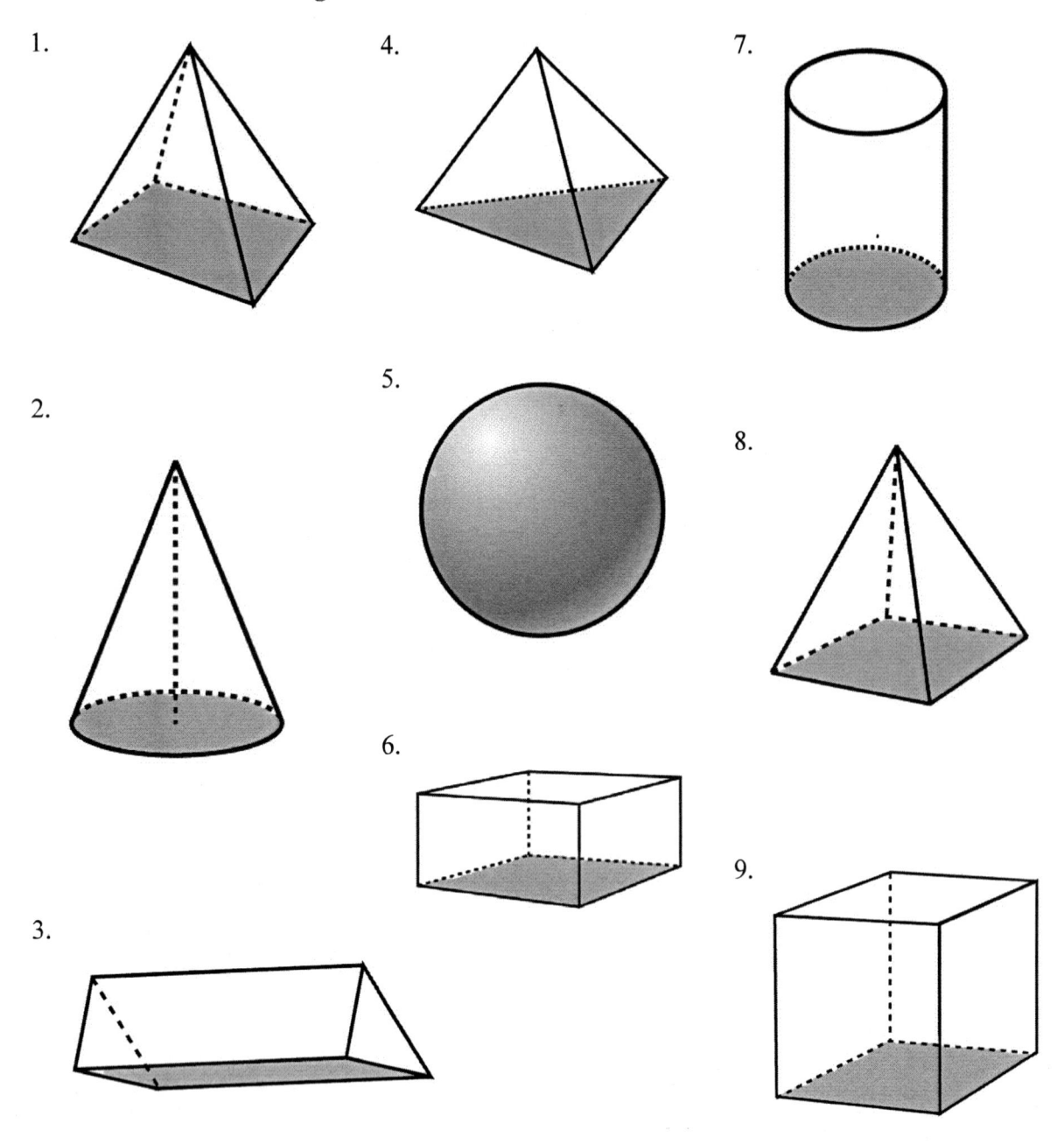

1.

2.

3.

4.

5.

6.

7.

8.

9.

7.2 Cross Sections (DOK 2)

Cross Sections of Cubes

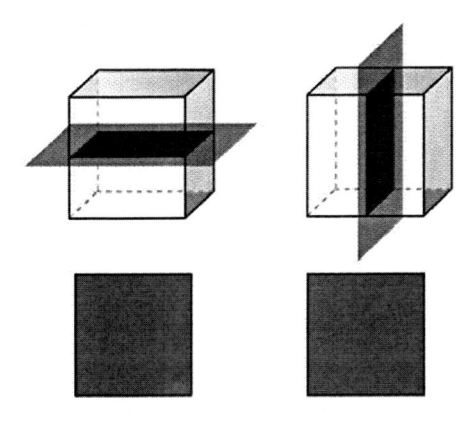

Cross Sections of Rectangular Prisms

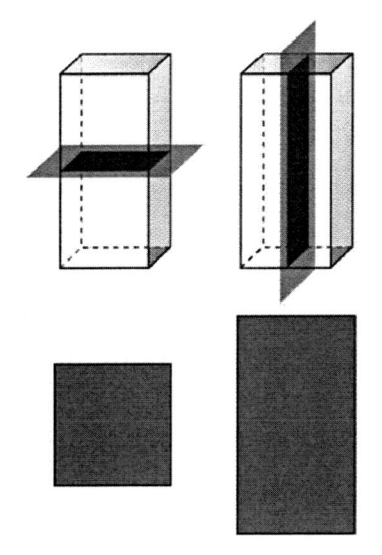

Cross Sections of Cones

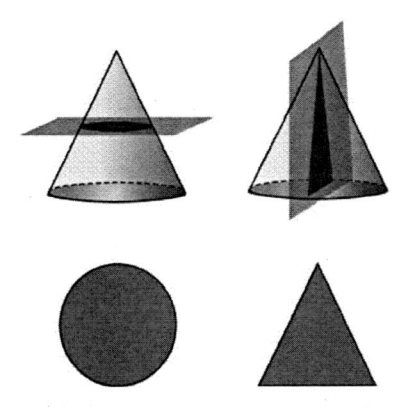

Chapter 7 Solid Geometry

Cross Sections of Cylinders

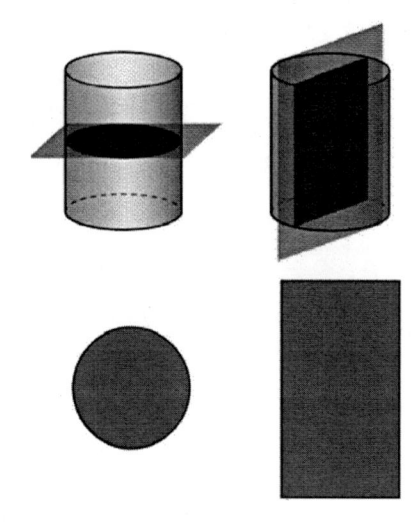

Cross Sections of Spheres

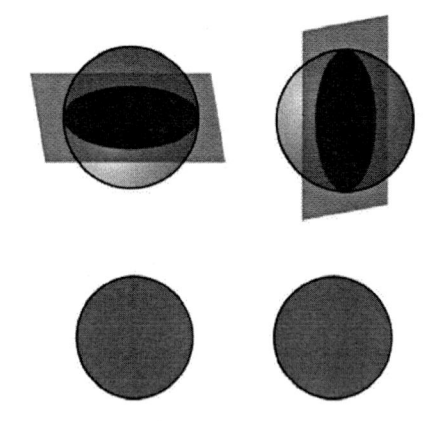

Cross Sections of Pyramids

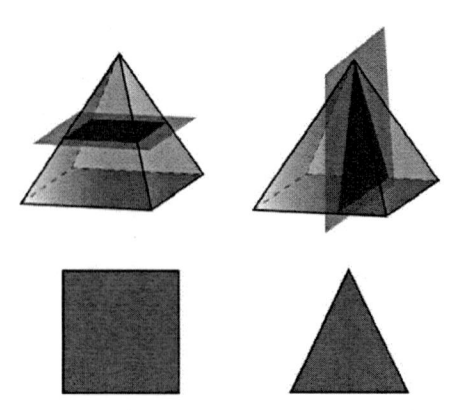

7.3 Formation of a Cube (DOK 2)

A cube is formed by translating a square in a straight line through space.

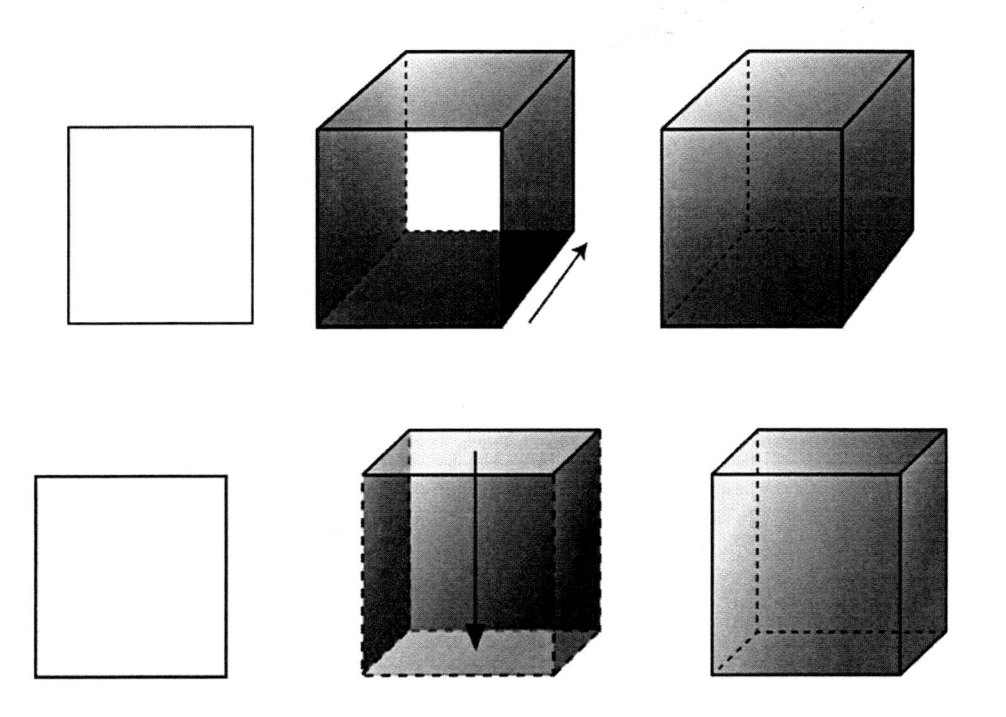

7.4 Formation of a Rectangular Prism (DOK 2)

A rectangular prism is formed by translating a rectangle or square in a straight line through space.

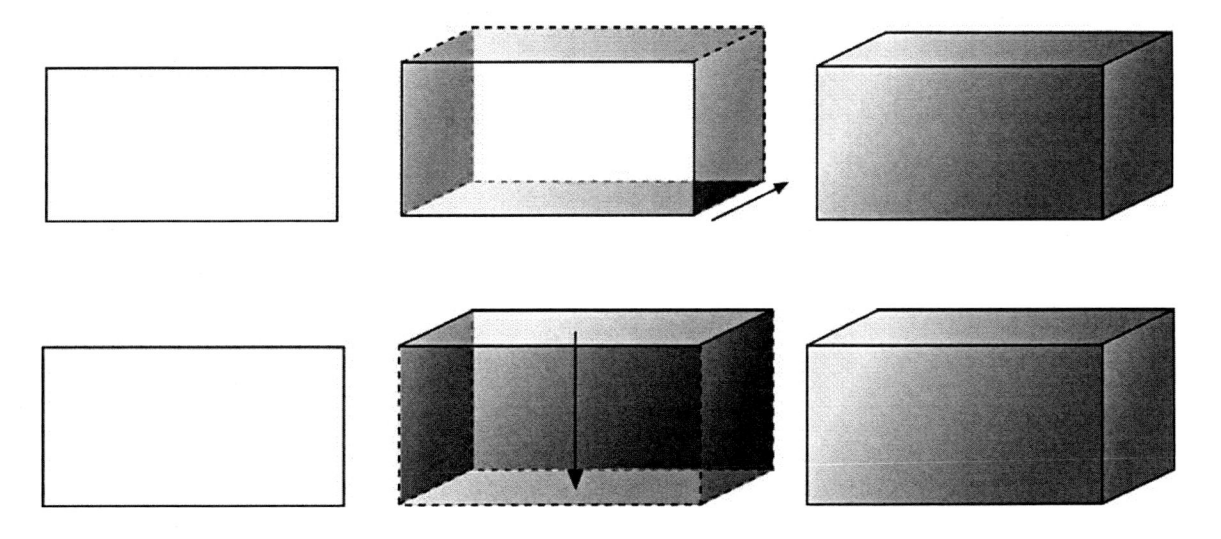

Chapter 7   Solid Geometry

7.5   Formation of a Cone (DOK 2)

A cone is formed by rotating a triangle 360° through space.

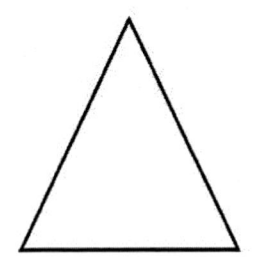

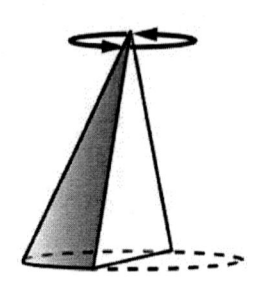

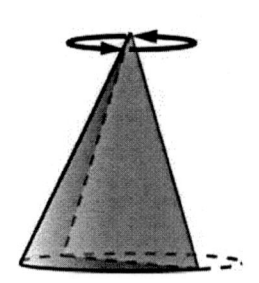

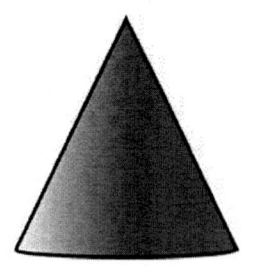

7.6   Formation of a Sphere (DOK 2)

A sphere is formed by rotating a circle 360° through space.

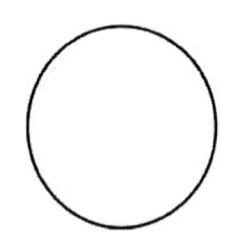

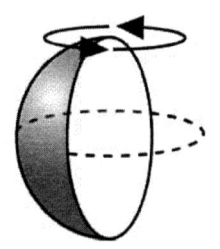

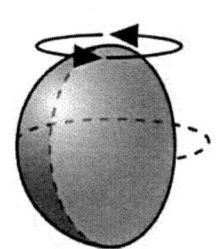

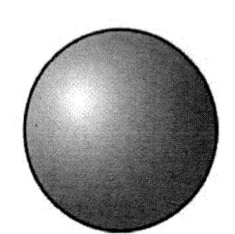

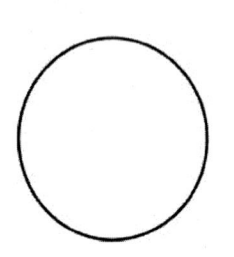

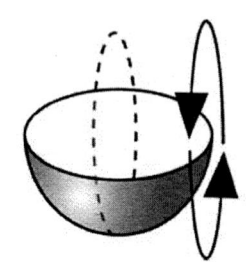

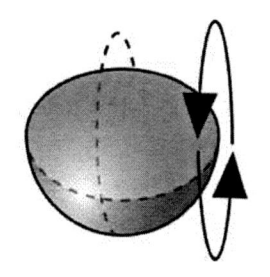

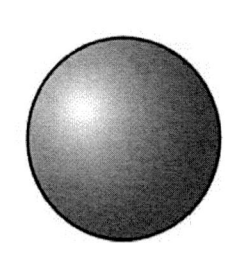

7.7 Formation of a Cylinder (DOK 2)

A cylinder is formed by either translating a circle in a straight line through space or rotating a rectangle 360° through space.

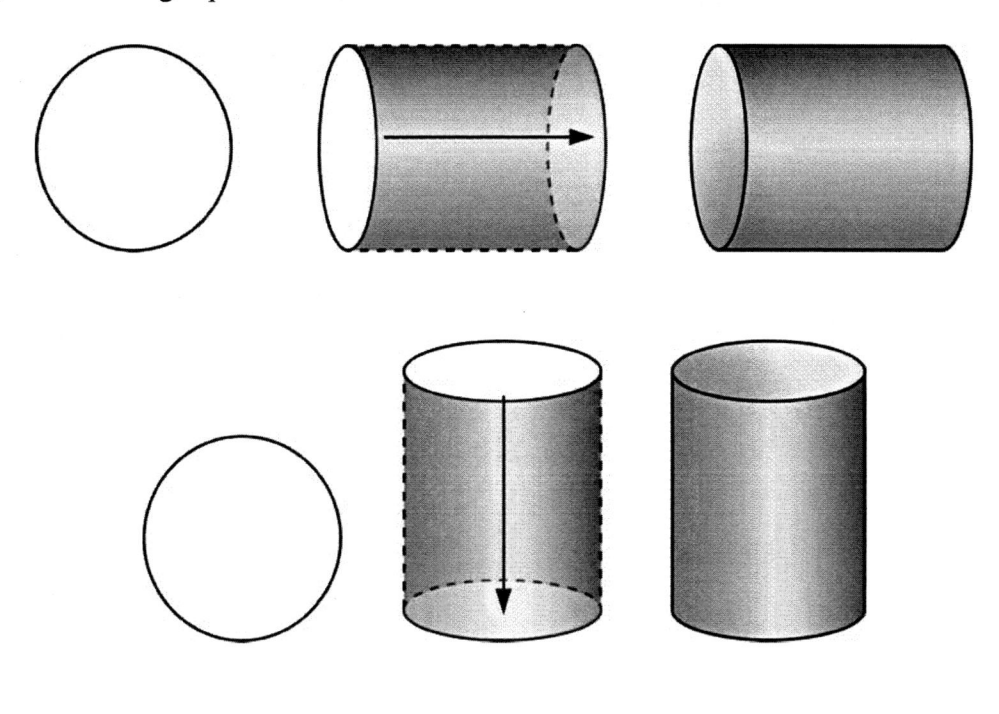

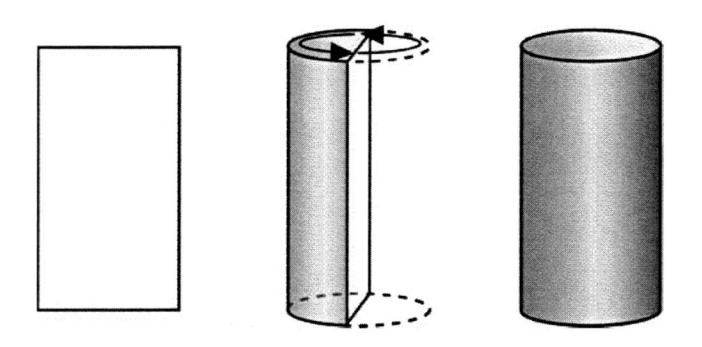

Chapter 7 Solid Geometry

7.8 Volume of Spheres, Cones, Cylinders, and Pyramids (DOK 2)

To find the volume of a solid, insert the measurements given for the solid into the correct formula and solve. Remember, volumes are expressed in cubic units such as in^3, ft^3, m^3, cm^3, or mm^3.

Cavalieri's Principle: Solids that have the same height and the same cross-sectional area at every level will have the same volume.

Volume of a Cylinder

We will take a rectangular prism of height 'h' for our proof.

A rectangular prism has volume = Area of base × height of prism. Because the cross-section of a rectangular prism as well as that of a cylinder is uniform throughout their height, we only need to calculate one cross-sectional area. If the bases and heights of the two solids are equal, then, according to Cavalier's Principle, the volumes are equal. Volume of the cylinder will also be = area of base × height.

The base of a cylinder is a circle, so the area of its base = πr^2.

Volume of the cylinder will therefore be = $\pi r^2 h$.

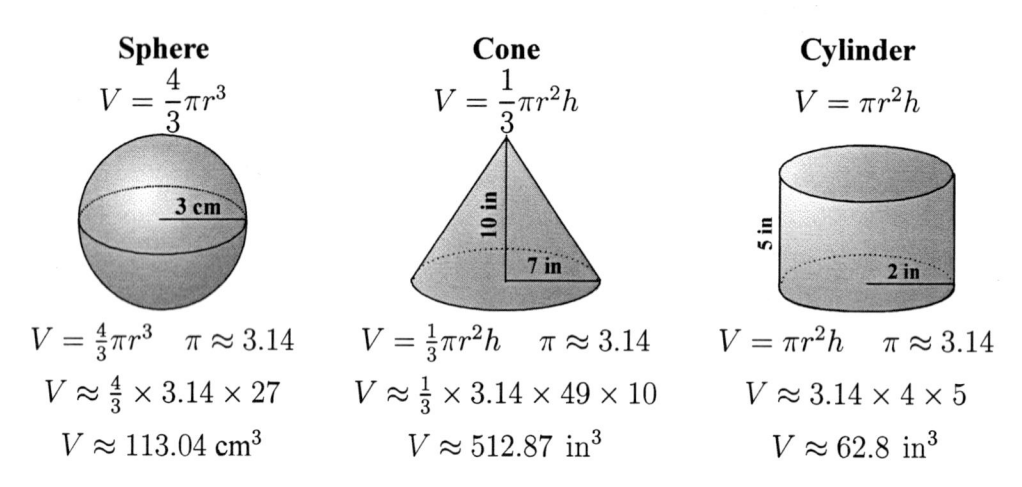

Sphere
$$V = \frac{4}{3}\pi r^3$$

3 cm

$V = \frac{4}{3}\pi r^3 \quad \pi \approx 3.14$

$V \approx \frac{4}{3} \times 3.14 \times 27$

$V \approx 113.04 \text{ cm}^3$

Cone
$$V = \frac{1}{3}\pi r^2 h$$

10 in

7 in

$V = \frac{1}{3}\pi r^2 h \quad \pi \approx 3.14$

$V \approx \frac{1}{3} \times 3.14 \times 49 \times 10$

$V \approx 512.87 \text{ in}^3$

Cylinder
$$V = \pi r^2 h$$

5 in

2 in

$V = \pi r^2 h \quad \pi \approx 3.14$

$V \approx 3.14 \times 4 \times 5$

$V \approx 62.8 \text{ in}^3$

Pyramids

$V = \frac{1}{3}Bh$ (B = area of rectangular base) $\qquad$ $V = \frac{1}{3}Bh$ (B = area of triangular base)

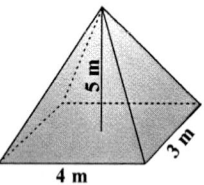

5 m

3 m

4 m

$V = \frac{1}{3}Bh \quad B = l \times w$

$V = \frac{1}{3} \times 4 \times 3 \times 5$

$V = 20 \text{ m}^3$

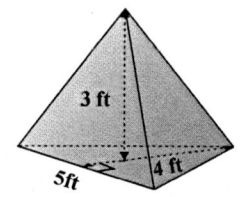

3 ft

5 ft   4 ft

$V = \frac{1}{3}Bh \quad B = \frac{1}{2} \times b \times h$

$B = \frac{1}{2} \times 5 \times 4 = 10 \text{ ft}^2$

$V = \frac{1}{3} \times 10 \times 3$

$V = 10 \text{ ft}^3$

7.8 Volume of Spheres, Cones, Cylinders, and Pyramids (DOK 2)

Find the volume of the following solids. Use $\pi \approx 3.14$.

1. $V = \pi r^2 h$

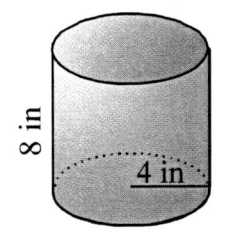

   (cylinder: 8 in height, 4 in radius)

2. $V = \frac{1}{3}Bh \quad B = \frac{1}{2}bh$

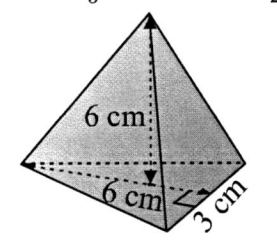

   (pyramid: 6 cm, 6 cm, 3 cm)

3. $V = \frac{4}{3}\pi r^3$

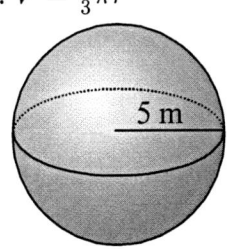

   (sphere: 5 m radius)

4. $V = \frac{1}{3}\pi r^2 h$

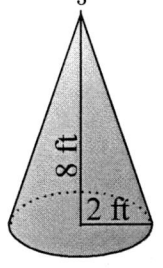

   (cone: 8 ft height, 2 ft radius)

5. $V = \frac{1}{3}Bh \quad B = lw$

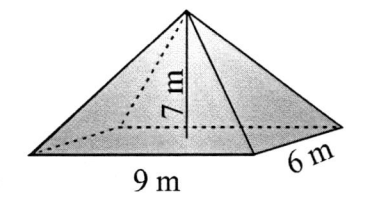

   (pyramid: 7 m height, 9 m, 6 m)

6. $V = \pi r^2 h$

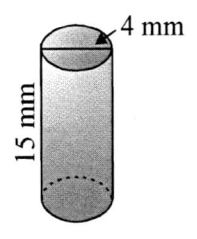

   (cylinder: 15 mm height, 4 mm radius)

7. $V = \frac{4}{3}\pi r^3$

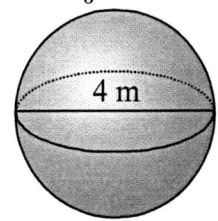

   (sphere: 4 m diameter)

8. $V = \frac{1}{3}Bh \quad B = lw$

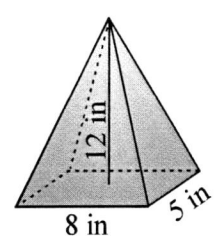

   (pyramid: 12 in height, 8 in, 5 in)

9. $V = \pi r^2 h$

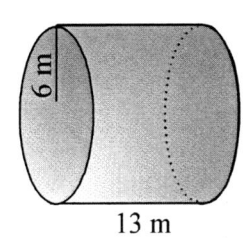

   (cylinder: 6 m radius, 13 m height)

10. $V = \frac{1}{3}Bh \quad B = \frac{1}{2}bh$

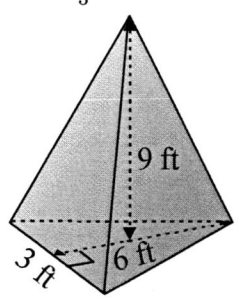

    (pyramid: 9 ft height, 3 ft, 6 ft)

7.9 Volume Word Problems (DOK 3)

1. A pyramid has a square base that measures 500 yards by 500 yards, and the pyramid stands 300 yards tall. What is the volume of the pyramid? The formula for volume of a pyramid is $V = \frac{1}{3}Bh$, where B is the area of the base.

2. Robert is using a barrel to collect rain water for his garden. The circular end has a radius of 1 ft. The barrel is 3 ft tall. How much water will the barrel hold? The formula for volume of a cylinder is $V = \pi r^2 h$. Use $\pi \approx 3.14$.

3. If a basketball measures 24 cm in diameter, what volume of air will it hold? The formula for volume of a sphere is $V = \frac{4}{3}\pi r^3$. Use $\pi \approx 3.14$.

4. What is the volume of a cone that is 2 inches in diameter and 5 inches tall? The formula for volume of a cone is $V = \frac{1}{3}\pi r^2 h$. Use $\pi \approx 3.14$.

5. Kelly has a fish aquarium that measures 24 inches wide, 12 inches long, and 18 inches tall. What is the maximum amount of water that the aquarium will hold? The formula for volume of a rectangular prism is $V = (lwh)$.

6. Jeff built a toy box for his son. Each side of the box measures 2 ft. How many cubic feet of toys will the box hold?

7. Three blocks are stacked on top of each other in the figure below. A rectangular hole measuring 1 cm on each side runs through the center of all the blocks. Find the volume of the figure below.

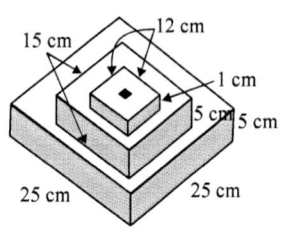

8. In the figure below, 3 cylinders are stacked on top of each other. The radii of the cylinders are 2 inches, 4 inches, and 6 inches. The height of each cylinder is one inch. Find the volume of the figure below. Use $\pi \approx 3.14$.

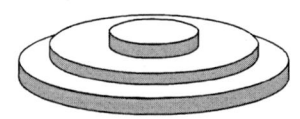

9. Samantha is building a model of the building she lives in. The model is shown below. Find the volume of the model.

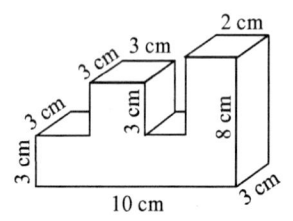

10. A hole, 1 m in diameter, has been cut through the cylinder in the figure below. Find the volume of the figure.

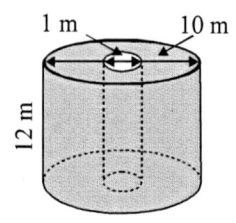

7.10 Geometric Relationships of Solids (DOK 3)

Now you will learn about the relationships between 3-dimensional figures. The formulas for finding the volumes of geometric solids are given below.

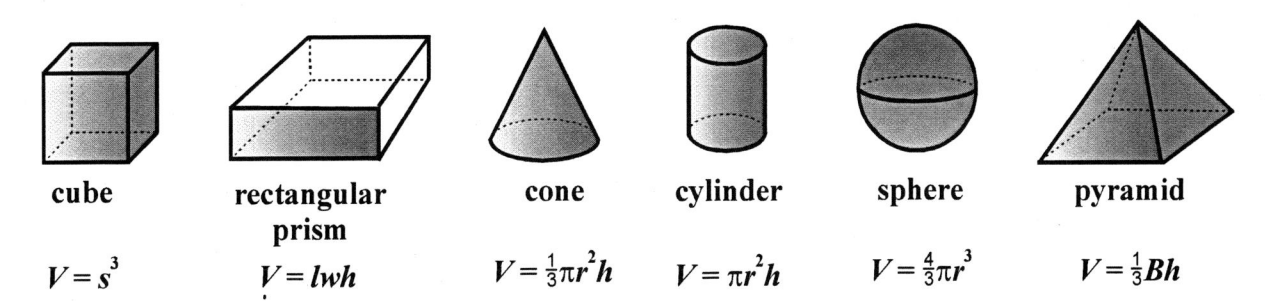

cube	rectangular prism	cone	cylinder	sphere	pyramid
$V = s^3$	$V = lwh$	$V = \frac{1}{3}\pi r^2 h$	$V = \pi r^2 h$	$V = \frac{4}{3}\pi r^3$	$V = \frac{1}{3}Bh$

By studying each formula and by comparing formulas between different solids, you can determine general relationships.

Example 6: How would doubling the radius of a sphere affect the volume?

The volume of a sphere is $V = \frac{4}{3}\pi r^3$. By doubling the radius, the volume would increase 8 times the original volume. A sphere with a radius of 2 would have a volume 8 times greater than a sphere with a radius of 1.

Example 7: A cylinder and a cone have the same radius and the same height. What is the difference between their volumes?

Compare the formulas for the volume of a cone and the volume of a cylinder. They are identical except that the cone is multiplied by $\frac{1}{3}$. Therefore, the volume of a cone with the same height and radius as a cylinder would be one-third less. Or, the volume of a cylinder with the same height and radius as a cone would be three times greater.

Example 8: If you double one dimension of a rectangular prism, how will the volume be affected? What if you double two dimensions? What if you double all three dimensions?

Doubling just one of the dimensions of a rectangular prism will also double the volume. Doubling two of the dimensions will cause the volume to increase 4 times the original volume. Doubling all three dimensions will cause the volume to increase to 8 times the original volume.

Example 9: A cylinder holds 100 cubic centimeters of water. If you triple the radius of the cylinder but keep the height the same, how much water would you need to fill the new cylinder?

Tripling the radius of a cylinder causes the volume to increase by 3^2 or 9 times the original volume. The volume of the new cylinder would hold 9×100 or 900 cubic centimeters of water.

Chapter 7 Solid Geometry

Answer the following questions by comparing the volumes of two solids that share some of the same dimensions.

1. If you have a cylinder with a height of 8 inches and a radius of 4 inches, and you have a cone with the same height and radius, how many times greater is the volume of the cylinder than the volume of the cone? Use the following formulas as needed: $V = \pi r^2 h$, $V = \frac{1}{3}\pi r^2 h$

2.

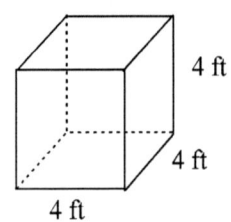

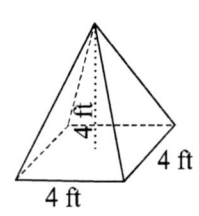

   In the two figures above, how many times greater is the volume of the cube than the volume of the pyramid? Use the following formulas as needed: $V = s^3$, $V = \frac{1}{3}Bh$

3. How many times greater is the volume of a cylinder if you double the radius? $V = \pi r^2 h$

4. How many times greater is the volume of a cylinder if you double the height? $V = \pi r^2 h$

5. In a rectangular prism, how many times greater is the volume if you double the length? $V = lwh$

6. In a rectangular prism, how many times greater is the volume if you double the length and the width? $V = lwh$

7. In a rectangular prism, how many times greater is the volume if you double the length and the width and the height? $V = lwh$

8. In the following two figures, how many cubes like Figure 1 will fit inside Figure 2? $V = s^3$

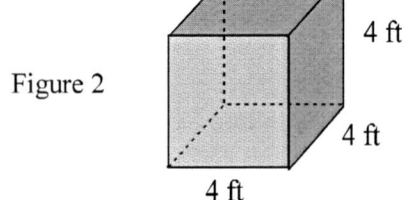

9. A sphere has a radius of 1. If the radius is increased to 3, how many times greater will the volume be? $V = \frac{4}{3}\pi r^3$

10. It takes 2 liters of water to fill cone A below. If the cone is stretched so the radius is doubled, but the height stays the same, how much water is needed to fill the new cone, B? $V = \frac{1}{3}\pi r^2 h$

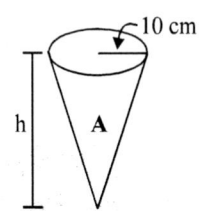

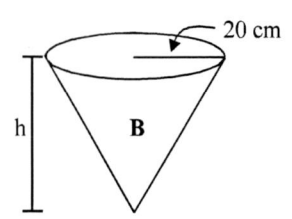

7.11 Applying Similar Solid Figures (DOK 3)

Solid figures are similar if they are the exact same shape but different sizes. Every cube is similar to every other cube, and every sphere is similar to every other sphere. The corresponding sides of similar figures are in **proportion**.

Models are similar to the original figure, and all the corresponding pieces are in the same proportion.

Example 10: If a similar cylinder were drawn with a radius of 8 inches, what would its height equal?

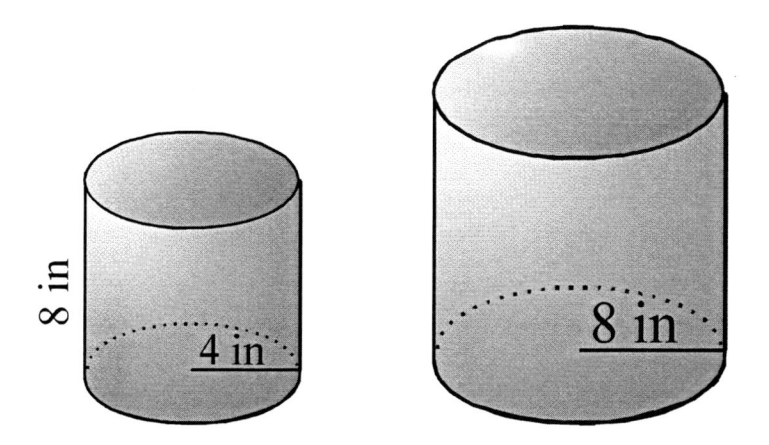

Since the radius of the similar cylinder is twice the radius of the original cylinder, the height of the similar cylinder would have to be in the same proportion. The original height is 8 inches, so the similar cylinder's height would be 8 times 2, or 16 inches.

Answer the following questions.

1. How would doubling the radius of a sphere affect the volume? $V = \frac{4}{3}\pi r^3$.

2. If you double one dimension of a rectangular prism, how will the volume be affected? What if you double two dimensions? What if you double all three dimensions? $V = lwh$

3. A cylinder holds 50 cubic centimeters of water. If you triple the radius and height of the cylinder, how much water would you need to fill the new cylinder? $V = \pi r^2 h$

Chapter 7 Review

Solve the following solid geometry problems. (DOK 2 & 3)

1. $V = \pi r^2 h$

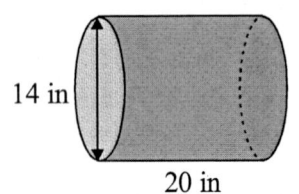

   14 in, 20 in

   Use $\pi \approx \frac{22}{7}$. $V \approx$ _____

2. $V = \frac{1}{3}Bh$ $B = lw$

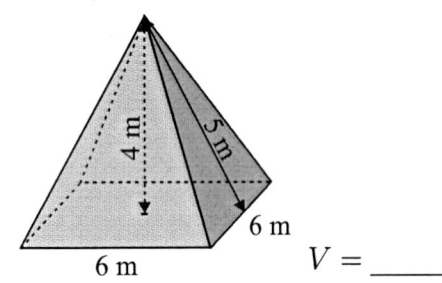

   4 m, 5 m, 6 m, 6 m

   $V =$ _____

3. $V = \frac{1}{3}\pi r^2 h$ Use $\pi \approx 3.14$.

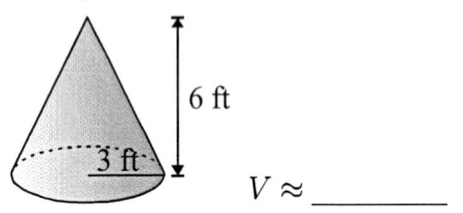

   6 ft, 3 ft

   $V \approx$ _____

4. $V = \frac{1}{3}Bh$ $B = \frac{1}{2}bh$

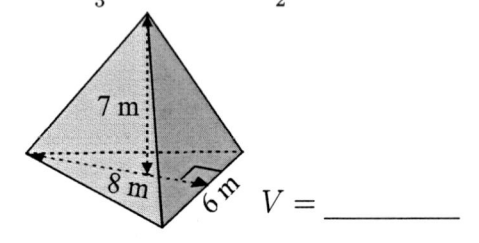

   7 m, 8 m, 6 m

   $V =$ _____

5. $V = \frac{4}{3}\pi r^3$

   7 in

   Use $\pi \approx \frac{22}{7}$.

   $V \approx$ _____

6. Estimate the volume of the figure below. Use the formulas, as needed: $V = \frac{4}{3}\pi r^3$, $V = \pi r^2 h$ Use $\pi \approx 3.14$.

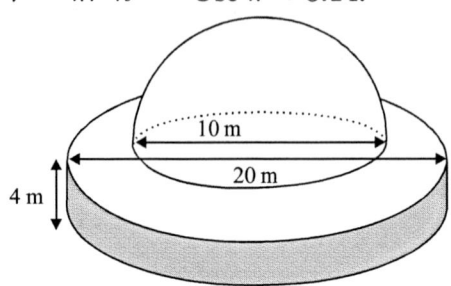

   10 m, 20 m, 4 m

7. Find the volume of the figure below. Use the following formulas as needed. $V = lwh$, $V = \frac{1}{3}Bh$

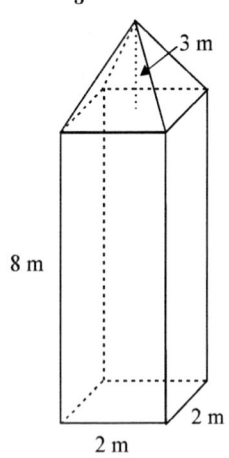

   3 m, 8 m, 2 m, 2 m

8. The sandbox at the local elementary school is 60 inches wide and 100 inches long. The sand in the box is 6 inches deep. How many cubic inches of sand are in the sandbox?

9. A grain silo is in the shape of a cylinder. If the silo has an inside diameter of 10 ft and a height of 35 ft, what is the maximum volume inside the silo? Use $\pi \approx \frac{22}{7}$.

10. A gigantic bronze sphere is being added to the top of a tall building downtown. The sphere will be 24 ft in diameter. What will be the approximate volume of the globe?

Chapter 7 Review

Answer the following questions about cross sections and formations of solid figures. (DOK 2)

11. What three solid figures have a circle as a horizontal cross section?

12. What two solid figures can be formed by moving a square in a straight line through space?

13. What two solid figures have a triangle as a cross section?

14. A sphere is formed when what plane figure moves through space?

15. If you rotate a triangle 360°, what solid figure is formed?

16. What three solids can have a rectangle as a cross section?

Answer the following geometry problems. (DOK 2 & 3)

17. Find the area of the shaded part of the image below. Use $\pi \approx 3.14$. $A = \pi r^2$

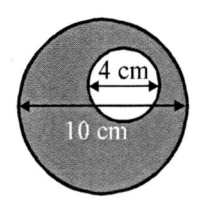

18. Calculate the approximate circumference and area of the following circle. Use the formulas $A = \pi r^2$ and $C = 2\pi r$. Use $\pi \approx \frac{22}{7}$.

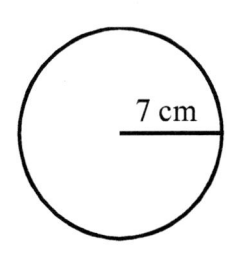

19. Calculate the approximate circumference and area of the following circle. Use the formulas $A = \pi r^2$ and $C = \pi d$. Use $\pi \approx 3.14$.

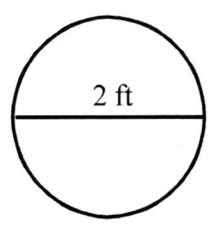

20. Calculate the circumference and the area of a circle with a radius of 2.4 cm. Use the formulas $A = \pi r^2$ and $C = \pi d$. Use $\pi \approx 3.14$.

Copyright © American Book Company

Chapter 7 Solid Geometry

21. Calculate the circumference and the area of a circle with a diameter of 4 ft. Use the formulas $A = \pi r^2$ and $C = \pi d$. Use $\pi \approx \frac{22}{7}$.

22. A parking lot has a population density of 0.75 cars/(parking spots). If there are 117 cars in the parking lot right now, how many parking spaces are not being used?

23. A block of graphite has the dimensions 4.5 cm by 5.2 cm by 6.0 cm. The block has a mass of 1587 g. What is the density of the block of graphite? Round your answer to the nearest tenth.

24. Alana is working on the lighting crew for a school play. The ceiling supports for the lighting trusses are 15 ft back from the center of the stage. The lights must form at least a 40° angle from the center of the stage or it is difficult for the actors to see. How high (in relation to the stage) should Alana hang the lighting trusses to achieve this angle? Round your answer to the nearest whole number.

25. A local restaurant offers a delivery service to their customers. The restaurant manager just received a call for two deliveries. The first delivery will need to go to a house that is located 8 miles north of the restaurant, and the second delivery will need to go to a house that is 15 miles east of the restaurant. It costs the restaurant $0.30/mile for deliveries. Is it more cost effective for the manager to send one employee to both houses and back, or two employees, one to each house and back?

26. How many 1-inch cubes will fit inside a larger 1 ft cube? (Figures are not drawn to scale.) $V = s^3$

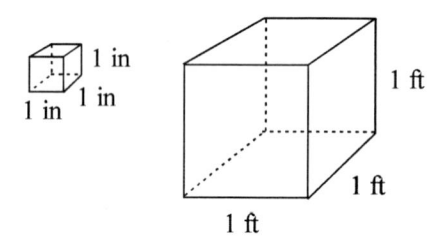

27. What is the ratio of the volumes of the following two spheres? $V = \frac{4}{3}\pi r^3$

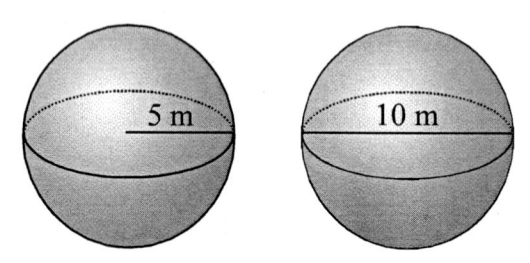

Chapter 7 Review

28. What is the ratio of the surface areas of the following cubes?

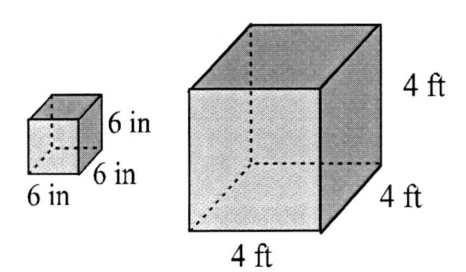

6 in, 6 in, 6 in

4 ft, 4 ft, 4 ft

29. What is the ratio of surface areas of the following similar cylinders?
$SA = 2\pi r^2 + 2\pi rh$

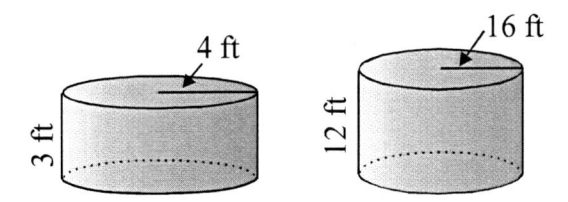

4 ft, 3 ft

16 ft, 12 ft

30. If a similar pyramid were drawn with a height of 8 m, what would the surface area equal?
$SA = B + \frac{1}{2}Pl$

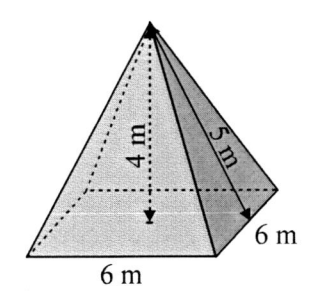

4 m, 5 m, 6 m, 6 m

Chapter 7 Solid Geometry

Chapter 7 Test

1 What is the volume, in cubic feet, of the square pyramid below? Use the formula $V = \frac{1}{3}Bh$.

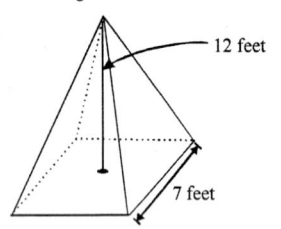

12 feet
7 feet

A 168 cubic feet
B 196 cubic feet
C 294 cubic feet
D 588 cubic feet   (DOK 2)

2 What is the volume of the following oil tank? Round your answer to the nearest hundredth. Use the formula $V = \pi r^2 h$, where $\pi \approx 3.14$.

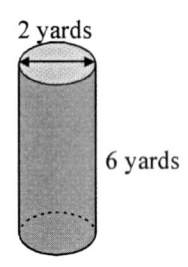

2 yards
6 yards

A 18.84 yd^3
B 37.68 yd^3
C 44.48 yd^3
D 75.36 yd^3 (DOK 2)

3 Find the approximate volume of the cone. Use the formula $V = \frac{1}{3}\pi r^2 h$, where $\pi \approx 3.14$.

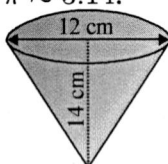

12 cm
14 cm

A 88 cm^3
B 176 cm^3
C 528 cm^3
D 2112 cm^3 (DOK 2)

4 What is volume of the box shown below? $V = lwh$

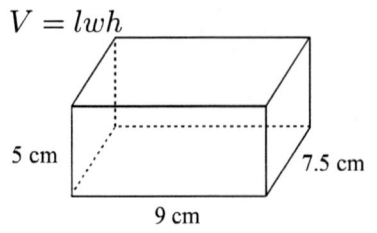

5 cm   7.5 cm
9 cm

A 22.5 cm^3
B 337.5 cm^3
C 1875 cm^3
D 2250 cm^3 (DOK 2)

5 Find the volume of the cube. Use the formula $V = s^3$.

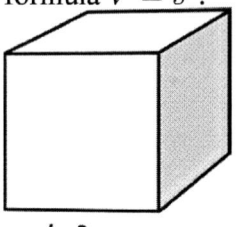

$l = 3$ cm

A 36 cm^3
B 9 cm^3
C 27 cm^3
D 12 cm^3 (DOK 2)

6 Find the volume of the figure below. Use the formula $V = lwh$.

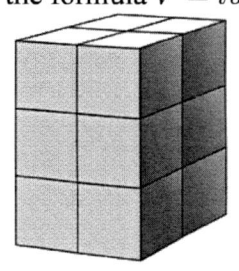

A 6 units^3
B 12 units^3
C 36 units^3
D 72 units^3 (DOK 2)

Chapter 7 Test

7 If a hole with a 3 inch diameter is cut through the cylinder, what is the volume? $V = \pi r^2 h$

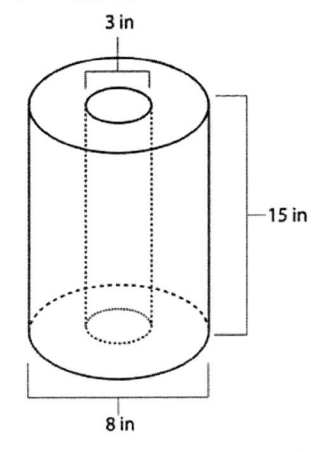

3 in
15 in
8 in

- **A** 1178 in³
- **B** 648 in³
- **C** 754 in³
- **D** 860 in³ (DOK 3)

8 If a sphere with a 6 m radius is cut out of a cube like the one shown below, what would the new volume of the cube be?

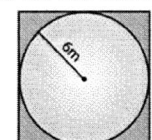

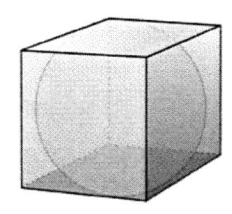

- **A** 823.22 m³
- **B** 1728 m³
- **C** 904.78 m³
- **D** 1441 m (DOK 3)

9 What shape is the vertical cross section of a cone?

- **A** triangle
- **B** circle
- **C** oval
- **D** square (DOK 2)

10 If you move a square through space in a straight line, what solid could it form?

- **A** cube
- **B** rectangular prism
- **C** pyramid
- **D** either A and B (DOK 2)

11 What shape does the horizontal cross section of the figure below create?

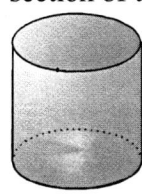

- **A** circle
- **B** cylinder
- **C** rectangle
- **D** oval (DOK 2)

12 What is the area of a circle with a radius of 9 cm? $A = \pi r^2$ (Round to the nearest whole number.)

- **A** 254 square cm
- **B** 196 square cm
- **C** 347 square cm
- **D** 616 square cm (DOK 2)

13 Find the approximate circumference. Use $\pi \approx 3.14$. $C = 2\pi r$

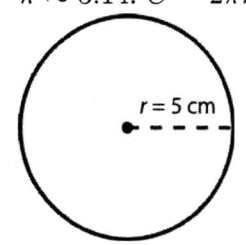

r = 5 cm

- **A** 15.7 cm
- **B** 62.8 cm
- **C** 31.4 cm
- **D** 0.314 cm (DOK 2)

Copyright © American Book Company

Chapter 7 Solid Geometry

14 Find the approximate area. Use $\pi \approx 3.14$.
$A = \pi r^2$

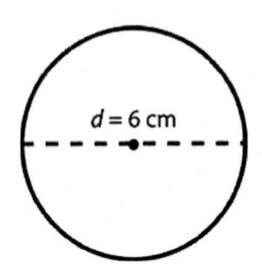

$d = 6$ cm

A 113.04 cm^2
B 28.26 cm^2
C 18.84 cm^2
D 188.4 cm^2 (DOK 2)

15 What is the circumference of a circle that has a diameter of 10 cm? $C = \pi d$

A 15.7 cm

B 31.4 cm

C 78.5 cm

D 310 cm (DOK 2)

16 An architect is designing a concert hall. He plans on building a 2nd floor balcony located 30 feet away from the stage and wants it to be at an angle of 40° to the vertical from the floor of the stage. If the stage is 2 feet higher than ground level, then how high should he build the balcony from the ground floor? Round your answer to the nearest hundredth.

A 35.75 feet
B 37.75 feet
C 39.75 feet
D 41.75 feet (DOK 3)

17 Goldfish care manuals recommend that you have only 12 goldfish per 10 gallons of water. If Alex owns a 40 gallon tank that has a population density of 1.75 fish/gallon, is Alex meeting the recommendations?

A No, Alex has more than 12 goldfish per 10 gallons.
B Yes, Alex has exactly 12 goldfish per 10 gallons.
C Yes, Alex has fewer than 12 goldfish per 10 gallons.
D There is no way to determine how many goldfish Alex actually has. (DOK 3)

18 Tim decided to paint the walls of his storage room which measures 7.5 ft by 10 ft. The room is 7.8 ft in height and has two 2 ft by 3 ft windows and a door that measures 3 ft by 7 ft. Tim needs to buy paint to cover how many square feet?

A 207 ft^2
B 219 ft^2
C 228 ft^2
D 240 ft^2 (DOK 3)

19 In science class, Jim is filling a graduated cylinder with ethyl alcohol. The container has a diameter of 2 inches and a height of 10 inches. When full, it holds 141.3 g of ethyl alcohol. Calculate the density of ethyl alcohol.

A 1.125 g/in^3
B 2.5 g/in^3
C 4.5 g/in^3
D 5.125 g/in^3 (DOK 3)

Chapter 7 Test

20 What is the reduced ratio of the surface areas of the following two spheres?
Use the formula $SA = 4\pi r^2$, where $\pi \approx 3.14$.

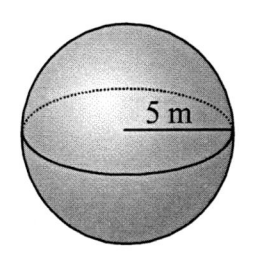

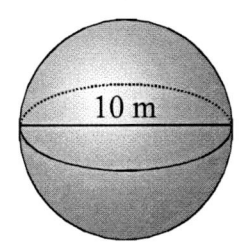

A 1:2
B 1:4
C 1:1
D 1:$\frac{1}{2}$ (DOK 2)

21 If the radius of a cylinder is doubled, how much larger will the volume be?
Use the formula $V = \pi r^2 h$, where $\pi \approx 3.14$.

A 12 times larger
B 8 times larger
C 4 times larger
D 2 times larger (DOK 3)

22 If the radius of a sphere is tripled, how much larger will the volume be?
Use the formula $V = \frac{4}{3}\pi r^3$, where $\pi \approx 3.14$.

A 81 times larger
B 9 times larger
C 3 times larger
D 27 times larger (DOK 3)

23 What is the reduced ratio of the volumes of the following cubes?
Use the formula $V = s^3$.

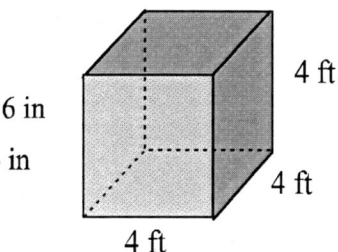

A 1:512
B 216:1728
C 6:4
D 6:48 (DOK 2)

24 The ratio of the radii of two circles is 1 to 3. What is the ratio of their areas?
Use the formula $A = \pi r^2$, where $\pi \approx 3.14$.

A $\frac{1}{2}$
B $\frac{1}{3}$
C $\frac{1}{9}$
D $\frac{1}{16}$ (DOK 2)

25 Terry built the box with measurements of 2 feet by 3 feet by 4 feet for his sister. He wants to build another box just like it, but instead he doubles each measurement. How many times larger will the volume of the new box be? Use the formula $V = lwh$.

A 2 times larger
B 4 times larger
C 8 times larger
D 16 times larger (DOK 3)

Copyright © American Book Company

Chapter 7 Solid Geometry

26 The first cylinder below has a height of 10 inches and a radius of 4 inches. For the second cylinder, the radius and height are doubled. Comparing the volumes, which of the following statements is true? $V = \pi r^2 h$.

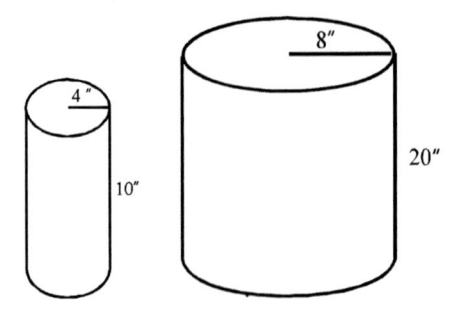

A The volume of the second cylinder is 4 times that of the first cylinder.

B The volume of the second cylinder is half that of the first cylinder.

C The volume of the second cylinder is 8 times that of the first cylinder.

D The volume of the second cylinder is double that of the first cylinder.

(DOK 3)

Chapter 8
Exponents

This chapter covers the following CCGPS standard(s):

| The Real Number System | N.RN.1, N.RN.2 |

8.1 Understanding Exponents (DOK 1)

Sometimes it is necessary to multiply a number by itself one or more times. For example, a math problem may need to multiply 3×3 or $5 \times 5 \times 5 \times 5$. In these situations, mathematicians have come up with a shorter way of writing out this kind of multiplication. Instead of writing 3×3, you can write 3^2, or instead of writing $5 \times 5 \times 5 \times 5$, 5^4 means the same thing. The first number is the **base**. The small, raised number is called the **exponent** or **power**. The exponent tells how many times the base should be multiplied by itself.

Example 1: 6^3 ← exponent (or power)
 ← base
This means multiply by 6 three times: $6 \times 6 \times 6$

Example 2: $4^1 = 4 \quad 10^1 = 10 \quad 25^1 = 25 \quad 4^0 = 1 \quad 10^0 = 1 \quad 25^0 = 1$

All rational numbers can have exponents.

Examples:
$$\left(\frac{1}{4}\right)^3 = \frac{1}{4} \times \frac{1}{4} \times \frac{1}{4} = \frac{1}{64}$$

$$\left(1\tfrac{1}{2}\right)^2 = \left(\frac{3}{2}\right)^2 = \frac{3}{2} \times \frac{3}{2} = \frac{9}{4}$$

$$0.2^3 = 0.2 \times 0.2 \times 0.2 = 0.008$$

$$\left(\frac{x}{y}\right)^2 = \frac{x}{y} \times \frac{x}{y} = \frac{x^2}{y^2}$$

Rewrite the following problems using exponents. (DOK 1)

Example 3: $2 \times 2 \times 2 = 2^3$

1. $7 \times 7 \times 7 \times 7$
2. 10×10
3. $12 \times 12 \times 12$
4. $4 \times 4 \times 4 \times 4$
5. $9 \times 9 \times 9$
6. 25×25

Copyright © American Book Company

147

Chapter 8   Exponents

Use your calculator to determine what product each number with an exponent represents. (DOK 1)

Example 4: $2^3 = 2 \times 2 \times 2 = 8$

7. 8^3
8. 12^2
9. 20^1
10. 5^4
11. 15^0
12. 16^2
13. 10^2
14. 3^5

Express each of the following numbers as a base with an exponent. (DOK 1)

Example 5: $4 = 2 \times 2 = 2^2$

15. 9
16. 16
17. 27
18. 36
19. 8
20. 32
21. 1000
22. 125

8.2   Multiplying Exponents with the Same Base (DOK 1)

To multiply two expressions with the same base, add the exponents together and keep the base the same.

Example 6: $2^3 \times 2^5 = (2 \times 2 \times 2)(2 \times 2 \times 2 \times 2 \times 2) = 2^8$

Example 7: $3a^2 \times 2a^3 = 6a^{2+3} = 6a^5$
   Notice that only the "a" is raised to a power and not the 3 or the 2.

Simplify each of the expressions below. (DOK 1)

1. $2^3 \times 2^5$
2. $x^5 \times x^3$
3. $2a^3 \times 3a^3$
4. $4^5 \times 4^3$
5. $2x^3 \times x^5$
6. $4b^3 \times 2b^4$
7. $10^5 \times 10^4$
8. $5^2 \times 5^4$
9. $3^3 \times 3^2$
10. $4x \times x^2$
11. $a^2 \times 3a^4$
12. $2^3 \times 2^4$

8.3   Multiplying Fractional Exponents with the Same Base (DOK 1)

To multiply two expressions with the same base, add the exponents together and keep the base the same. Numbers with **fractional exponents** follow the same rules as numbers with whole numbers as the exponent.

Example 8: $(4)^{\frac{1}{2}} \times (4)^{\frac{3}{2}} = 4^{\frac{1}{2}+\frac{3}{2}} = 4^{\frac{4}{2}} = 4^2$

Example 9: $2x^{\frac{3}{4}} \times 5x^{\frac{1}{5}} = 10x^{\frac{3}{4}+\frac{1}{5}} = 10x^{\frac{15}{20}+\frac{4}{20}} = 10x^{\frac{19}{20}}$
   Notice that only the "x" is raised to a power and not the 2 or the 5.

8.5 Expressions Raised to a Power (DOK 1)

Simplify each of the expressions below. (DOK 1)

1. $y^{\frac{1}{3}} \times y^{\frac{2}{3}}$
2. $(2)^{\frac{5}{9}} \times (2)^{\frac{1}{3}}$
3. $(10)^{\frac{4}{7}} \times (10)^{\frac{5}{7}}$
4. $b^{\frac{1}{2}} \times b^{\frac{1}{2}}$
5. $x^{\frac{1}{6}} \times x^{\frac{2}{3}}$
6. $(6)^{\frac{1}{12}} \times (6)^{\frac{3}{4}}$
7. $(3)^{\frac{5}{12}} \times (3)^{\frac{7}{2}}$
8. $(12)^{\frac{13}{25}} \times (12)^{\frac{1}{5}}$
9. $2a^{\frac{7}{9}} \times a^{\frac{4}{9}}$
10. $(8)^{\frac{2}{5}} \times (8)^{\frac{1}{10}}$
11. $4a^{\frac{6}{7}} \times 4a^{\frac{4}{5}}$
12. $3(x)^{\frac{4}{13}} \times 5(x)^{\frac{1}{2}}$

8.4 Multiplying Exponents Raised to an Exponent (DOK 1)

If a power is raised to another power, multiply the exponents together and keep the base the same.

Example 10: $(2^3)^2 = (2^3)(2^3) = (2 \times 2 \times 2)(2 \times 2 \times 2) = 2^{3 \times 2} = 2^6$

Example 11: $(y^4)^3 = y^{4 \times 3} = y^{12}$

Simplify each of the expressions below. (DOK 1)

1. $(5^3)^2$
2. $(x^5)^2$
3. $(6^2)^5$
4. $(3^4)^2$
5. $(3^2)^4$
6. $(y^2)^3$
7. $(3^3)^2$
8. $(9^2)^2$
9. $(7^2)^2$
10. $(a^4)^2$
11. $(5^4)^5$
12. $(x^3)^2$

8.5 Expressions Raised to a Power (DOK 1)

If an expression is raised to a power, do not raise each term to the power, but rather consider the expression as a whole and raise it to the power.

Example 12: $(2+3)^2 = (2+3)(2+3) = (5)(5) = 25$

Example 13: $(2-5)^3 = (2-5)(2-5)(2-5) = (-3)(-3)(-3) = -27$

Simplify each of the expressions below. (DOK 1)

1. $(2+6)^2$
2. $(1-2)^2$
3. $(3+2)^2$
4. $(4+5)^2$
5. $(5+1)^3$
6. $(2+3)^3$
7. $(4-6)^2$
8. $(2+1)^4$
9. $(25+15)^0$
10. $(4+1+7)^2$
11. $(5-9+2)^3$
12. $(2+1+4)^2$
13. $(-7+-7+6)^3$
14. $(2+2-5+4)^2$
15. $(20+45+40)^0$

Copyright © American Book Company

Chapter 8   Exponents

8.6   Fractions Raised to a Power (DOK 1)

A fraction can also be raised to a power.

Example 14: $\left(\dfrac{3}{4}\right)^3 = \dfrac{3^3}{4^3} = \dfrac{27}{64}$

Simplify the following fractions. (DOK 1)

1. $\left(\dfrac{2}{3}\right)^2$
2. $\left(\dfrac{7}{8}\right)^3$
3. $\left(\dfrac{1}{2}\right)^2$
4. $\left(\dfrac{1}{4}\right)^2$
5. $\left(\dfrac{2}{3}\right)^3$
6. $\left(\dfrac{3}{4}\right)^2$
7. $\left(\dfrac{1}{2}\right)^3$
8. $\left(\dfrac{5}{7}\right)^2$
9. $\left(\dfrac{2}{3}\right)^4$
10. $\left(\dfrac{3}{10}\right)^2$
11. $\left(\dfrac{4}{5}\right)^2$
12. $\left(\dfrac{1}{10}\right)^4$

8.7   More Multiplying Exponents (DOK 1)

If a product in parentheses is raised to a power, then each factor is raised to the power when parentheses are eliminated.

Example 15: $(2 \times 4)^2 = 2^2 \times 4^2 = 4 \times 16 = 64$

Example 16: $(3a)^3 = 3^3 \times a^3 = 27a^3$

Example 17: $(7b^5)^2 = 7^2 b^{10} = 49b^{10}$

Simplify each of the following. (DOK 1)

1. $(2^3)^2$
2. $(7a^5)^2$
3. $(6b^2)^2$
4. $(3^2)^2$
5. $(3 \times 5)^2$
6. $(3x^4)^2$
7. $(6y^7)^2$
8. $(11w^3)^2$
9. $(3^3)^2$
10. $(3 \times 3)^2$
11. $(2a)^4$
12. $(2^2)^3$
13. $(3 \times 2)^3$
14. $(5^3)^2$
15. $(4r^7)^3$
16. $(2m^3)^2$
17. $(6 \times 4)^2$
18. $(9a^5)^2$
19. $(5x^4)^2$
20. $(9^2)^2$
21. 4×4^3
22. $(3a)^2$
23. $(2 \times 3)^3$
24. $(5p^4)^3$
25. $(4y^4)^2$
26. $(2b^3)^4$
27. $(5a^2)^2$
28. $(8a^3)^2$
29. $(2 \times 6)^2$
30. $(7^2)^2$

8.8 More Multiplying Fractional Exponents (DOK 2)

If a product in parentheses is raised to a power, then each factor is raised to the power when the parentheses are eliminated. This rule applies to fractional exponents as well.

Example 18: $(2x)^{\frac{2}{3}} = 2^{\frac{2}{3}} \times x^{\frac{2}{3}} = 2^{\frac{2}{3}} x^{\frac{2}{3}}$

Example 19: $(3a^2)^{\frac{1}{4}} = 3^{\frac{1}{4}} a^{\frac{2}{4}} = 3^{\frac{1}{4}} a^{\frac{1}{2}}$

Simplify each of the following. (DOK 2)

1. $(5)^{\frac{1}{2}}$
2. $(7x)^{\frac{6}{7}}$
3. $(14x^2)^{\frac{2}{9}}$
4. $\left(x^{\frac{5}{7}}\right)^{\frac{1}{5}}$
5. $(x^3)^{\frac{1}{3}}$
6. $(4x^5)^{\frac{4}{25}}$
7. $(23y)^{\frac{1}{2}}$
8. $\left(3x^{\frac{1}{2}}\right)^{\frac{2}{5}}$
9. $(16)^{\frac{1}{2}}$
10. $(11x^2)^{\frac{5}{2}}$
11. $(8)^{\frac{6}{13}}$
12. $\left(4a^{\frac{7}{2}}\right)^{\frac{1}{2}}$
13. $(5t)^{\frac{9}{11}}$
14. $(z^9)^{\frac{1}{9}}$
15. $(17)^{\frac{3}{4}}$
16. $\left(9y^{\frac{1}{2}}\right)^{\frac{1}{3}}$
17. $(8)^{\frac{1}{4}}$
18. $(33x)^{\frac{4}{7}}$
19. $(x^4)^{\frac{3}{8}}$
20. $\left(12x^{\frac{7}{9}}\right)^{\frac{3}{7}}$
21. $(4w)^{\frac{1}{6}}$
22. $(z^2)^{\frac{2}{13}}$
23. $(15)^{\frac{1}{15}}$
24. $\left(5a^{\frac{2}{9}}\right)^{\frac{3}{10}}$

Chapter 8 Exponents

8.9 Negative Exponents (DOK 1)

Expressions can also have negative exponents. Negative exponents do not indicate negative numbers. They indicate **reciprocals**. The **reciprocal** of a number is 1 divided by that number. For example, the reciprocal of 2 is $\frac{1}{2}$. (A number multiplied by its reciprocal is equal to 1.) If the negative exponent is in the bottom of a fraction (denominator), the reciprocal will put the expression on the top of the fraction (numerator) without a negative sign.

Example 20: $2^{-3} = \dfrac{1}{2^3} = \dfrac{1}{8}$

Example 21: $3a^{-5} = 3 \times \dfrac{1}{a^5} = \dfrac{3}{a^5}$ Notice that the 3 is not raised to the -5 power, only the a.

Example 22: $\dfrac{6}{5x^{-2}} = \dfrac{6x^2}{5}$ The 5 is not raised to the -2 power, only the x.

Rewrite using only positive exponents. (DOK 1)

1. $5m^{-6}$
2. $\dfrac{5x^{-4}}{7}$
3. $14z^{-8}$
4. $\dfrac{1}{5s^{-4}}$
5. $14h^{-5}$
6. $\dfrac{h^{-3}}{5}$
7. $\dfrac{2y^{-3}}{4}$
8. x^{-4}
9. $-2y^{-2}$
10. $5y^{-5}$
11. $\dfrac{x^{-3}}{5}$
12. $10z^{-7}$
13. $7x^{-3}$
14. r^{-2}
15. $\dfrac{m^{-4}}{6}$

8.10 Multiplying with Negative Exponents (DOK 2)

Multiplying with negative exponents follows the same rules as multiplying with positive exponents.

Example 23: $6^2 \cdot 6^{-3} = 6^{2+(-3)} = 6^{-1} = \dfrac{1}{6}$

Example 24: $(5a \times 2)^{-3} = (10a)^{-3} = \dfrac{1}{(10a)^3} = \dfrac{1}{1000a^3}$

Example 25: $(7a^2)^{-3} = 7^{-3}a^{-6} = \dfrac{1}{7^3 a^6}$

Simplify the following. Answers should <u>not</u> have any negative exponents. (DOK 2)

1. $5^{-2} \cdot 5^5$
2. $(6^3 \cdot 6^{-2})^{-2}$
3. $10^{-4} \cdot 10^2$
4. $11^{-5} \cdot 11^7$
5. $4^7 \cdot 4^{-10}$
6. $20^8 \cdot 20^{-6}$
7. $5^{-8} \cdot 5^4$
8. $(2^{-2} \cdot 2^3)^{-4}$
9. $7^{-2} \cdot 7^{-1}$
10. $(3x^4)^{-3}$
11. $12^{-10} \cdot 12^8$
12. $(10^8 \cdot 10^{-10})^2$
13. $3^{-2} \cdot 2^{-2}$
14. $(8x^5)^{-4}$
15. $(6b^3)^{-6}$
16. $(9y)^{-2}$

8.11 Dividing with Exponents (DOK 1)

Exponents that have the same base can also be divided.

Example 26: $\dfrac{3^5}{3^3}$ This problem means $3^5 \div 3^3$. Let us look at 2 ways to solve this problem.

Solution 1: $\dfrac{3^5}{3^3} = \dfrac{3 \cdot 3 \cdot 3 \cdot 3 \cdot 3}{3 \cdot 3 \cdot 3} = 3 \cdot 3 = 9$ First, rewrite the fraction with the exponents in expanded form, and then multiply.

Solution 2: $\dfrac{3^5}{3^3} = 3^{5-3} = 3^2 = 9$ A quick way to simplify this same problem is to subtract the exponents. **When dividing exponents with the same base, subtract the exponents.**

Example 27: $\dfrac{(4x)^{-3}}{2x^4}$

Step 1: $(4x)^{-3} = \dfrac{1}{(4x)^3} = \dfrac{1}{4^3 x^3}$ Remove the parentheses from the top of the fraction.

Step 2: $\dfrac{1}{4^3 x^3 \cdot 2x^4} = \dfrac{1}{128 x^7}$ The bottom of the fraction remains the same, so put the two together and simplify.

Simplify the problems below. You may be able to cancel. Be sure to follow order of operations. Remove parentheses before canceling. (DOK 1)

1. $\dfrac{5^5}{5^3}$
2. $\dfrac{x^2}{x^3}$
3. $\dfrac{(10^2)^4}{10^5}$
4. $\dfrac{3^5}{3^2}$
5. $\dfrac{8^{10}}{8^8}$
6. $\dfrac{5^2}{5}$

7. $\dfrac{(7^2)^3}{7^5}$
8. $\dfrac{(x^3)^4}{x^6}$
9. $\dfrac{4^3}{4^2}$
10. $\dfrac{2}{(2^2)^2}$
11. $\dfrac{(3x)^{-2}}{9x^5}$
12. $\dfrac{(11^4)^4}{(11^7)^2}$

13. $\dfrac{x^3}{(x^2)^3}$
14. $\dfrac{2^2}{2^7}$
15. $\dfrac{6^2}{6}$
16. $\dfrac{9^{11}}{9^9}$
17. $\dfrac{(15)^5}{15^6}$
18. $\dfrac{(x^3)^{-2}}{(x^2)^5}$

19. $\dfrac{12^{-4}}{12^{-2}}$
20. $\dfrac{6^{12}}{6^9}$
21. $\dfrac{8^8}{8^{10}}$
22. $\dfrac{3(x^{-3})^{-2}}{3x^7}$
23. $\dfrac{7^3}{7^5}$
24. $\dfrac{10^3}{10^{-1}}$

Chapter 8   Exponents

8.12   Dividing with Fractional Exponents (DOK 2)

Fractional exponents that have the same base can also be divided.

Example 28: $\dfrac{5^{\frac{3}{4}}}{5^{\frac{1}{2}}}$ This problem means $5^{\frac{3}{4}} \div 5^{\frac{1}{2}}$.

Solution: $\dfrac{5^{\frac{3}{4}}}{5^{\frac{1}{2}}} = 5^{\frac{3}{4} - \frac{1}{2}} = 5^{\frac{3}{4} - \frac{2}{4}} = 5^{\frac{1}{4}}$

A quick way to simplify this problem is to subtract the exponents. **When dividing exponents with the same base, subtract the exponents.**

Simplify the problems below. (DOK 2)

1. $\dfrac{x^{\frac{7}{9}}}{x^{\frac{1}{9}}}$

2. $\dfrac{2^{\frac{3}{4}}}{2^{\frac{3}{5}}}$

3. $\dfrac{12^{\frac{6}{7}}}{12^{\frac{1}{2}}}$

4. $\dfrac{a^{\frac{1}{3}}}{a^{\frac{1}{6}}}$

5. $\dfrac{6^{\frac{3}{5}}}{6^{\frac{1}{7}}}$

6. $\dfrac{x^{\frac{13}{25}}}{x^{\frac{2}{5}}}$

7. $\dfrac{y^{\frac{8}{9}}}{y^{\frac{2}{3}}}$

8. $\dfrac{8^{\frac{3}{4}}}{8^{\frac{1}{5}}}$

9. $\dfrac{2^{\frac{14}{15}}}{2^{\frac{4}{15}}}$

10. $\dfrac{7^{\frac{6}{7}}}{7^{\frac{2}{7}}}$

11. $\dfrac{b^{\frac{1}{2}}}{b^{\frac{1}{6}}}$

12. $\dfrac{5^{\frac{9}{10}}}{5^{\frac{2}{5}}}$

13. $\dfrac{4^{\frac{9}{17}}}{4^{\frac{1}{2}}}$

14. $\dfrac{9^{\frac{13}{15}}}{9^{\frac{2}{3}}}$

15. $\dfrac{17^{\frac{2}{5}}}{17^{\frac{1}{7}}}$

16. $\dfrac{3^{\frac{2}{3}}}{3^{\frac{1}{12}}}$

17. $\dfrac{x^{\frac{3}{4}}}{x^{\frac{4}{9}}}$

18. $\dfrac{2^{\frac{3}{2}}}{2^{\frac{9}{10}}}$

19. $\dfrac{16^{\frac{5}{3}}}{16^{\frac{7}{8}}}$

20. $\dfrac{10^{\frac{3}{24}}}{10^{\frac{1}{24}}}$

8.13 Order of Operations (DOK 2)

In long math problems with +, −, ×, ÷, (), and exponents in them, you have to know what to do first. Without following the same rules, you could get different answers. If you will memorize the silly sentence, Please Excuse My Dear Aunt Sally, you can memorize the order you must follow.

Please "P" stands for parentheses. You must get rid of parentheses first.
Examples: $3(1+4) = 3(5) = 15$
$6(10-6) = 6(4) = 24$

Excuse "E" stands for exponents. You must eliminate exponents next.
Example: $4^2 = 4 \times 4 = 16$

My Dear "M" stands for multiply. "D" stands for divide. Start on the left of the equation and perform all multiplications and divisions in the order in which they appear.

Aunt Sally "A" stands for add. "S" stands for subtract. Start on the left and perform all additions and subtractions in the order they appear.

Example 29: $12 \div 2(6-3) + 3^2 - 1$

Please	Eliminate **parentheses**. $6-3=3$ so now we have	$12 \div 2 \times 3 + 3^2 - 1$
Excuse	Eliminate **exponents**. $3^2 = 9$ so now we have	$12 \div 2 \times 3 + 9 - 1$
My Dear	**Multiply** and **divide** next in order from left to right.	$12 \div 2 = 6$ then $6 \times 3 = 18$
Aunt Sally	Last, we **add** and **subtract** in order from left to right.	$18 + 9 - 1 = 26$

Simplify the following problems. (DOK 2)

1. $6 + 9 \times 2 - 4$
2. $3(4+2) - 6^2$
3. $3(6-3) - 2^3$
4. $49 \div 7 - 3 \times 3$
5. $10 \times 4 - (7-2)$
6. $2 \times 3 \div 6 \times 4$
7. $4^3 \div 8(4+2)$
8. $7 + 8(14-6) \div 4$
9. $(2 + 8 - 12) \times 4$
10. $4(8-13) \times 4$

11. $8 + 4^2 \times 2 - 6$
12. $3^2(4+6) + 3$
13. $(12-6) + 27 \div 3^2$
14. $8^20 - 1 + 4 \div 2^2$
15. $1 - (2-3) + 8$
16. $-4\{18 - (4 + 2 \times 6)\}$
17. $18 \div (6+3) - 12$
18. $10^2 + 3^3 - 2 \times 3$
19. $4^2 + (7+2) \div 3$
20. $7 \times 4 - 9 \div 3$

21. $\dfrac{4 - (2+7)}{13 + (6-9)}$

22. $\dfrac{5(3-8) - 2^2}{7 - 3(6+1)}$

23. $\dfrac{3(3-8) + 5}{8^2 - (5+9)}$

24. $\dfrac{6^2 - 4(7+3)}{8 + (9-3)}$

Chapter 8 Exponents

Chapter 8 Review

Put all the expressions below into simplest terms. Make all exponents positive. (DOK 1)

1. $5^2 \times 5^3$
2. $(4^4)^5$
3. $(4y^3)^3$
4. $6x^{-3}$
5. $(3a^2)^{-2}$
6. $(b^3)^{-4}$
7. $\dfrac{4^6}{4^4}$
8. $\left(\dfrac{3}{5}\right)^2$
9. $(2x)^{-4}$
10. $3^3 \times 3^2$
11. $(2^4)^2$
12. $\dfrac{(3a^2)^3}{a^3}$
13. $(4^2)^{-2}$
14. $\dfrac{(2^3)^2}{2^4}$
15. $(6x)^{-3}$
16. $(4d^5)^{-3}$

Put all the expressions below into simplest terms. Make all exponents positive. (DOK 2)

17. $x^3 \cdot x^{-7}$
18. $5^7 \times 5^{-4}$
19. $(5^{-9} \times 5^7)^{-2}$
20. $\dfrac{y^{-2}}{3y^4}$
21. $\dfrac{x^{\frac{6}{5}}}{x^{\frac{7}{15}}}$
22. $(5)^{\frac{3}{4}} \times (5)^{\frac{1}{3}}$
23. $\left(3^{\frac{1}{4}}\right)^{\frac{3}{5}}$
24. $(6)^{\frac{1}{12}} \times (6)^{\frac{1}{4}}$
25. $(3a^2)^{\frac{3}{2}}$
26. $\dfrac{8^{\frac{8}{9}}}{8^{\frac{2}{9}}}$
27. $(19)^{\frac{6}{13}}$
28. $(z)^{\frac{5}{7}} \times (z)^{\frac{5}{14}}$
29. $\left(5^{\frac{9}{13}}\right)^{\frac{1}{3}}$
30. $\dfrac{10^{\frac{3}{4}}}{10^{\frac{1}{12}}}$

Write using exponents. (DOK 1)

31. $3 \times 3 \times 3 \times 3$
32. $6 \times 6 \times 6 \times 6 \times 6 \times 6$
33. $11 \times 11 \times 11$
34. $2 \times 2 \times 2 \times 2 \times 2 \times 2 \times 2 \times 2$

Simplify the following problems using the correct order of operations. (DOK 2)

35. $10 \div (-1 - 4) + 2$
36. $5 + (2)(4 - 1) \div 3$
37. $5 - 5^2 + (2 - 4)$
38. $(8 - 10) \times (5 + 3) - 10$
39. $\dfrac{10 + 5^2 - 3}{2^2 + 2(5 - 3)}$
40. $1 - (9 - 1) \div 2$
41. $\dfrac{5(3 - 6) + 3^2}{4(2 + 1) - 6}$
42. $-4(6 + 4) \div (-2) + 1$
43. $12 \div (7 - 4) - 2$
44. $1 + 4^2 \div (3 + 1)$

Chapter 8 Test

1 Simplify the expression shown below:

$$\frac{8x^4}{2x^2}$$

A $2x^4$

B $4x^2$

C $\frac{1}{4x^2}$

D $\frac{4x^2}{x}$

(DOK 1)

2 Simplify the following:

$5 \cdot x^4 \cdot y^5 \cdot z^{-3}$

A $\frac{5x^4y^5}{z^3}$

B $(5xyz)^6$

C $\frac{625x^4y^5}{z^3}$

D $x^{20}y^{25}z^{-15}$

(DOK 2)

3 What is the solution to $2(5-2)^2 - 15 \div 5$?

A $-\frac{3}{5}$

B $\frac{3}{5}$

C 15

D $4\frac{1}{5}$

(DOK 2)

4 Simplify: $(6)^{\frac{5}{6}} \times (6)^{\frac{1}{3}}$

A $(6)^{\frac{7}{6}}$

B $(6)^{\frac{5}{18}}$

C $(6)^{\frac{6}{9}}$

D $(36)^{\frac{7}{6}}$

(DOK 2)

5 Simplify: $3^2 + 4 \times 18 \div 9$

A 4

B 14

C 17

D 26

(DOK 2)

6 $4 \times 18 - 9 \div 3^2 =$

A 4

B 7

C 68

D 71

(DOK 2)

7 Simplify: $\frac{(4x)^{-3}}{6x^4}$

A $\frac{2}{3x^{12}}$

B $\frac{2}{3x^7}$

C $\frac{1}{384x^7}$

D $-\frac{32}{3x}$

(DOK 2)

8 Simplify: $(6x^4)^{-2}$

A $\frac{1}{36x^8}$

B $-36x^8$

C $-12x^{-8}$

D $6x^{-8}$

(DOK 1)

9 $(3a^2)^{\frac{3}{4}} =$

A $3^{\frac{3}{4}}a^{\frac{5}{4}}$

B $3^{\frac{3}{4}}a^{\frac{3}{2}}$

C $3a^{\frac{11}{4}}$

D $3a^{\frac{3}{2}}$

(DOK 1)

Chapter 8 Exponents

10 $x^2 \cdot x^4 =$

  A x^8
  B $8x$
  C x^6
  D $6x$

(DOK 1)

11 Write using exponents: $4a \times 4a \times 4a$

  A $4a^3$
  B $64a^3$
  C $3(4a)$
  D $(4+a)^3$

(DOK 1)

12 Simplify: $(3^4)^2$

  A 3^8
  B 3^6
  C 12^2
  D 7^2

(DOK 1)

13 $(5y)^{-4} =$

  A $\dfrac{1}{(5y)^4}$
  B $\dfrac{1}{5y^4}$
  C $-20 - 4y$
  D $\dfrac{5}{y^4}$

(DOK 1)

14 $\dfrac{12^{\frac{7}{8}}}{12^{\frac{1}{4}}} =$

  A $1^{\frac{5}{8}}$
  B $12^{\frac{7}{2}}$
  C $12^{\frac{5}{8}}$
  D $12^{\frac{9}{8}}$

(DOK 2)

15 $3^3 x^2 \cdot 4x^5 =$

  A $108x^{10}$
  B $108x^5$
  C $108x^7$
  D $12x^{10}$

(DOK 1)

16 $3^2 \cdot 3^{-3} =$

  A 9^{-6}
  B $\frac{1}{3}$
  C $\frac{1}{9}$
  D 3

(DOK 2)

17 $8y^{-2} =$

  A $\dfrac{1}{8y^2}$
  B $6y$
  C $-16y$
  D $\dfrac{8}{y^2}$

(DOK 1)

18 $(4+2)^3 =$

  A 18
  B 72
  C 216
  D 54

(DOK 1)

19 $\dfrac{(2+3)^{-2}}{11(2+3)^{-2}} =$

  A 11
  B $\frac{1}{11}$
  C $\frac{1}{5}$
  D $\frac{1}{55}$

(DOK 2)

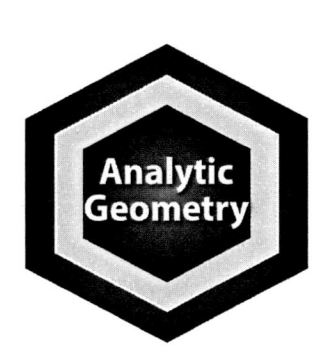

Chapter 9
Roots

Analytic Geometry

This chapter covers the following CCGPS standard(s):

| The Real Number System | N.RN.1, N.RN.2, N.RN.3 |

9.1 Square Root (DOK 1)

Just as working with exponents is related to multiplication, finding square roots is related to division. In fact, the sign for finding the square root of a number looks similar to a division sign. The best way to learn about square roots is to look at examples.

Example: This is a square root problem: $\sqrt{64}$
It is asking, "What is the square root of 64?"
It means, "What number multiplied by itself equals 64?"
The answer is 8. $8 \times 8 = 64$.

Find the square roots of the following numbers. (DOK 1)

1. $\sqrt{49}$
2. $\sqrt{81}$
3. $\sqrt{25}$
4. $\sqrt{16}$
5. $\sqrt{121}$
6. $\sqrt{625}$
7. $\sqrt{100}$
8. $\sqrt{289}$
9. $\sqrt{196}$
10. $\sqrt{36}$
11. $\sqrt{4}$
12. $\sqrt{900}$
13. $\sqrt{64}$
14. $\sqrt{9}$

9.2 Simplifying Square Roots Using Factors (DOK 1)

Square roots can sometimes be simplified even if the number under the square root is not a perfect square. One of the rules of roots is that if a and b are two positive real numbers, then it is always true that $\sqrt{a \cdot b} = \sqrt{a} \cdot \sqrt{b}$. You can use this rule to simplify square roots.

Example 1: $\sqrt{100} = \sqrt{4 \cdot 25} = \sqrt{4} \cdot \sqrt{25} = 2 \cdot 5 = 10$

Example 2: $\sqrt{200} = \sqrt{100 \cdot 2} = 10\sqrt{2}$ ⟵ Means 10 multiplied by the square root of 2

Example 3: $\sqrt{160} = \sqrt{10 \cdot 16} = 4\sqrt{10}$

Simplify. (DOK 1)

1. $\sqrt{98}$
2. $\sqrt{600}$
3. $\sqrt{50}$
4. $\sqrt{27}$
5. $\sqrt{8}$
6. $\sqrt{63}$
7. $\sqrt{48}$
8. $\sqrt{75}$
9. $\sqrt{54}$
10. $\sqrt{40}$
11. $\sqrt{72}$
12. $\sqrt{80}$
13. $\sqrt{90}$
14. $\sqrt{175}$
15. $\sqrt{18}$
16. $\sqrt{20}$

Copyright © American Book Company

Chapter 9   Roots

9.3   Adding, Subtracting, and Simplifying Square Roots (DOK 2)

You can add and subtract terms with square roots only if the number under the square root sign is the same.

Example 4: $2\sqrt{2} + 3\sqrt{2} = 5\sqrt{2}$

Example 5: $12\sqrt{7} - 3\sqrt{7} - 9\sqrt{7}$

Or, look at the following examples where you can simplify the square roots and then add or subtract.

Example 6: $2\sqrt{25} + \sqrt{36}$

   Step 1:   Simplify. You know that $\sqrt{25} = 5$, and $\sqrt{36} = 6$ so the problem simplifies to $2(5) + 6$

   Step 2:   Solve: $2(5) + 6 = 10 + 6 = 16$

Example 7: $2\sqrt{72} - 3\sqrt{2}$

   Step 1:   Simplify what you know. $\sqrt{72} = \sqrt{36 \cdot 2} = 6\sqrt{2}$

   Step 2:   Substitute $6\sqrt{2}$ for $\sqrt{72}$ and simplify.
   $2(6)\sqrt{2} - 3\sqrt{2} = 12\sqrt{2} - 3\sqrt{2} = 9\sqrt{2}$

Simplify the following addition and subtraction problems. (DOK 2)

1. $3\sqrt{5} + 9\sqrt{5}$
2. $3\sqrt{25} + 4\sqrt{16}$
3. $4\sqrt{8} + 2\sqrt{2}$
4. $3\sqrt{32} - 2\sqrt{2}$
5. $\sqrt{25} - \sqrt{49}$
6. $2\sqrt{5} + 4\sqrt{20}$
7. $5\sqrt{8} - 3\sqrt{72}$
8. $\sqrt{27} + 3\sqrt{27}$

9. $3\sqrt{20} - 4\sqrt{45}$
10. $4\sqrt{45} - \sqrt{75}$
11. $2\sqrt{28} + 2\sqrt{7}$
12. $\sqrt{64} + \sqrt{81}$
13. $5\sqrt{54} - 2\sqrt{24}$
14. $\sqrt{32} + 2\sqrt{50}$
15. $2\sqrt{7} + 4\sqrt{63}$
16. $8\sqrt{2} + \sqrt{8}$

17. $2\sqrt{8} - 4\sqrt{32}$
18. $\sqrt{36} + \sqrt{100}$
19. $\sqrt{9} + \sqrt{25}$
20. $\sqrt{64} - \sqrt{36}$
21. $\sqrt{75} + \sqrt{108}$
22. $\sqrt{81} + \sqrt{100}$
23. $\sqrt{192} - \sqrt{75}$
24. $3\sqrt{5} + \sqrt{245}$

9.4 Multiplying and Simplifying Square Roots (DOK 2)

You can also multiply square roots. To multiply square roots, you just multiply the numbers under the square root sign and then simplify. Look at the examples below.

Example 8: $\sqrt{2} \times \sqrt{6}$

Step 1: $\sqrt{2} \times \sqrt{6} = \sqrt{2 \times 6} = \sqrt{12}$ Multiply the numbers under the square root sign.

Step 2: $\sqrt{12} = \sqrt{4 \times 3} = 2\sqrt{3}$ Simplify.

Example 9: $3\sqrt{3} \times 5\sqrt{6}$

Step 1: $(3 \times 5)\sqrt{3 \times 6} = 15\sqrt{18}$ Multiply the numbers in front of the square root, and multiply the numbers under the square root sign.

Step 2: $15\sqrt{18} = 15\sqrt{2 \times 9}$ Simplify.
$15 \times 3\sqrt{2} = 45\sqrt{2}$

Example 10: $\sqrt{14} \times \sqrt{42}$ For this more complicated multiplication problem, use the rule of roots that you learned on page 159, $\sqrt{a \cdot b} = \sqrt{a} \cdot \sqrt{b}$.

Step 1: $\sqrt{14} = \sqrt{7} \times \sqrt{2}$ and   Instead of multiplying 14 by 42, divide these
$\sqrt{42} = \sqrt{2} \times \sqrt{3} \times \sqrt{7}$ numbers into their roots.

$\sqrt{14} \times \sqrt{42} = \sqrt{7} \times \sqrt{2} \times \sqrt{2} \times \sqrt{3} \times \sqrt{7}$

Step 2: Since you know that $\sqrt{7} \times \sqrt{7} = 7$ and $\sqrt{2} \times \sqrt{2} = 2$, the problem simplifies to $(7 \times 2)\sqrt{3} = 14\sqrt{3}$.

Simplify the following multiplication problems. (DOK 2)

1. $\sqrt{5} \times \sqrt{7}$
2. $\sqrt{32} \times \sqrt{2}$
3. $\sqrt{10} \times \sqrt{14}$
4. $2\sqrt{3} \times 3\sqrt{6}$
5. $4\sqrt{2} \times 2\sqrt{10}$

6. $\sqrt{5} \times 3\sqrt{15}$
7. $\sqrt{45} \times \sqrt{27}$
8. $5\sqrt{21} \times \sqrt{7}$
9. $\sqrt{42} \times \sqrt{21}$
10. $4\sqrt{3} \times 2\sqrt{12}$

11. $\sqrt{56} \times \sqrt{24}$
12. $\sqrt{11} \times 2\sqrt{33}$
13. $\sqrt{13} \times \sqrt{26}$
14. $2\sqrt{2} \times 5\sqrt{5}$
15. $\sqrt{6} \times \sqrt{12}$

Chapter 9   Roots

9.5   Dividing and Simplifying Square Roots (DOK 2)

When dividing a number or a square root by another square root, you cannot leave the square root sign in the denominator (the bottom number) of a fraction. You must simplify the problem so that the square root is not in the denominator. This is also called rationalizing the denominator. Look at the examples below.

Example 11: $\dfrac{\sqrt{2}}{\sqrt{5}}$

Step 1: $\dfrac{\sqrt{2}}{\sqrt{5}} \times \dfrac{\sqrt{5}}{\sqrt{5}}$ ← The fraction $\dfrac{\sqrt{5}}{\sqrt{5}}$ is equal to 1, and multiplying by 1 does not change the value of a number.

Step 2: $\dfrac{\sqrt{2 \times 5}}{5} = \dfrac{\sqrt{10}}{5}$ Multiply and simplify. Since $\sqrt{5} \times \sqrt{5}$ equals 5, you no longer have a square root in the denominator.

Example 12: $\dfrac{6\sqrt{2}}{2\sqrt{10}}$ In this problem, the numbers outside of the square root will also simplify.

Step 1: $\dfrac{6}{2} = 3$ so you have $\dfrac{3\sqrt{2}}{\sqrt{10}}$

Step 2: $\dfrac{3\sqrt{2}}{\sqrt{10}} \times \dfrac{\sqrt{10}}{\sqrt{10}} = \dfrac{3\sqrt{2 \times 10}}{10} = \dfrac{3\sqrt{20}}{10}$

Step 3: $\dfrac{3\sqrt{20}}{10}$ will further simplify because $\sqrt{20} = 2\sqrt{5}$, so you then have $\dfrac{3 \times 2\sqrt{5}}{10}$ which reduces to $\dfrac{3\sqrt{5}}{5}$.

Simplify the following division problems. Show your work. (DOK 2)

1. $\dfrac{9\sqrt{3}}{\sqrt{5}}$

2. $\dfrac{16}{\sqrt{8}}$

3. $\dfrac{24\sqrt{10}}{12\sqrt{3}}$

4. $\dfrac{\sqrt{121}}{\sqrt{6}}$

5. $\dfrac{\sqrt{40}}{\sqrt{90}}$

6. $\dfrac{33\sqrt{15}}{11\sqrt{2}}$

7. $\dfrac{\sqrt{32}}{\sqrt{12}}$

8. $\dfrac{\sqrt{11}}{\sqrt{5}}$

9. $\dfrac{\sqrt{2}}{\sqrt{6}}$

10. $\dfrac{2\sqrt{7}}{\sqrt{14}}$

11. $\dfrac{5\sqrt{2}}{4\sqrt{8}}$

12. $\dfrac{4\sqrt{21}}{7\sqrt{7}}$

13. $\dfrac{9\sqrt{22}}{2\sqrt{2}}$

14. $\dfrac{\sqrt{35}}{2\sqrt{14}}$

15. $\dfrac{\sqrt{40}}{\sqrt{15}}$

16. $\dfrac{\sqrt{3}}{\sqrt{12}}$

9.6 Cube Roots (DOK 1)

Cube roots look like square roots, except that there is a "3" raised in front of the root sign:

Square root of 64: $\sqrt{64}$

Cube root of 64: $\sqrt[3]{64}$

In fact, they function very much like square roots, with one important difference. Recall asking, "What is the square root of 64?" means:

"What number multiplied by itself equals 64?"

Asking "What is the cube root of 64?" means:

"What number multiplied 3 times ('cubed') by itself equals 64?"

The answer is 4. $4 \times 4 \times 4 = 64$.

Find the cube root of the following numbers.

Examples: $\sqrt[3]{27}$ $3 \times 3 \times 3 = 27$ so $\sqrt[3]{27} = 3$

$\sqrt[3]{1000}$ $10 \times 10 \times 10 = 1000$ so $\sqrt[3]{1000} = 10$

Find the cube roots of the following numbers. (DOK 1)

1. $\sqrt[3]{1}$

2. $\sqrt[3]{8}$

3. $\sqrt[3]{64}$

4. $\sqrt[3]{125}$

5. $\sqrt[3]{27}$

6. $\sqrt[3]{\frac{64}{27}}$

7. $\sqrt[3]{1000}$

8. $\sqrt[3]{\frac{125}{1000}}$

9.7 Rational Exponents (DOK 1)

Sometimes exponents can be rational numbers.

Example 13: $4^{\frac{3}{2}}$ ← the exponent of the number under the radical
← the root you need to take

$$4^{\frac{3}{2}} = \sqrt[2]{4^3} = \sqrt{4^3} = \sqrt{64} = 8$$

Note: To solve these problems you can raise the base to the exponent *before* or *after* you take the root.

Example 14:
$$\sqrt{2} = 2^{\frac{1}{2}}$$
$$\sqrt[3]{10} = 10^{\frac{1}{3}}$$
$$\sqrt[3]{3^2} = 3^{\frac{2}{3}}$$
$$\sqrt[5]{4^2} = 4^{\frac{2}{5}}$$

Example 15: Evaluate $\sqrt[10]{7^{20}}$ using rational exponents.

$$\sqrt[10]{7^{20}} = 7^{\frac{20}{10}} = 7^2 = 7 \times 7 = 49$$

Change the following to an expression using a radical and find its integer equivalent. (DOK 1)

1. $25^{\frac{3}{2}}$
2. $16^{\frac{5}{2}}$
3. $4^{\frac{5}{2}}$
4. $36^{\frac{3}{2}}$
5. $8^{\frac{2}{3}}$
6. $125^{\frac{1}{3}}$
7. $32^{\frac{3}{5}}$
8. $4^{\frac{1}{2}}$
9. $25^{\frac{5}{2}}$
10. $64^{\frac{2}{3}}$
11. $27^{\frac{2}{3}}$
12. $81^{\frac{3}{4}}$

Change the following expressions to a whole number base with a rational exponent. (DOK 1)

13. $\sqrt{5}$
14. $\sqrt[3]{19}$
15. $\sqrt[4]{2^3}$
16. $\sqrt[5]{9^2}$
17. $\sqrt[3]{20}$
18. $\sqrt[4]{11^3}$
19. $\sqrt{5^3}$
20. $\sqrt[4]{7^3}$
21. $\sqrt[5]{2^3}$
22. $\sqrt{50}$
23. $\sqrt[3]{8^2}$
24. $\sqrt[5]{9^3}$
25. $\sqrt[10]{5^{30}}$
26. $\sqrt{2^6}$
27. $\sqrt[4]{7^5}$
28. $\sqrt[5]{12^2}$

Find the value of each expression. (DOK 1)

29. $\left(\sqrt{10}\right)^4$
30. $\sqrt[9]{10^{27}}$
31. $\sqrt{1,307,428^2}$
32. $\sqrt{3^4}$

9.8 Real Numbers (DOK 1)

Real numbers include all positive and negative numbers and zero. Included in the set of real numbers are positive and negative fractions, decimals, and rational and irrational numbers. Rational numbers are all repeating and terminating decimals, like $\frac{1}{3} = 0.\overline{3}$ and 1.9856. If a number cannot be expressed as a fraction, then it is not rational. It is irrational. Irrational numbers are all numbers that are nonrepeating and nonterminating decimals, like $\sqrt{26} \approx 5.0990195...$

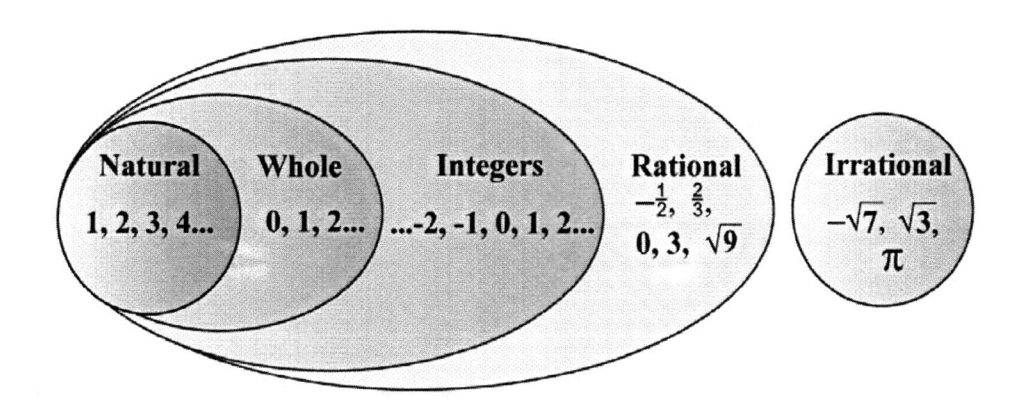

Use the diagram above and your calculator to answer the following questions. (DOK 1)

1. Using your calculator, find the square root of 7. Does it repeat? Does it end? Is it a rational or an irrational number?

2. Find $\sqrt{25}$. Is it rational or irrational? Is it an integer?

3. Is an integer an irrational number?

4. Is an integer a real number?

5. Is $\frac{1}{8}$ a real number? Is it rational or irrational?

6. Is zero a natural number?

Identify the following numbers as rational (R) or irrational (I). (DOK 1)

7. 5π

8. $\sqrt{8}$

9. $\frac{1}{3}$

10. -7.2

11. $-\frac{3}{4}$

12. $\frac{\sqrt{2}}{2}$

13. $9 + \pi$

14. 1.0004

15. $-\frac{4}{5}$

16. $1.1\overline{8}$

17. $\sqrt{81}$

18. $\frac{\pi}{4}$

19. $-\sqrt{36}$

20. $17\frac{1}{2}$

21. $-\frac{5}{3}$

22. 0

23. -7

24. $\sqrt{11}$

Chapter 9   Roots

9.9   Computing with Real Numbers (DOK 1)

When you add, subtract, multiply, or divide with different combinations of rational and irrational numbers, the answer can be either rational or irrational. A few rules apply; otherwise, you will need to perform the operation and decide whether or not the answer is rational.

Rules for Addition:	Examples:
Rational + Rational = Rational	$\frac{2}{3}+\frac{3}{5}=1\frac{4}{15}$
Irrational + Irrational = Irrational	$2\sqrt{7}+4\sqrt{7}=6\sqrt{7}$
Rational + Irrational = Irrational	$7+\sqrt{5}=7+\sqrt{5}$

Rules for Multiplication:	Examples:
Rational × Rational = Rational	$\frac{2}{3}\times\frac{3}{5}=\frac{2}{5}$
Rational × Irrational = Irrational	$4\times\pi=4\pi$

Perform the operation and label your answer as either rational (R) or irrational (I). (DOK 1)

1. $(\sqrt{7})(\sqrt{5}) =$
2. $4 + \pi =$
3. $4 + \frac{2}{3} =$
4. $7(\sqrt{5}) =$
5. $(\sqrt{7})(\pi) =$
6. $2\pi \cdot \pi =$
7. $\sqrt{7} \div \sqrt{5} =$
8. $4 - \pi =$
9. $4 - \frac{2}{3} =$
10. $9 \div \sqrt{5} =$
11. $\sqrt{3} \div \pi =$
12. $2\pi \div \pi =$

9.10   Going Deeper into Roots (DOK 3)

Solve the problems below. Show your work and provide an explanation. (DOK 3)

1. Jason gave an incorrect answer for the problem shown below. The steps he made to solve the problem are also shown below. How should Jason have solved the problem to get the correct solution? Give the correct solution.

Original Problem	Step 1	Step 2	Step 3	Final Solution
$\frac{\sqrt{9}+\sqrt{24}}{\sqrt{6}}$	$\frac{3+2\sqrt{6}}{\sqrt{6}}$	$\frac{3+2\sqrt{\cancel{6}}}{\sqrt{\cancel{6}}}$	$3+2$	5

2. Kristen's teacher gave the class 5 problems to do. The answers of Kristen and the teacher are shown below.

	#1	#2	#3	#4	#5
Teacher's Answers:	$\sqrt[5]{2^7}=2^{\frac{7}{5}}$	$\sqrt[3]{5^9}=5^{\frac{9}{3}}$	$\sqrt[14]{3^7}=3^{\frac{1}{2}}$	$\sqrt[6]{8^3}=8^{\frac{1}{2}}$	$\sqrt{24}=24^{\frac{1}{2}}$
Kristen's Answers:	$\sqrt[5]{2^7}=2\sqrt[5]{4}$	$\sqrt[3]{5^9}=125$	$\sqrt[14]{3^7}=9$	$\sqrt[6]{8^3}=2\sqrt[6]{8}$	$\sqrt{24}=2\sqrt{6}$

Which questions did Kristen solve incorrectly? Explain how you know.

3. Given the expression $\sqrt{x+11}$, find 3 values for x that would make the expression rational and 3 values of x that would make the expression irrational. Explain how you know the values of x chosen would result in the answer you are looking for.

Chapter 9 Review

Simplify the following square root expressions. (DOK 1)

1. $\sqrt{50}$
2. $\sqrt{44}$
3. $\sqrt{12}$
4. $\sqrt{18}$
5. $\sqrt{8}$
6. $\sqrt{48}$
7. $\sqrt{75}$
8. $\sqrt{200}$
9. $\sqrt{32}$
10. $\sqrt{20}$
11. $\sqrt{63}$
12. $\sqrt{80}$

Simplify the following square root problems. (DOK 2)

13. $5\sqrt{27} + 7\sqrt{3}$
14. $\sqrt{40} - \sqrt{10}$
15. $\sqrt{64} + \sqrt{81}$
16. $8\sqrt{50} - 3\sqrt{32}$
17. $14\sqrt{5} + 8\sqrt{80}$
18. $\sqrt{63} \times \sqrt{28}$
19. $\dfrac{\sqrt{56}}{\sqrt{35}}$
20. $\sqrt{8} \times \sqrt{50}$
21. $\dfrac{\sqrt{20}}{\sqrt{45}}$
22. $5\sqrt{40} \times 3\sqrt{20}$
23. $2\sqrt{48} - \sqrt{12}$
24. $\dfrac{2\sqrt{5}}{\sqrt{30}}$
25. $\dfrac{3\sqrt{22}}{2\sqrt{3}}$
26. $\sqrt{72} \times 3\sqrt{27}$
27. $4\sqrt{5} + 8\sqrt{45}$

Rewrite each of the following with rational (fractional) exponents. (DOK 1)

28. $\sqrt[3]{2^4}$
29. $\sqrt[5]{1^2}$
30. $\sqrt[4]{8^3}$
31. $\sqrt[3]{6^2}$
32. $\sqrt[6]{2^{12}}$
33. $\sqrt[5]{9^3}$

Find the cube roots of the following numbers. (DOK 1)

34. $\sqrt[3]{512}$
35. $\sqrt[3]{\dfrac{125}{8}}$

Find the value of each expression. (DOK 2)

36. $9^{\frac{3}{2}}$
37. $100^{\frac{3}{2}}$
38. $4^{\frac{1}{2}}$
39. $\left(\sqrt[3]{3}\right)^6$

Perform the operation and label your answer as either rational (R) or irrational (I). (DOK 1)

40. $7 + 1\frac{2}{3} =$
41. $\sqrt{14} \div \sqrt{7} =$
42. $2\pi - \pi =$
43. $3\pi(\pi) =$

Identify the following numbers as whole (W), natural (N), rational (R), or irrational (I). Some numbers may fit more than one category. (DOK 1)

44. $\sqrt{6}$
45. 0
46. $-\dfrac{7}{3}$

Chapter 9 Test

1 Simplify: $\sqrt{135}$

A $3\sqrt{15}$

B $\sqrt{72}$

C $9\sqrt{15}$

D $\sqrt{9} \times \sqrt{15}$

(DOK 1)

2 Express $\dfrac{\sqrt{20}}{\sqrt{35}}$ in simplest form.

A $\dfrac{2\sqrt{7}}{7}$

B $\dfrac{2}{\sqrt{7}}$

C $\dfrac{2\sqrt{5}}{\sqrt{7}}$

D $\dfrac{4}{7}$

(DOK 2)

3 Simplify: $\dfrac{3\sqrt{12}}{2\sqrt{3}}$

A 3

B $\dfrac{3\sqrt{4}}{2}$

C $\dfrac{6}{\sqrt{6}}$

D $3\sqrt{3}$

(DOK 2)

4 Simplify: $\sqrt{44} \cdot 2\sqrt{33}$

A $2\sqrt{77}$

B $44\sqrt{3}$

C $22\sqrt{7}$

D $22\sqrt{12}$

(DOK 2)

5 Rewrite with rational exponents:

$\sqrt[5]{3^2}$

A 3^{10}

B $5^{\frac{2}{3}}$

C $3^{\frac{5}{2}}$

D $3^{\frac{2}{5}}$

(DOK 1)

6 Simplify: $\sqrt{45} \times \sqrt{27}$

A $3\sqrt{15}$

B $\sqrt{72}$

C $9\sqrt{15}$

D $\sqrt{9} \times \sqrt{15}$

(DOK 2)

7 Simplify the following by rationalizing the denominator.

$\dfrac{\sqrt{3}}{\sqrt{15}}$

A $\dfrac{\sqrt{5}}{5}$

B $\dfrac{1}{\sqrt{5}}$

C $\dfrac{\sqrt{45}}{15}$

D $\dfrac{3\sqrt{5}}{15}$

(DOK 2)

8 Simplify: $\sqrt[3]{40}$

A $8\sqrt[3]{5}$

B $\sqrt[3]{8} \times \sqrt[3]{5}$

C $2\sqrt[3]{5}$

D You cannot take the cube root of 40.

(DOK 1)

9 Which is equivalent to $\sqrt[9]{2^3}$?

A $\frac{8}{9}$

B $2^{\frac{1}{3}}$

C 1.2

D 27

(DOK 1)

10 Which is equivalent to $5^{\frac{2}{9}}$?

A $\sqrt[9]{25}$

B $\frac{25}{9}$

C $\sqrt[3]{125}$

D $(\sqrt{5})^9$

(DOK 1)

11 Which of the following is a rational number?

A $\sqrt{6}$

B 2π

C $\sqrt{196}$

D $\sqrt{8}$

(DOK 1)

12 Perform the operation, then determine if the answer is rational or irrational.
$6 \div \frac{2}{3} =$

A 4, rational

B 9, rational

C 9, irrational

D 4, irrational

(DOK 1)

13 Which of the following illustrates that a rational number multiplied by a rational number is rational?

A $14 \times 0.76 = 10.64$

B $14 \times 2\pi = 28\pi$

C $14 \times \sqrt{2} = 14\sqrt{2}$

D $14 + 17 = 31$

(DOK 2)

14 Erin solved the problem: $\dfrac{\sqrt{32} + \sqrt{16}}{\sqrt{8}}$ using the following steps.

Step #1	Step #2	Step #4	Answer
$\dfrac{\sqrt{4 \cdot 8} + \sqrt{4 \cdot 4}}{\sqrt{8}}$	$\sqrt{4} + 4$	$2 + 4$	6

What mistake did Erin make when solving this problem?

A Erin made no mistake in this problem.

B You must divide all the terms by $\sqrt{8}$, but Erin only divided 2 of the 3 terms by the $\sqrt{8}$.

C You must add the numerators together, (i.e. $\sqrt{48}$), before dividing by $\sqrt{8}$.

D You must simplify each radical prior to canceling out anything.

(DOK 3)

15 Which part of the following statement is false, and what should it be so the following sentence is true?

Every <u>rational</u> number can be written as a fraction, but every <u>integer</u> cannot be expressed as a fraction.

A The word rational should be changed to irrational.

B The word integer should be changed to irrational numbers.

C The word integer should be changed to natural numbers.

D The word integer should be changed to rational numbers.

(DOK 3)

Chapter 10
Complex Numbers

This chapter covers the following CCGPS standard(s):

| Complex Numbers | N.CN.1, N.CN.2, N.CN.3 |

10.1 Complex Numbers Defined (DOK 1)

Complex numbers are usually written in the form $a + bi$, where a and b are real numbers and i is defined as $\sqrt{-1}$. Because $\sqrt{-1}$ does not exist in the set of real numbers, i is referred to as the imaginary unit.

When talking about a complex number, $a + bi$, the real number a is called the real part, and the real number b is called the imaginary part.

If the real part, a, is zero, then the complex number $a + bi$ is just bi, so it is imaginary.

If the real part, b, is zero, then the complex number $a + bi$ is just a, so it is real.

Example 1: What is the real part of the complex number $9 + 16i$?

Solution: The complex number $9 + 16i$ is written in the form $a + bi$. Here $a = 9$ and $b = 16$. The real part of the complex number $9 + 16i$ is 9.

Example 2: What is the imaginary part of the complex number $23 - 6i$?

Solution: The complex number $23 - 6i$ is written in the form $a + bi$. Here $a = 23$ and $b = -6$. The imaginary part of the complex number $a + bi$ is b. The imaginary part of the complex number $23 - 6i$ is -6.

Name the real part of each of the following complex numbers.

1. $-\frac{4}{5} - 3i$
2. $7 + 2i$
3. $20 - 11i$
4. $\frac{2}{9}$
5. $15i$
6. $-13 + \frac{i}{4}$
7. $12 + 5i$
8. $\frac{-9 + 2i}{25}$

Name the imaginary part of each of the following complex numbers.

9. $4 - i$
10. $\frac{6i}{5}$
11. $5 + \frac{2}{3}i$
12. $-9 + 8i$
13. 18
14. $\frac{1 - 3i}{2}$
15. $51 - 2i$
16. $14 + i$

10.2 Imaginary Numbers (DOK 1)

The square root of a negative number is an imaginary number. You know that $\sqrt{-1} = i$. Therefore, $i^2 = -1$.

Where n is some natural number $(1, 2, 3...)$, then $\sqrt{-n} = \sqrt{(-1) \times n} = \sqrt{-1} \times \sqrt{n} = i\sqrt{n}$. To remove the negative number from under the radical, just take i out. Remember $\sqrt{-n} = i\sqrt{n}$.

Example 3: Simplify: $\sqrt{-450}$

Step 1: Factor -450 and rewrite: $\sqrt{-450} = \sqrt{225 \times 2 \times (-1)}$.

Step 2: By root laws, $\sqrt{225 \times 2 \times (-1)} = \sqrt{225} \times \sqrt{2} \times \sqrt{-1}$.

Step 3: Since $\sqrt{225} = 15$ and $\sqrt{-1} = i$, we have $\sqrt{225} \times \sqrt{2} \times \sqrt{-1} = 15 \times \sqrt{2} \times i$.

Step 4: Write in standard form as $15i\sqrt{2}$.

Example 4: Multiply: $5i \times 2i$

Step 1: Using the basic rules of multiplication, we know $5i \times 2i = 5 \times i \times 2 \times i$.

Step 2: Use the commutative property of multiplication, $5 \times i \times 2 \times i = 5 \times 2 \times i \times i$.

Step 3: Simplify: $5 \times 2 \times i \times i = 10 \times i^2$.

Step 4: Since $i^2 = -1$, then $10 \times i^2$ can be simplified to $10 \times -1 = -10$.

Find the square root of each of the following numbers.

1. -8
2. $-\dfrac{4}{49}$
3. -441
4. $-\dfrac{81}{16}$
5. -44
6. -0.0121
7. -144
8. -64

Use what you know about the imaginary number i to solve the following problems.

9. $-3i + \sqrt{-3}$
10. $7i - 8i$
11. $\sqrt{-4} \times \sqrt{-9}$
12. $2i \times (-4i)$
13. $\left(\sqrt{-16}\right) \div (2i)$
14. $14i + i$
15. $\sqrt{-25} - 3i$
16. $(12i) \div (3i)$

Chapter 10 Complex Numbers

10.3 Adding and Subtracting Complex Numbers (DOK 2)

Complex numbers, written as $a + bi$ or $c + di$ may be added and subtracted. The real parts are added or subtracted together and the imaginary parts are added or subtracted together. So, where a, c are the real parts of two complex numbers and b, d are the corresponding imaginary parts.

$$(a + bi) + (c + di) = (a + c) + (b + d)i$$

Example 5: What is $(6 + i) - (5 - 7i)$?

Solution: Collect the real parts and the imaginary parts and do the arithmetic.
$(6 + i) - (5 - 7i) = 6 + i - 5 + 7i = (6 - 5) + (i + 7i) = 1 + 8i$

Example 6: What is $(10 + 3i) + (-7 - 6i) + (18 - 5i)$?

Step 1: Group the real and imaginary parts.
$(10 + 3i) + (-7 - 6i) + (18 - 5i) = (10 - 7 + 18) + (3 - 6 - 5)i$

Step 2: Add or subtract the real and imaginary parts separately.
$(10 - 7 + 18) + (3 - 6 - 5)i = 21 - 8i$

Example 7: z and v are complex numbers. $z = -2 + 3i$ and $v = 3 - 2i$. Compute $z - v$.

Solution: $z - v = (-2 + 3i) - (3 - 2i) = (-2 - 3) + (3i + 2i) = -5 + 5i$

Add.

1. $(5 + 2i) + (-3 - 11i)$
2. $(1 - 5i) + \left(4 + \frac{4i}{3}\right) + (20 - 7i)$
3. $(12 + 2i) + \left(\frac{4}{5} - i\right)$
4. $(3 + 8i) + (4 + 9i) + (2 - 3i)$
5. $(-13 + 4i) + \left(9 - \frac{i}{5}\right)$
6. $22 + (3 - 7i) + (-1 + 20i)$
7. $(16 - 10i) + (-3 - 4i) + 19i$
8. $(2 - 6i) + \left(5 + \frac{9i}{2}\right)$
9. $(30 + 7i) + (-23 - 15i) + (8 + 6i)$
10. $(-4 + i) + (21 - 18i) + 10$
11. $\left(\frac{9}{10} - 17i\right) + (9 + 13i)$
12. $(12 + 2i) + (2 - 12i) + 23i$

Subtract.

13. $(17 + 10i) - (6 + 12i)$
14. $(11 - i) - \left(2 + \frac{2}{11}i\right)$
15. $(3 - 2i) - (-4 + 5i) - (2 - 16i)$
16. $\left(9 + \frac{i}{8}\right) - (20 - 22i)$
17. $(40 + 4i) - (2 + 13i) - i$
18. $(5 - 4i) - (4 + 3i) - (18 + 3i)$
19. $(7 - i) - (10 + 35i) - 8$
20. $(12 + 8i) - (5 - 8i) - (1 + i)$
21. $(6 + 3i) - (5 + 4i) - (11 - i)$
22. $\left(13 - \frac{7i}{6}\right) - \left(31 + \frac{6i}{7}\right)$
23. $(2 + 20i) - (-1 + 6i) - (3 + 14i)$
24. $(-11 - 5i) - (5 - 11i) - 11i$

10.4 Multiplying Complex Numbers (DOK 2)

Multiplying two complex numbers, $a+bi$ and $c+di$, should remind you of the FOIL (First Outside Inside Last) method for multiplying two binomials like $(x+2)(x+3) = x^2 + 2x + 3x + 6 = x^2 + 5x + 6$. Generally, multiplying two complex numbers is the same:

$$(a+bi)(c+di) = ac + adi + bci + bdi^2$$

Simplify:

Remember that $i^2 = \left(\sqrt{-1}\right)^2 = -1$, so $ac + adi + bci + bdi^2 = ac + (ad+bc)i - bd$

Example 8: Multiply $1 + 2i$ and $-4 + 3i$.

Step 1: Use the FOIL method to multiply:
$(1+2i)(-4+3i) = (1)(-4) + (1)(3i) + (2i)(-4) + (2i)(3i)$

Step 2: Simplify:
$(1)(-4) + (1)(3i) + (2i)(-4) + (2i)(3i) = -4 + 3i - 8i + 6i^2 = -4 - 5i + 6i^2 = -4 - 5i + 6(-1)$

Step 3: Combine like terms:
$-4 - 5i + 6(-1) = -10 - 5i$

Example 9: What is $-7i \times -4i$?

Solution: $-7i \times -4i = 28i^2 = -28$

Example 10: Compute $(4 - 8i)(6 + 2i)$.

Solution: Use the FOIL method and add the results.
$(4-8i)(6+2i) = 24 + 8i - 48i - 16i^2 = 40 - 40i$

Multiply.

1. $(4 + 7i) \times (1 - i)$
2. $(5 - 2i) \times (6 + 3i) \times 2$
3. $(8 + 4i) \times (2 + 5i) \times (4 + i)$
4. $(10 - i) \times (2 - 3i) \times 12i$
5. $(25 + 7i) \times (25 - 7i)$
6. $(3 - i)^3$
7. $\left(\frac{1}{4} + 4i\right) \times (4 - 8i)$
8. $(7 + 5i) \times (7 + 5i)$
9. $(1 - 8i) \times (1 - 8i) \times (1 + 8i)$
10. $(6 - i) \times (7 - i) \times i$
11. $(11 - 9i) \times (11 + 9i)$
12. $(3 + 4i) \times (1 + 10i) \times 3$

Chapter 10 Complex Numbers

10.5 Dividing Complex Numbers (DOK 2)

There are three steps to remember when dividing complex numbers.

1. Write the complex number as a fraction.
2. Multiply the numerator and denominator of the fraction by the complex conjugate of the denominator. The **complex conjugate** of $a + bi$ is $a - bi$.
3. Simplify.

Example 11: What is $(6 + 3i) \div (2 + i)$?

   Step 1: Write the expression as a fraction:
   $$\frac{6 + 3i}{2 + i}$$

   Step 2: Multiply the top (numerator) and the bottom (denominator) by $2 - i$, the complex conjugate of the denominator.
   $$\frac{6 + 3i}{2 + i} \times \frac{2 - i}{2 - i}$$

   Step 3: Simplify.
   $$\frac{6 + 3i}{2 + i} \times \frac{2 - i}{2 - i} = \frac{12 - 6i + 6i - 3i^2}{4 - 2i + 2i - i^2} = \frac{12 - 3(-1)}{4 - (-1)} = \frac{15}{5} = 3$$

Example 12: What is $(5 - 4i) \div (7 - 8i)$?

   Step 1: Write the expression as a fraction, $\dfrac{5 - 4i}{7 - 8i}$.

   Step 2: Multiply by $\dfrac{7 + 8i}{7 + 8i}$.

   Step 3: Simplify:
   $$\frac{5 - 4i}{7 - 8i} \times \frac{7 + 8i}{7 + 8i} = \frac{35 + 40i - 28i - 32i^2}{49 + 56i - 56i - 64i^2} = \frac{35 + 12i - 32(-1)}{49 - 64(-1)}$$
   $$= \frac{67 + 12i}{113} = \frac{67}{113} + \frac{12}{113}i.$$

Divide.

1. $(7 + i) \div (4 - 3i)$
2. $25 \div (-1 + 5i)$
3. $(10 - 2i) \div (3 + 6i)$
4. $(-9 - 4i) \div (9 - 4i)$
5. $(5 + 3i) \div (-2 + i)$
6. $(8 - 7i) \div (1 - 8i)$
7. $(-11 - i) \div (7 + 3i)$
8. $(20 + i) \div 9i$
9. $(3 - 8i) \div (-8 + 3i)$
10. $(6 + 2i) \div (10 + i)$
11. $(-1 + 5i) \div (1 - 5i)$
12. $(15 + 2i) \div (3 - 7i)$

10.6 Simplifying Complex Numbers (DOK 2)

The order of operations for expressions with complex numbers is the same as the order of operations for real number expressions.

The **absolute value** of $a + bi$ is $\sqrt{a^2 + b^2}$.

Example 13: Simplify the expression $12 \times [(-1 + 4i)^2 + (2 - 3i)] \div i$.

Step 1: First, $-1 + 4i$ is squared.
$(-1 + 4i)^2 = (-1 + 4i)(-1 + 4i) = 1 - 4i - 4i + 16i^2 = 1 - 8i + 16(-1) = -15 - 8i$

Step 2: Plug the result into the original expression.
$12 \times [(-15 - 8i) + (2 - 3i)] \div i$

Step 3: Add $-15 - 8i$ and $2 - 3i$:
$(-15 - 8i) + (2 - 3i) = -15 - 8i + 2 - 3i = (-15 + 2) + (-8 - 3)i = -13 - 11i$

Step 4: Plug the result from step 3 into equation in step 2:
$12 \times (-13 - 11i) \div i$.

Step 5: Multiply first:
$12 \times (-13 - 11i) = (12)(-13) + (12)(-11i) = -156 - 132i$

The equation in step 4 can now be written as $\dfrac{-156 - 132i}{i}$.

Step 6: Finally, $-156 - 132i$ is divided by i as follows:
$$\frac{-156 - 132i}{i} = \frac{(-156 - 132i)(-i)}{(i)(-i)} = \frac{(-156)(-i) + (-132i)(-i)}{(i)(-i)}$$
$$= \frac{156i + 132i^2}{-i^2} = \frac{156i + 132(-1)}{-(-1)} = \frac{156i - 132}{1} = -132 + 156i$$

Simplify each of the following expressions.

1. $(-10 + i) + (5 - 4i) \times (1 + 13i)$

2. $(4 - 5i) - (7 + 2i) \div (2 - i)^2$

3. $((9 - 3i) + (3 + 9i)) \times (6i + (3 + i))$

4. $((17 - 2i) + (-15 + 4i))^2$

5. $((8 + 5i)^2 - (33 + 75i))^2$

6. $(2 + 11i) \div ((1 - 7i) - (8 - 8i))$

7. $\dfrac{3 + i}{3 - i} + \dfrac{2 + i}{4 + 3i}$

8. $((39 - 36i) - (4 - 5i)^2)^2$

9. What is the absolute value of $5 - 2i$?

10. What is the absolute value of $-3 + 9i$?

Chapter 10 Complex Numbers

Chapter 10 Review

Name the real part of each of the following complex numbers. (DOK 1)

1. $-38 + 17i$

2. $\dfrac{8}{5} - \dfrac{13}{5}i$

Name the imaginary part of each of the following complex numbers. (DOK 1)

3. $11 - 16i$

4. $225 + 725i$

Use what you know about the imaginary number i to solve the following problems. (DOK 2)

5. $-5i + 18i$

6. $\sqrt{-12} \times \sqrt{3}$

Find the square root of each of the following numbers. (DOK 2)

7. -64

8. -361

Add. (DOK 2)

9. $(-14 - 3i) + (7 + 9i)$

10. $\left(\dfrac{3}{10} - 20i\right) + \left(-8 + \dfrac{1}{5}i\right)$

11. $(4 - 15i) + (12 + 19i)$

12. $(-21 + 20i) + (32 - 5i)$

Subtract. (DOK 2)

13. $(10 - 4i) - (42 + 7i)$

14. $(-13 + 23i) - (18 - 6i)$

15. $(76 - 52i) - (43 + 27i)$

16. $\left(26 + \dfrac{4}{9}i\right) - \left(31 - \dfrac{2}{3}i\right)$

Multiply. (DOK 2)

17. $(1 - 6i) \times (5 + 5i)$

18. $(4 + 2i) \times (10 - 15i)$

19. $(8 + i) \times (-4 - 3i)$

20. $(13 - 4i) \times (13 + 4i)$

Divide. (DOK 2)

21. $(5 + 4i) \div (-3 - 2i)$

22. $(14 - 7i) \div (7 + i)$

23. $(6 + 3i) \div (-8 - 4i)$

24. $(-2 + 8i) \div (-10 + 2i)$

25. $104 \div (10 - 2i)$

26. $89 \div (-5 + 8i)$

Chapter 10 Test

1 What is $\dfrac{-9-2i}{3-i} + \dfrac{2+7i}{1-4i}$?

A $\quad -\dfrac{137}{34} - \dfrac{21}{34}i$

B $\quad -\dfrac{137}{34} + \dfrac{21}{34}i$

C $\quad \dfrac{137}{34} - \dfrac{21}{34}i$

D $\quad \dfrac{137}{34} + \dfrac{21}{34}i$

(DOK 2)

2 Which of the following statements is true?

A Every real number is a complex number with an imaginary part of -1.

B Every real number is a complex number with an imaginary part of 0.

C Every real number is a complex number with an imaginary part of 1.

D Every complex number is a real number with an imaginary part of 1.

(DOK 2)

3 What is the complex conjugate of $\frac{1}{10} - \frac{3}{10}i$?

A $\quad -\frac{1}{10} - \frac{3}{10}i$

B $\quad -\frac{1}{10} + \frac{3}{10}i$

C $\quad \frac{1}{10} - \frac{3}{10}i$

D $\quad \frac{1}{10} + \frac{3}{10}i$

(DOK 2)

4 What is $(8 - 18i) - (-3 - 13i)$?

A $\quad 5 - 31i$

B $\quad 5 - 5i$

C $\quad 11 - 31i$

D $\quad 11 - 5i$

(DOK 2)

5 What is $(4-i) - \left(8 + \frac{1}{2}i\right) \times (2+6i)^2$?

A $\quad -272 - 177i$

B $\quad -264 - 175i$

C $\quad 272 - 177i$

D $\quad 264 - 175i$

(DOK 2)

6 What is $(2 - 9i) \div (-4 + 6i)$?

A $\quad -\dfrac{31}{26} - \dfrac{6}{13}i$

B $\quad -\dfrac{31}{26} + \dfrac{6}{13}i$

C $\quad \dfrac{31}{26} - \dfrac{6}{13}i$

D $\quad \dfrac{31}{26} + \dfrac{6}{13}i$

(DOK 2)

Chapter 10   Complex Numbers

7 What is $\left(-\frac{5}{6} - \frac{1}{3}i\right) + \left(\frac{1}{2} + \frac{1}{6}i\right)$?

A $-\frac{1}{3} - \frac{1}{6}i$

B $-\frac{1}{3} + \frac{1}{6}i$

C $\frac{1}{3} - \frac{1}{6}i$

D $\frac{1}{3} + \frac{1}{6}i$

(DOK 2)

8 What is the imaginary part of the complex number $\frac{76}{3} - \frac{32}{3}i$?

A i

B $-\frac{32}{3}$

C $\frac{32}{3}$

D $\frac{76}{3}$

(DOK 2)

9 When multiplying two complex numbers that both have a real part and an imaginary part not equal to 0, which of the following statements is true regarding the result?

A It will always have a real part equal to 0.

B It will always have an imaginary part equal to 0.

C It will sometimes have an imaginary part equal to 0.

D It will never have an imaginary part equal to 0.

(DOK 2)

10 What is the simplest form of the expression $(-2 - i)(4 + i)$?

A -7

B 7

C $-7 - 6i$

D $-7 + 6i$

(DOK 2)

11 What is the simplest form of the expression $(2 - 4i)(-6 + 4i)$?

A $4 + 32i$

B $4 - 32i$

C $-12 + 16i$

D $-12 - 16i$

(DOK 2)

12 What is the simplest form of the expression $(-3 + 2i)(-6 - 8i)$?

A $34 - 12i$

B $34 + 12i$

C $18 + 16i$

D $18 - 16i$

(DOK 2)

13 What is the simplest form of the expression $(8 - 3i)^2$?

A $55 - 48i$

B $55 + 48i$

C $64 - 9i$

D $64 - 9i$

(DOK 2)

Chapter 11
Polynomials

This chapter covers the following CCGPS standard(s):

Arithmetic Operations on Polynomials	A.APR.1
Structure of Expressions	A.SSE.1a, A.SSE.1b, A.SSE.2, A.SSE.3a, A.SSE.3b

11.1 Parts of an Expression (DOK 1)

Constant: a number that stands alone.

    Example: 5

Variable: a symbol that represents an unknown value.

    Example: x

Coefficient: a number used to multiply a variable.

    Example: $5x$

Term: a constant or variable, or a coefficient and its variable (separated by addition and subtraction operation signs)

    Example: $\underline{5} + \underline{2}x$

Exponent: a number or variable that denotes how often a number or variable should be multiplied to itself.

    Example: 5^3

Factors: two or more parts of an expression which are multiplied together.

    Example: $(5 \cdot x)(x + 8)$

Copyright © American Book Company

Chapter 11 Polynomials

Identify the parts of the following expression for problems numbers 1–5.

$$(7x + 8)(x + 2)$$

1. What are the terms in this expression?
2. What are the factors in this expression?
3. What are the coefficients in this expression?
4. What are the variables in this expression?
5. What are the constants in this expression?

Identify the parts of the following expression for problems numbers 6–10.

$$6x^5 + y^3 + 9z + 5$$

6. What are the terms in this expression?
7. What are the coefficients in this expression?
8. What are the variables in this expression?
9. What are the constants in this expression?
10. What are the exponents in this expression?

Identify the parts of the following expression for problems numbers 11–15.

$$\frac{4x^6 + 3}{7y}$$

11. What are the terms in this expression?
12. What are the coefficients in this expression?
13. What are the variables in this expression?
14. What are the constants in this expression?
15. What are the factors in this expression?

Identify the parts of the following expression for problems numbers 16–20.

$$5x^2(12t + 11) + 8$$

16. What are the terms in this expression?
17. What are the factors in this expression?
18. What are the exponents in this expression?
19. What are the constants in this expression?
20. What are the variables in this expression?

11.2 Rewriting Expressions (DOK 2)

Rewriting expressions can be useful, and usually makes the problem easier to solve. However, knowing how to simplify the problem can be different based on the structure of the problem.

Example 1: Given $x^4 - y^4$, rewrite the expression by factoring.

Step 1: Note that this expression looks much like the difference of two squares, which we know is factored in this way: $x^2 - y^2 = (x+y)(x-y)$

Step 2: Remember that because this problem has a higher exponent that we will reduce this to: $x^4 - y^4 = (x^2 + y^2)(x^2 - y^2)$

Step 3: There is another possible factorization beyond this point. Once again, we have a difference of squares, and therefore must continue to simplify.
$(x^2 + y^2)(x^2 - y^2) = (x^2 + y^2)(x+y)(x-y)$

Step 4: Confirm that this is the most simplified expression by checking to see that there are no other common factors.

Answer: $(x^2 + y^2)(x+y)(x-y)$

Example 2: Given $6x^2 + 3x + 9$, rewrite the expression by factoring.

Step 1: Look at the coefficients and constants of the expression and see if they have any common factors. 6, 3, and 9 are each divisible by 3.

Step 2: Factor out a 3 from the expression, and rewrite it.
$6x^2 + 3x + 9 = 3(2x^2 + x + 3)$

Step 3: Check to see if it is possible to factor the expression any further. In this particular case, there are no further factorizations.

Answer: $3(2x^2 + x + 3)$

Chapter 11 Polynomials

Directions: Rewrite the following expressions by factoring. Reduce to the simplest form.

1. $18x^3 - 12x^2 + 4x$
2. $49x^2 - 144$
3. $16x^2 - 24x + 9$
4. $x^2 - 36$
5. $20x^3 + 8x^2 + 10x + 4$
6. $x^2 - 14x + 45$
7. $8x^2 - 16x + 8$
8. $49x^4 - 81$
9. $4x^4 - 17x^2$
10. $18x^5 - 24x^3$
11. $2x^2 - 20x + 32$
12. $25x^2 - 100$
13. $7x^2 + 42x + 63$
14. $25x^2 - 90xy + 81y^2$
15. $37x^4 - 259x^3 + 444x^2$

Polynomials are algebraic expressions which include **monomials** containing one term, **binomials** which contain two terms, and **trinomials**, which contain three terms. Expressions with more than three terms are called **polynomials**. **Terms** are separated by plus and minus signs.

EXAMPLES

Monomials	Binomials	Trinomials	Polynomials
$5f$	$5t + 20$	$x^2 + 4x + 3$	$x^3 - 3x^4 + 3x - 20$
$3x^3$	$20 - 8g$	$7x^4 - 6x - 2$	$p^5 + 4p^3 + p^4 + 20p - 7$
$5g^4$	$7x^4 + 8x$	$y^5 + 27y^4 + 200$	
4	$6x^3 - 9x$		

11.3 Adding Polynomials (DOK 2)

When adding **polynomials,** make sure the exponents and variables are the same on the terms you are combining. The easiest way is to put the terms in columns with **like exponents** under each other. Each column is added as a separate problem. Fill in the blank spots with zeros if it helps you keep the columns straight. You never carry to the next column when adding polynomials.

Example 3: Add $3x^4 + 25$ and $7x^4 + 4x$

$$\begin{array}{r} 3x^4 + 0x + 25 \\ (+)\ 7x^4 + 4x + 0 \\ \hline 10x^4 + 4x + 25 \end{array}$$

Example 4: $(5x^3 - 4x) + (-x^3 - 5)$

$$\begin{array}{r} 5x^3 - 4x + 0 \\ (+)\ -x^3 + 0x - 5 \\ \hline 4x^3 - 4x - 5 \end{array}$$

Add the following polynomials.

1. $y^4 + 3y + 4$ and $4y^4 + 5$
2. $(7y^4 + 5y - 6) + (4y^4 - 7y + 9)$
3. $-4x^4 + 7x^3 + 5x - 2$ and $3x^4 - x + 4$
4. $-p + 5$ and $7p^4 - 4p + 4$
5. $(w - 4) + (w^4 + 4)$
6. $5t^4 - 7t - 8$ and $9t + 4$
7. $t^5 + t + 9$ and $4t^3 + 5t - 5$
8. $(s^4 + 3s^3 - 4) + (-4s^3 + 5)$
9. $(-v^4 + 8v - 9) + (5v^3 - 6v + 5)$
10. $6m^4 - 4m + 20$ and $m^4 - m - 9$
11. $-x + 5$ and $3x^4 + x - 4$
12. $(9t^4 + 3t) + (-8t^4 - t + 5)$

11.4 Subtracting Polynomials (DOK 2)

When you subtract polynomials, it is important to remember to change all the signs in the subtracted polynomial (the subtrahend) and then add.

Example 5: $(5y^4 + 9y + 20) - (4y^4 + 6y - 5)$

Step 1: Copy the subtraction problem into vertical form. Make sure you line up the terms with like exponents under each other just like you did for adding polynomials.

$$\begin{array}{r} 5y^4 + 9y + 20 \\ (-)\ 4y^4 + 6y - 5 \\ \hline \end{array}$$

Step 2: Change the subtraction sign to addition and all the signs of the subtracted polynomial to the opposite sign. The bottom polynomial in the problem becomes $-4y^4 - 6y + 5$.

Step 3: Add:
$$\begin{array}{r} 5y^4 + 9y + 20 \\ (+)\ -4y^4 - 6y + 5 \\ \hline y^4 + 3y + 25 \end{array}$$

Chapter 11 Polynomials

Subtract the following polynomials.

1. $(4x^4 + 7x + 4) - (x^4 + 3x + 2)$
2. $(9y - 5) - (5y + 3)$
3. $(-5t^4 + 22t^3 + 3) - (5t^4 - t^3 - 7)$
4. $(-3w^4 + 20w - 7) - (-7w^4 - 7)$
5. $(6a^7 - a^3 + a) - (8a^7 + a^4 - 3a)$
6. $(25c^5 + 40c^4 + 20) - (8c^5 + 7c^4 + 24)$
7. $(7x^4 - 20x) - (-8x^4 + 5x + 9)$
8. $(-9y^4 + 24y^3 - 20) - (3y^3 + y + 20)$
9. $(-3h^4 - 8h + 8) - (7h^4 + 5h + 20)$
10. $(20k^3 - 9) - (k^4 - 5k^3 + 7)$
11. $(x^4 - 7x + 20) - (6x^4 - 7x + 8)$
12. $(24p^4 + 5p) - (20p - 4)$
13. $(-4m - 9) - (6m + 4)$
14. $(4y^4 + 23y^3 - 9y) - (5y^4 + 4y^3 - 8y)$
15. $(8g + 3) - (g^4 + 5g - 7)$
16. $(-9w^3 + 5w) - (-5w^4 - 20w^3 - w)$
17. $(x^4 + 24x^3 - 20) - (4x^4 + 3x^3 + 2)$
18. $(4a^4 + 4a + 4) - (-a^4 + 3a + 3)$
19. $(c + 220) - (3c^4 - 8c + 4)$
20. $(-6v^4 + 24v) - (3v^4 + 4v + 6)$
21. $(3b^4 + 5b^3 + 7) - (8b^3 - 9)$
22. $(7x^4 + 27x^3 - 5) - (-5x^4 + 5x^3)$
23. $(9y^4 - 4) - (22y^4 - 4y - 3)$
24. $(-z^4 - 7z - 9) - (3z^4 - 7z + 7)$

11.5 Multiplying Monomials (DOK 2)

When two monomials have the **same variable**, you can multiply them. Then, add the **exponents** together. If the variable has no exponent, it is understood that the exponent is 1.

Example 6: $\quad 5x^5 \times 3x^4 = 15x^9 \qquad\qquad 4y \times 7y^4 = 28y^5$

Multiply the following monomials.

1. $6a \times 20a^7$
2. $4x^6 \times 7x^3$
3. $5y^3 \times 3y^4$
4. $20t^4 \times 4t^4$
5. $4p^7 \times 5p^4$
6. $20b^4 \times 9b$
7. $-8s^5 \times 7s^3$
8. $-6a \times -20a^7$
9. $5x \times -x$
10. $-3y^4 \times -y^3$
11. $-7b^4 \times 3b^7$
12. $20c^5 \times -4c$

11.6 Multiplying Monomials with Different Variables (DOK 2)

Warning: You cannot add the exponents of variables that are different.

Example 7: $\qquad (-5wx)(6w^3x^4)$

To work this problem, first multiply the whole numbers: $-5 \times 6 = -30$. Then multiply the w's: $w \times w^3 = w^4$. Last, multiply the x's: $x \times x^4 = x^5$. The answer is $-30w^4x^5$.

Multiply the following monomials.

1. $(4x^4y^4)(-5xy^3) =$
2. $(20p^3q^5)(4p^4q) =$
3. $(-3t^5v^4)(t^4v) =$
4. $(8w^3z^4)(3wz) =$
5. $(-4st^6)(-9s^4t) =$
6. $(xy^3)(5x^4y^4) =$
7. $(3st)(5s^3t^4)(4s^4t^5) =$
8. $(xy)(x^4y^4)(4x^3y^4) =$
9. $(4a^4b^4)(a^3b^3)(4ab) =$
10. $(5y^4z^5)(4y^3)(4z^4) =$
11. $(7cd^3)(3c^4d^4)(d^4) =$
12. $(4w^4x^3)(3x^5)(4w^3) =$

Chapter 11   Polynomials

11.7   Multiplying Monomials by Polynomials (DOK 2)

Example 8: $-7t\left(4t^4 - 8t + 20\right)$

Step 1:   Multiply $-7t \times 4t^4 = \mathbf{-28t^5}$

Step 2:   Multiply $-7t \times -8t = \mathbf{56t^2}$

Step 3:   Multiply $-7t \times 20 = \mathbf{-140t}$

Step 4:   Arrange the answers horizontally in order: $\mathbf{-28t^5 + 56t^2 - 140t}$

Remove parentheses in the following problems.

1. $3x\left(3x^4 + 5x - 2\right)$
2. $5y\left(y^3 - 8\right)$
3. $8a^4\left(4a^4 + 3a + 4\right)$
4. $-7d^3\left(d^4 - 7d\right)$
5. $4w\left(-5w^4 + 3w - 9\right)$
6. $9p\left(p^3 - 6p + 7\right)$
7. $-20b^4\left(-4b + 7\right)$
8. $4t\left(t^4 - 5t - 20\right)$
9. $20c\left(5c^4 + 3c - 8\right)$
10. $6z\left(4z^5 - 7z^4 - 5\right)$

11. $-20t^4\left(3t^4 + 7t + 6\right)$
12. $c\left(-3c - 7\right)$
13. $3p\left(-p^4 + p^3 - 20\right)$
14. $-k^4\left(4k + 5\right)$
15. $-3\left(5m^4 - 7m + 9\right)$
16. $6x\left(-8x^3 + 20\right)$
17. $-w\left(w^4 - 5w + 8\right)$
18. $4y\left(7y^4 - y\right)$
19. $3d\left(d^7 - 8d^3 + 5\right)$
20. $-7t\left(-5t^4 - 9t + 2\right)$

21. $8\left(4w^4 - 20w + 5\right)$
22. $3y^4\left(y^4 - 22\right)$
23. $v^4\left(v^4 + 3v + 3\right)$
24. $9x\left(4x^3 + 3x + 2\right)$
25. $-7d\left(5d^4 + 8d - 4\right)$
26. $-k^4\left(-3k + 6\right)$
27. $3x\left(-x^4 - 7x + 7\right)$
28. $5z\left(5z^5 - z - 8\right)$
29. $-7y\left(20y^3 - 3\right)$
30. $4b^4\left(8b^4 + 5b + 5\right)$

11.8 Dividing Polynomials by Monomials (DOK 2)

Example 9: $\dfrac{-8wx + 6x^2 - 16wx^2}{2wx}$

Step 1: Rewrite the problem. Divide each term from the top by the denominator, $2wx$.
$$\dfrac{-8wx}{2wx} + \dfrac{6x^2}{2wx} + \dfrac{-16wx^2}{2wx} = \dfrac{-4}{1} + \dfrac{3x}{w} + \dfrac{-8x}{1}$$

Step 2: Simplify each term in the problem. Then combine like terms.
$$-4 + \dfrac{3x}{w} - 8x$$

Simplify each of the following.

1. $\dfrac{bc^4 - 9bc - 4b^4c^4}{4bc}$

2. $\dfrac{3jk^4 + 24k + 20j^4k}{3jk}$

3. $\dfrac{7x^4y - 9xy^4 + 4y^3}{4xy}$

4. $\dfrac{26st^4 + st - 24s}{5st}$

5. $\dfrac{5wx^4 + 6wx - 24w^3}{4wx}$

6. $\dfrac{cd^4 + 20cd^3 + 26c^4}{4cd}$

7. $\dfrac{y^4z^3 - 4yz - 9z^4}{-4yz^4}$

8. $\dfrac{a^4b + 4ab^4 - 25ab^3}{4a^4}$

9. $\dfrac{pr^4 + 6pr + 9p^4r^4}{4pr^4}$

10. $\dfrac{6xy^4 - 3xy + 29x^4}{-3xy}$

11. $\dfrac{6x^4y + 24xy - 45y^4}{6xy}$

12. $\dfrac{7m^4n - 20mn - 47n^4}{7mn}$

13. $\dfrac{st^4 - 20st - 26s^4t^4}{4st}$

14. $\dfrac{8jk^4 - 25jk - 63j^4}{8jk}$

Chapter 11 Polynomials

11.9 Dividing Polynomials Using Synthetic Division(DOK 2)

Synthetic division is a way of dividing polynomials without the variables. Remember, terms are to be in descending order. Missing terms are to be replaced with "0".

Example 10: Divide $(x^2 - 8x - 20)$ by $(x + 2)$ using synthetic division.

Step 1: Set the polynomial you are dividing by equal to zero and solve for x.

If $x + 2 = 0$, then $x = -2$.

Step 2: Take the value found in Step 1 and write it to off to the left, then write only the coefficients of the devidend to the right.

$$-2] \quad 1 \quad -8 \quad -20$$

Step 3: Bring the first coefficient under the line.

$$\begin{array}{r} -2] \quad 1 \quad -8 \quad -20 \\ \downarrow \\ \hline 1 \end{array}$$

Step 4: Multiply the number on the bottom and the number in the bracket and place it below the next coefficient. Then add the two terms and bring the result down.

$$\begin{array}{r} -2] \quad 1 \quad -8 \quad -20 \\ -2 \\ \hline 1 \quad -10 \end{array}$$

Step 5: Repeat step 4 for the remaining coefficients.

$$\begin{array}{r} -2] \quad 1 \quad -8 \quad -20 \\ -2 \\ \hline 1 \quad -10 \quad 0 \end{array}$$

Step 6: The last number is the remainder. The other numbers become the coefficients of the answer. Place variables and corresponding exponents behind the new coefficients.
1 and -10 correspond to the equation, $1x - 10$, with a remainder of 0.

11.10 Removing Parentheses and Simplifying (DOK 2)

In the following problem, you must multiply each term inside the parentheses by the numbers and variables outside the parentheses, and then add the polynomials to simplify the expressions.

Example 11: $9x(4x^4 - 7x + 8) - 3x(5x^4 + 3x - 9)$

Step 1: Multiply to remove the first set of parentheses.

$9x(4x^4 - 7x + 8) = 36x^5 - 63x^2 + 72x$

Step 2: Multiply to remove the second set of parentheses.

$-3x(5x^4 + 3x - 9) = -15x^5 - 9x^2 + 27x$

Step 3: Copy each polynomial in columns, making sure the terms with the same variable and exponent are under each other. Add to simplify.

$$\begin{array}{r} 36x^5 - 63x^2 + 72x \\ (+) -15x^5 - 9x^2 + 27x \\ \hline 21x^5 - 72x^2 + 99x \end{array}$$

Remove the parentheses and simplify the following problems.

1. $5t(t + 8) + 7t(4t^4 - 5t + 2)$

2. $-7y(3y^4 - 7y + 3) - 6y(y^4 - 5y - 5)$

3. $-3(3x^4 + 5x) + 7x(x^4 + 3x + 4)$

4. $4b(7b^4 - 9b - 2) - 3b(5b + 3)$

5. $9d^4(3d + 5) - 8d(3d^4 + 5d + 7)$

6. $7a(3a^4 + 3a + 2) - (-4a^4 + 7a - 5)$

7. $3m(m + 8) + 9(5m^4 + m + 5)$

8. $5c^4(-6c^4 - 3c + 4) - 8c(7c^3 + 4c)$

9. $-9w(-w + 2) - 5w(3w - 7)$

10. $6p(4p^4 - 5p - 6) + 3p(p^4 + 6p + 20)$

Chapter 11 Review

Simplify. (DOK 2)

1. $3a^4 + 20a^4$
2. $(8x^4y^5)(20xy^7)$
3. $-6z^4(z+3)$
4. $(5b^4)(7b^3)$
5. $8x^4 - 20x^4$
6. $(7p-5) - (3p+4)$
7. $-7t(3t+20)^2$
8. $(3w^3y^4)(5wy^7)$
9. $3(4g+3)^2$
10. $25d^5 - 20d^5$
11. $(8w-5)(w-9)$
12. $27t^4 + 5t^4$
13. $(8c^5)(20c^4)$
14. $(20x+4)(x+7)$
15. $5y(5y^4 - 20y + 4)$
16. $(9a^5b)(4ab^3)(ab)$
17. $(7w^6)(20w^{20})$
18. $9x^3 + 24x^3$
19. $27p^7 - 22p^7$
20. $(3s^5t^4)(5st^3)$
21. $(5d+20)(4d+8)$
22. $5w(-3w^4 + 8w - 7)$
23. $45z^6 - 20z^6$
24. $-8y^3 - 9y^3$
25. $(8x^5)(8x^7)$
26. $28p^4 + 20p^4$
27. $(a^4v)(4av)(a^3v^6)$
28. $5(6y-7)^2$
29. $(3c^4)(6c^9)$
30. $(5x^7y^3)(4xy^3)$
31. Add $4x^4 + 20x$ and $7x^4 - 9x + 4$
32. $5t(6t^4 + 5t - 6) + 9t(3t+3)$
33. Subtract $y^4 + 5y - 6$ from $3y^4 + 8$
34. $4x(5x^4 + 6x - 3) + 5x(x+3)$
35. $(6t-5) - (6t^4 + t - 4)$
36. $(5x+6) + (8x^4 - 4x + 3)$
37. Subtract $7a - 4$ from $a + 20$
38. $(-4y+5) + (5y-6)$
39. $4t(t+6) - 7t(4t+8)$
40. Add $3c - 5$ and $c^4 - 3c - 4$
41. $4b(b-5) - (b^4 + 4b + 2)$
42. $(6k^4 + 7k) + (k^4 + k + 20)$
43. $(q^4r^3)(3qr^4)(4q^5r)$
44. $(7df)(d^5f^4)(4df)$
45. $(8g^4h^3)(g^3h^6)(6gh^3)$
46. $(9v^4x^3)(3v^6x^4)(4v^5x^5)$
47. $(3n^4m^4)(20n^4m)(n^3m^8)$
48. $(22t^4a^4)(5t^3a^9)(4t^6a)$
49. $\dfrac{24(4a^3)b}{3a^4b^{-4}}$
50. $\dfrac{8(g^3h^3)}{5(g^4h)^{-4}}$
51. $\dfrac{26(m^4n^3)^4}{5(m^4n)^{-4}}$
52. $\dfrac{25p^3q^3}{4p^4q}$
53. $\dfrac{9(e^5h^{-4})^{-4}}{36e^4h^7}$
54. $\dfrac{44x^3y^5}{154(x^{-3}y^8)^4}$

Chapter 11 Test

1 $2x^2 + 5x^2 =$

A $10x^4$

B $7x^4$

C $7x^2$

D $10x^2$

(DOK 2)

2 $-8m^3 + m^3 =$

A $-8m^6$

B $-8m^9$

C $-9m^6$

D $-7m^3$

(DOK 2)

3 $(6x^3 + x^2 - 5) + (-3x^3 - 2x^2 + 4) =$

A $3x^3 - x^2 - 1$

B $3x^3 - 3x^2 - 1$

C $3x^3 - 3x^2 - 9$

D $-3x^3 - 3x^2 - 1$

(DOK 2)

4 $(-7c^2 + 5c + 3) + (-c^2 - 7c + 2) =$

A $-3x^3 - 3x^2 - 1$

B $-8c^2 - 2c + 5$

C $-6c^2 - 12c + 5$

D $-8c^2 - 12c + 5$

(DOK 2)

5 $(5x^3 - 4x^2 + 5) - (-2x^3 - 3x^2) =$

A $3x^3 + x^2 + 5$

B $3x^3 - 7x^2 + 5$

C $7x^3 - x^2 + 5$

D $7x^3 - 7x^2 + 5$

(DOK 2)

6 $(-z^3 - 4z^2 - 6) - (3z^3 - 6z + 5) =$

A $-4z^3 - 4z^2 + 6z - 11$

B $-2z^3 - 10z - 1$

C $-4z^3 - 10z^2 - 1$

D $-2z^2 + 2z - 11$

(DOK 2)

7 $(-7d^5)(-3d^2) =$

A $-21d^7$

B $21d^{10}$

C $21d^7$

D $-21d^{10}$

(DOK 2)

8 $(-5c^3d)(3c^5d^3)(2cd^4) =$

A $30c^{15}d^8$

B $15c^8d^{12}$

C $-17c^{15}d^{12}$

D $-30c^9d^8$

(DOK 2)

Chapter 11   Polynomials

9 $-11j^2 \times -j^4 =$

A $11j^6$

B $11j^8$

C $-11j^6$

D $-11j^8$

(DOK 2)

10 $-6m^2(7m^2 + 5m - 6) =$

A $-42m^2 + 30m^3 - 36$

B $-42m^4 - 30m^3 + 36m^2$

C $-13m^4 - m^2 + 36m^2$

D $42m^4 - 30m^3 - 36m^2$

(DOK 2)

11 $-h^2(-4h + 5) =$

A $-4h^3 - 5h^2$

B $4h^3 - 5h^2$

C $-5h^2 - 5h^2$

D $-5h^3 - 5h^2$

(DOK 2)

12 $\dfrac{4xy^2 - 6xy + 8x^2y}{2xy} =$

A $2xy - 3 + 4x$

B $2y - 3 + 4xy$

C $2y - 3 + 4x$

D $2xy - 3 + 4x^2$

(DOK 2)

13 $\dfrac{3cd^3 + 6c^2d - 12cd}{3cd} =$

A $cd + 3c - 4cd$

B $d^2 + 2c - 4cd$

C $cd^2 + 2c - 4cd$

D $d^2 + 2c - 4$

(DOK 2)

14 $4m(m - 5) + 3m(2m^2 - 6m + 4) =$

A $6m^3 - 14m^2 - 8m$

B $-8m^2 - 8m - 1$

C $7m - 14m^2 - 1$

D $10m^2 - 26m - 20$

(DOK 2)

15 $2h(3h^2 - 5h - 2) + 4h(h^2 + 6h + 8) =$

A $6h^3 + 19h^2 + 28h$

B $-8m^2 - 8m - 1$

C $7m - 14m^2 - 1$

D $10h^3 + 14h^2 + 28h$

(DOK 2)

16 Multiply the following binomial and simplify. $(x - 3)(x + 3)$

A $x^2 - 3x + 3x - 9$

B $x^2 - 9$

C $x^2 + 9$

D $x^2 + 6x + 9$

(DOK 2)

Chapter 11 Test

17 Multiply the following binomial and simplify. $(x+9)(x+1)$

   A $x^2 + 10x + 9$

   B $x^2 + 10x + 10$

   C $x^2 + 9x + 9$

   D $x^2 + 9x + x + 9$

(DOK 2)

18 Multiply the following binomial and simplify. $(x-2)^2$

   A $x^2 - 4x - 4$

   B $x^2 - 2x + 4$

   C $x^2 - 2x - 4$

   D $x^2 - 4x + 4$

(DOK 2)

19 $(x+4)^2 =$

   A $x^2 + 4$

   B $x^2 + 16$

   C $x^2 + 16x + 8$

   D $x^2 + 8x + 16$

(DOK 2)

20 How many terms are in the expression $5x^2(8x + 6y + 7)$?

   A 4

   B 3

   C 2

   D 1

(DOK 2)

21 How many factors are in the expression $4x^3(8x + 9)(3x^2 + y)$?

   A 4

   B 3

   C 2

   D 1

(DOK 2)

22 What is the exponent in the expression $12x^2 + 6y + 9$?

   A 4

   B 3

   C 2

   D 1

(DOK 2)

23 What is the y coefficient in the expression $6x^2 + 8y^4 + 3z + 17$?

   A 6

   B 8

   C 3

   D 17

(DOK 2)

Chapter 12
Solving Multi-Step Equations and Inequalities

This chapter covers the following CCGPS standard(s):

| Arithmetic Operations | A.APR.1 |
| The Real Number System | N.RN.1, N.RN.2 |

12.1 Two-Step Algebra Problems (DOK 2)

In the following two-step algebra problems, **addition** and **subtraction** are performed first and then **multiplication** and **division**.

Example 1: Solve for x: $-4x + 7 = 31$

Step 1: Subtract 7 from both sides.

$$\begin{array}{r} -4x + 7 = 31 \\ -7 -7 \\ \hline -4x = 24 \end{array}$$

Step 2: Divide both sides by -4.

$$\frac{-4x}{-4} = \frac{24}{-4}$$

Answer: $x = -6$

Example 2: Solve for y: $-8 - y = 12$

Step 1: Add 8 to both sides.

$$\begin{array}{r} -8 - y = 12 \\ +8 +8 \\ \hline -y = 20 \end{array}$$

Step 2: To finish solving a problem with a negative sign in front of the variable, multiply both sides by -1. The variable needs to be positive in the answer.

$$(-1)(-y) = (-1)(20)$$

Answer: $y = -20$

12.2 Multi-Step Algebra Problems (DOK 2)

Solve the two-step algebra problems below.

1. $6x - 4 = -34$
2. $5y - 3 = 32$
3. $8 - t = 1$
4. $10p - 6 = -36$
5. $11 - 9m = -70$
6. $4x - 12 = 24$
7. $3x - 17 = -41$
8. $9d - 5 = 49$
9. $10h + 8 = 78$
10. $-6b - 8 = 10$
11. $-g - 24 = -17$
12. $-7k - 12 = 30$
13. $9 - 5r = 64$
14. $6y - 14 = 34$
15. $12f + 15 = 51$
16. $21t + 17 = 80$
17. $20y + 9 = 149$
18. $15p - 27 = 33$
19. $22h + 9 = 97$
20. $-5 + 36w = 175$

12.2 Multi-Step Algebra Problems (DOK 2)

You can now use what you know about removing parentheses, combining like terms, and solving simple algebra problems to solve problems that involve three or more steps. Study the examples below to see how easy it is to solve multi-step problems.

Example 3: $3(x + 6) = 5x - 2$

Step 1:	Use the distributive property to remove parentheses.	$3x + 18 = 5x - 2$
Step 2:	Subtract $5x$ from each side to move the terms with variables to the left side of the equation.	$\dfrac{-5x \qquad -5x}{-2x + 18 = -2}$
Step 3:	Subtract 18 from each side to move the integers to the right side of the equation.	$\dfrac{-18 \qquad -18}{\dfrac{-2x}{-2} = \dfrac{-20}{-2}}$
Step 4:	Divide both sides by -2 to solve for x.	$x = 10$

Answer: $x = 10$

Example 4: Solve for x: $\dfrac{3(x-3)}{2} = 9$

Step 1:	Use the distributive property to remove parentheses.	$\dfrac{3x - 9}{2} = 9$
Step 2:	Multiply both sides by 2 to eliminate the fraction.	$\dfrac{2(3x-9)}{2} = 2(9)$
Step 3:	Add 9 to both sides, and combine like terms.	$3x - 9 = 18$
Step 4:	Divide both sides by 3 to solve for x.	$\dfrac{+9 \quad +9}{\dfrac{3x}{3} = \dfrac{27}{3}}$
		$x = 9$

Answer: $x = 9$

Example 5: Solve for x: $4(x - 2) = 4x - 8$

Step 1: Use the distributive property to remove parentheses.
$4x - 8 = 4x - 8$

Chapter 12   Solving Multi-Step Equations and Inequalities

Step 2:   Both sides of the equation are the same, so when you finish solving you will get $0 = 0$. In this specific case, any real number would make this equation true. Therefore, all real numbers are solutions for x.

Answer: All real numbers

Example 6: $x - 12 = x + 2$

Step 1:   Subtract x from both sides.
$x - x - 12 = x - x + 2$
$-12 = 2$

Step 2:   The result $-12 = 2$ is false. This means the equation has no solution. Therefore, there are no solutions for x or $x = \emptyset$.

Answer: No solutions, $x = \emptyset$

Solve the following multi-step algebra problems.

1. $2(y - 3) = 4y + 6$

2. $\dfrac{2(a+4)}{2} = 12$

3. $\dfrac{10(x-2)}{5} = 14$

4. $\dfrac{12y - 18}{6} = 4y + 3$

5. $2x + 3x = 30 - x$

6. $\dfrac{2a+1}{3} = a + 5$

7. $5(b - 4) = 10b + 5$

8. $-8(y + 4) = 10y + 4$

9. $\dfrac{x+4}{-3} = 6 - x$

10. $\dfrac{4(n+3)}{4} = n + 3$

11. $3(2x - 5) = 8x - 9$

12. $7 - 9a = 9 - 9a$

13. $7 - 5x = 10 - (6x + 7)$

14. $4(x - 3) - x = x - 6$

15. $4a + 4 = 3a - 4$

16. $-3(x - 4) + 5 = -2x - 2$

17. $5b - 11 = 13 - b$

18. $\dfrac{-4x+3}{2x} = \dfrac{7}{2x}$

19. $-(x + 1) = -2(5 - x)$

20. $4(2c + 3) - 7 = 13$

21. $6 - 3a = 9 - 2(2a + 5)$

22. $-5x + 9 = -3x + 11$

23. $3y + 2 - 2y - 5 = 4y + 3$

24. $3y - 10 = 4 - 4y$

25. $-(a + 3) = -(a + 1) - 2$

26. $5m - 2(m + 1) = m - 10$

27. $\dfrac{1}{2}(b - 2) = 5$

28. $-3(b - 4) = -2b$

29. $4x + 12 = -2(x + 3)$

30. $\dfrac{7x+4}{3} = 2x - 1$

31. $9x - 5 = 8x - 7$

32. $4x - 5 = 4x + 10$

33. $\dfrac{4x+8}{2} = 6$

34. $2(c + 4) + 8 = 10$

35. $y - (y + 3) = y + 6$

36. $4 + x - 2(x - 6) = 8$

12.3 Solving Radical Equations (DOK 2)

Some multi-step equations contain radicals. An example of a radical is a square root, $\sqrt{a}$.

Example 7: Solve the following equation for x. $\sqrt{4x-3} + 2 = 5$

Step 1: The first step is to get the constants that are not under the radical on one side. Subtract 2 from both sides of the equation.
$\sqrt{4x-3} + 2 - 2 = 5 - 2$
$\sqrt{4x-3} = 3$

Step 2: Next, you must eliminate the radical sign by squaring both sides of the equation.
$\left(\sqrt{4x-3}\right)^2 = \left((4x-3)^{1/2}\right)^2 = (4x-3)^{(1/2) \times 2} = (4x-3)^1 = 4x - 3$
$4x - 3 = (3)^2$
$4x - 3 = 9$

Step 3: Add 3 to both sides of the equation to get the constants on just one side of the equation.
$4x - 3 + 3 = 9 + 3$
$4x = 12$

Step 4: Last, isolate x by dividing both sides by 4.
$\dfrac{4x}{4} = \dfrac{12}{4}$
$x = 3$

Solve the following equations.

1. $\sqrt{x+3} - 13 = -8$
2. $3 + \sqrt{7t-3} = 5$
3. $\sqrt{3q+12} - 4 = 5$
4. $\sqrt{11f+3} + 2 = 8$
5. $5 = \sqrt{6g-5} + (-2)$
6. $2 = \sqrt{x-3}$
7. $\sqrt{-8t} - 3 = 1$
8. $\sqrt{-d+1} - 9 = -6$
9. $10 - \sqrt{8x+2} = 9$
10. $\sqrt{15y+4} + 4 = 12$
11. $\sqrt{r+14} - 1 = 9$
12. $3 - \sqrt{2q-1} = 0$
13. $\sqrt{5t+16} + 4 = 13$
14. $17 = \sqrt{23-f} + 15$
15. $19 - \sqrt{7x-5} = 16$

Chapter 12 Solving Multi-Step Equations and Inequalities

12.4 More Equations (DOK 2)

Some multi-step equations may have rational exponents. Rational exponents are often fractional exponents. The fractional exponents represent two parts, an integer exponent (power) and a root.

$\dfrac{x}{y} = \dfrac{\text{power}}{\text{root}}$ Example: $a^{\frac{x}{y}} = \sqrt[y]{a^x}$

Example 8:   Solve the following equation for x.
$5(x+10)^{\frac{7}{2}} = 390,625$

Step 1:   Isolate the parentheses and exponent by dividing both sides of the equation by 5.
$$\dfrac{5(x+10)^{\frac{7}{2}}}{5} = \dfrac{390,625}{5}$$
$(x+10)^{\frac{7}{2}} = 78,125$

Step 2:   Eliminate the exponent by raising each side of the equation by the reciprocal. Remember that raising an exponent by an exponent means to multiply the two exponents together.
$\left((x+10)^{\frac{7}{2}}\right)^{\frac{2}{7}} = (78,125)^{\frac{2}{7}}$
$x + 10 = 25$

Step 3:   Isolate the variable by subtracting 10 from each side of the equation.
$x + 10 - 10 = 25 - 10$
$x = 15$

Answer: $x = 15$

Check your work by substituting 15 in for x.
$5(15+10)^{\frac{7}{2}} = 390,625$
$5(25)^{\frac{7}{2}} = 390,625$
$5 \times 78,125 = 390,625$
$390,625 = 390,625$

Solve the following equations. Check your work.

1. $8(6x+9)^{\frac{2}{3}} = 72$

2. $15(x-6)^{\frac{2}{7}} = 15$

3. $5(4x+3)^{\frac{2}{3}} = 20$

4. $\dfrac{(x+3)^{\frac{5}{2}}}{8} = 4$

5. $-54 = 10 - (x-10)^{\frac{3}{2}}$

6. $-3 + (8-2x)^{\frac{5}{4}} = 29$

7. $(x-27)^{\frac{3}{2}} = 64$

8. $-5126 = -6 - 5(3x+22)^{\frac{5}{3}}$

12.5 Equation Word Problems (DOK 3)

Example 9: You want to buy either a shirt or sweater as a gift for 12 people. The shirts cost $10 each, and the sweaters cost $20 each. Write an equation for the number of shirts and sweaters you can buy if you want to spend a total of $190. Let n represent the number of shirts and sweaters bought.

Step 1: Set up all the information you know.

Sweater price = $20   Shirt price = $10
No. of sweaters = n No. of shirts = $(12 - n)$

*Note: Since you are buying 12 shirts and sweaters altogether, if you buy n sweaters, then you will need $12 - n$ shirts.

Step 2: Set up the equation.

$20n + 10(12 - n) = 190$

Step 3: Solve for n.

$20n + 10(12 - n) = 190$
$20n + 120 - 10n = 190$
$10n + 120 = 190$
$10n = 70$
$n = 7$

Step 4: Since we found the value of n, we can find $12 - n$.

Sweaters = n Shirts = $12 - n$
Sweaters = 7   Shirts = $12 - 7 = 5$

Answer: With $190, you can buy 7 sweaters and 5 shirts.

Solve the following problems.

1. The population of Big War Creek, TN for 2000 through 2011 can be modeled by $15.6t + 1399$, where t is the numbers of years since 1980. What was the population of Big War Creek in 2000? What was the population increase from 2000 to 2011?

2. An airline ticket to India costs $1,675$ for an adult and $1,000$ for toddlers who are less than three years old. The airline charges a percentage of the ticket price as tax. What would it cost for three adults and one toddler to travel to India of the tax is 12%?

3. The length of a standard nail can be modeled by $l = 54d^{\frac{3}{2}}$, where d is the diameter of the nail. What is the diameter of a standard nail that is 2 inches long?

4. Your salary is $1,500$ per week, and you receive 11% commission on your sales each week. What are the possible amounts (in dollars) that you can sell each week to earn at least $2,000$ per week? Write an expression and solve. Assume x is the amount of sales.

Chapter 12 Solving Multi-Step Equations and Inequalities

12.6 Multi-Step Inequalities (DOK 2)

Remember that adding and subtracting with inequalities follow the same rules as equations. When you multiply or divide both sides of an inequality by the same positive number, the rules are also the same as for equations. However, when you multiply or divide both sides of an inequality by a **negative** number, you must **reverse** the inequality symbol.

Example 10:
$-x > 4$
$(-1)(-x) < (-1)(4)$
$x < -4$

Example 11:
$-4x < 2$
$\dfrac{-4x}{-4} > \dfrac{2}{-4}$
$x > -\dfrac{1}{2}$

Reverse the symbol when you multiply or divide by a negative number.

When solving multi-step inequalities, first add and subtract to isolate the term with the variable. Then multiply and divide.

Example 12: $2x - 8 > 4x + 1$

Step 1: Add 8 to both sides.

$2x - 8 + 8 > 4x + 1 + 8$
$2x > 4x + 9$

Step 2: Subtract $4x$ from both sides.

$2x - 4x > 4x + 9 - 4x$
$-2x > 9$

Step 3: Divide by -2. Remember to change the direction of the inequality sign.

$\dfrac{-2x}{-2} < \dfrac{9}{-2}$

Answer: $x < -\dfrac{9}{2}$

12.6 Multi-Step Inequalities (DOK 2)

Solve each of the following inequalities.

1. $8 - 3x \leq 7x - 2$

2. $3(2x - 5) \geq 8x - 5$

3. $\frac{1}{3}b - 2 > 5$

4. $7 + 3y > 2y - 5$

5. $3a + 5 < 2a - 6$

6. $3(a - 2) > -5a - 2(3 - a)$

7. $2x - 7 \geq 4(x - 3) + 3x$

8. $6x - 2 \leq 5x + 5$

9. $-\frac{x}{4} > 12$

10. $-\frac{2x}{3} \leq 6$

11. $3b + 5 < 2b - 8$

12. $4x - 5 \leq 7x + 13$

13. $4x + 5 \leq -2$

14. $2y - 5 > 7$

15. $4 + 2(3 - 2y) \leq 6y - 20$

16. $-4c + 6 \leq 8$

17. $-\frac{1}{2}x + 2 > 9$

18. $\frac{1}{4}y - 3 \leq 1$

19. $-3x + 4 > 5$

20. $\frac{y}{2} - 2 \geq 10$

21. $7 + 4c < -2$

22. $2 - \frac{a}{2} > 1$

23. $10 + 4b \leq -2$

24. $-\frac{1}{2}x + 3 > 4$

Chapter 12 Solving Multi-Step Equations and Inequalities

12.7 Solving Equations and Inequalities with Absolute Values (DOK 2)

When solving equations and inequalities which involve variables placed in absolute values, remember that there will be two or more numbers that will work as correct answers. This is because the absolute value variable will signify both positive and negative numbers as answers.

Example 13: $5 + 3|k| = 8$ Solve as you would any equation.

  Step 1: $3|k| = 3$ Subtract 5 from each side.

  Step 2: $|k| = 1$ Divide by 3 on each side.

  Step 3: $k = 1$ or $k = -1$ Because k is an absolute value, the answer can be 1 or -1.

Example 14: $2|x| - 3 < 7$ Solve as you would any inequality.

  Step 1: $2|x| < 10$ Add 3 to both sides.

  Step 2: $|x| < 5$ Divide by 2 on each side.

  Step 3: $x < 5$ or $x > -5$ or $-5 < x < 5$ Because x is an absolute value, the answer is a set of both positive and negative numbers.

Read each problem, and write the number or set of numbers which solves each equation or inequality.

1. $7 + 2|y| = 15$
2. $4|x| - 9 < 3$
3. $6|k| + 2 = 14$
4. $10 - 4|n| > -14$
5. $-3 = 5|z| + 12$
6. $-4 + 7|m| < 10$
7. $5|x| - 12 > 13$

8. $21|g| + 7 = 49$
9. $-9 + 6|x| = 15$
10. $12 - 6|w| > -12$
11. $31 > 13 + 9|r|$
12. $-30 = 21 - 3|t|$
13. $9|x| - 19 < 35$
14. $-13|c| + 21 \geq -31$

15. $5 - 11|k| < -17$
16. $-42 + 14|p| = 14$
17. $15 < 3|b| + 6$
18. $9 + 5|q| = 29$
19. $-14|y| - 38 < -45$
20. $36 = 4|s| + 20$
21. $20 \leq -60 + 8|e|$

12.8 More Solving Equations and Inequalities with Absolute Values (DOK 2)

Now, look at the following examples in which numbers and variables are added or subtracted within the absolute value symbols ($||$).

Example 15: $|3x - 5| = 10$ Remember an equation with absolute value symbols has two solutions.

Step 1:
$3x - 5 = 10$
$3x - 5 + 5 = 10 + 5$
$\dfrac{3x}{3} = \dfrac{15}{3}$
$x = 5$

To find the first solution, remove the absolute value symbol and solve the equation.

Step 2:
$-(3x - 5) = 10$
$-3x + 5 = 10$
$-3x + 5 - 5 = 10 - 5$
$-3x = 5$
$x = -\dfrac{5}{3}$

To find the second solution, solve the equation for the negative of the expression in absolute value symbols.

Answer: $x = \left\{5, -\dfrac{5}{3}\right\}$

Example 16: $|5z - 10| < 20$ Remove the absolute value symbols and solve the inequality.

Step 1:
$5z - 10 < 20$
$5z - 10 + 10 < 20 + 10$
$\dfrac{5z}{5} < \dfrac{30}{5}$
$z < 6$

Step 2:
$-(5z - 10) < 20$
$-5z + 10 < 20$
$-5z + 10 - 10 < 20 - 10$
$\dfrac{-5z}{5} < \dfrac{10}{5}$
$-z < 2$
$z > -2$

Next, solve the equation for the negative of the expression in the absolute value symbols.

Answer: $-2 < z < 6$

Chapter 12   Solving Multi-Step Equations and Inequalities

Example 17: $|4y + 7| - 5 > 18$

Step 1: $4y + [7 - 5 + 5] > 18 + 5$ Remove the absolute value symbols and solve the
$4y + 7 > 23$ inequality.
$4y + 7 - 7 > 23 - 7$
$4y > 16$
$y > 4$

Step 2: $-(4y + 7) - 5 > 18$ Solve the equation for the negative of the
$-4y - 7 - 5 + 5 > 18 + 5$ expression in the absolute value symbols.
$-4y - 7 + 7 > 23 + 7$
$-4y > 30$
$y < -7\frac{1}{2}$

Answer: $y > 4$ or $y < -\frac{15}{2}$

Solve the following equations and inequalities below.

1. $-4 + |2x + 4| = 14$
2. $|4b - 7| + 3 > 12$
3. $6 + |12e + 3| < 39$
4. $-15 + |8f - 14| > 35$
5. $|-9b + 13| - 12 = 10$
6. $-25 + |7b + 11| < 35$
7. $|7w + 2| - 60 > 30$
8. $63 + |3d - 12| = 21$
9. $|-23 + 8x| - 12 > +37$
10. $|61 + 20x| + 32 > 51$

11. $|4a + 13| + 31 = 50$
12. $4 + |4k - 32| < 51$
13. $8 + |4x + 3| = 21$
14. $|28 + 7v| - 28 < 77$
15. $|62p + 31| + 43 = 136$
16. $18 - |6v + 22| < 22$
17. $12 = 4 + |42 + 10m|$
18. $53 < 18 + |12e + 31|$
19. $38 > -39 + |7j + 14|$
20. $9 = |14 + 15u| + 7$

21. $11 - |2j + 50| > 45$
22. $|35 + 6i| - 3 = 14$
23. $|26 - 8r| - 9 > 41$
24. $|25 + 6z| - 21 = 28$
25. $12 < |2t + 6| - 14$
26. $50 > |9q - 10| + 6$
27. $12 + |8v - 18| > 26$
28. $-38 + |16i - 33| = 41$
29. $|-14 + 6p| - 9 < 7$
30. $28 > |25 - 5f| - 12$

12.9 Inequality Word Problems (DOK 3)

Inequality word problems involve staying under a limit or having a minimum goal one must meet.

Example 18: A contestant on a popular game show must earn a minimum of 800 points by answering a series of questions worth 40 points each per category in order to win the game. The contestant will answer questions from each of four categories. Her results for the first three categories are as follows: 160 points, 200 points, and 240 points. Write an inequality which describes how many points, (p), the contestant will need on the last category in order to win.

Step 1: Add to find out how many points she already has.
$160 + 200 + 240 = 600$

Step 2: Subtract the points she already has from the minimum points she needs. $800 - 600 = 200$. She must get at least 200 points in the last category to win. If she gets more than 200 points, that is okay, too. To express the number of points she needs, use the following inequality statement:

$p \geq 200$ The points she needs must be greater than or equal to 200.

Solve each of the following problems using inequalities.

1. Stella wants to place her money in a high interest money market account. However, she needs at least $1,000 to open an account. Each month, she set aside some of her earnings in a savings account. In January through June, she added the following amounts to her savings: $121, $206, $138, $212, $109, and $134. Write an inequality which describes the amount of money she can set aside in July to qualify for the money market account.

2. A high school band program will receive $2,000.00 for selling $10,000.00 worth of coupon books. Six band classes participate in the sales drive. Classes 1–5 collect the following amounts of money: $1,400, $2,600, $1,800, $2,450, and $1,550. Write an inequality which describes the amount of money the sixth class must collect so that the band will receive $2,000.

3. A small elevator has a maximum capacity of 1,000 pounds before the cable holding it in place snaps. Six people get on the elevator. Five of their weights follow: 146, 180, 130, 262, and 135. Write an inequality which describes the amount the sixth person can weigh without snapping the cable.

4. A small high school class of 9 students were told they would receive a pizza party if their class average was 92% or higher on the next exam. Students 1–8 scored the following on the exam: 86, 91, 98, 83, 97, 89, 99, and 96. Write an inequality which describes the score the ninth student must make for the class to qualify for the pizza party.

5. Raymond wants to spend his entire credit limit on his credit card. His credit limit is $2,000. He purchases items costing $600, $800, $50, $168, and $3. Write an inequality which describes the amounts Raymond can put on his credit card for his next purchases.

Chapter 12 Solving Multi-Step Equations and Inequalities

12.10 Recognizing Errors in Problems (DOK 3)

Example 19: David and Esther solved an equation $|4x + 2| = 18$ as shown below.

David's work

$$|4x + 2| = 18 \quad | \quad |4x + 2| = 18$$
$$4x + 2 = 18 \quad | \quad -4x + 2 = 18$$
$$4x = 16 \quad | \quad -4x = 16$$
$$x = 4 \quad | \quad x = -4$$

Esther's work

$$|4x + 2| = 18 \quad | \quad |4x + 2| = 18$$
$$4x + 2 = 18 \quad | \quad -4x - 2 = 18$$
$$4x = 16 \quad | \quad -4x = 20$$
$$x = 4 \quad | \quad x = -5$$

Looking at their work, we can see that Esther solved the equation correctly.

Decide whether the statement is true or false.

1. Multiplying both sides of an inequality by the same number sometimes produces an equivalent inequality.

2. Four is a solution for the inequality $4x + 3 < 11$.

3. Three is a solution for the inequality $-4 < 2x \leq 8$.

4. Five is a solution for the inequality $-10 < x - 11 < -7$.

5. Three is a solution for the inequality $-0.2 \leq 4.8x - 1.8 < 7.8$.

6. The inequality $|ax + b| > c$, where $c > 0$, means that $ax + b$ is beyond $-c$ and c. This is equivalent to $ax + b > c$ or $ax + b > -c$.

7. The inequality $|3x - 2| \geq 8$ is equivalent to $-8 \leq 3x - 2 \leq 8$.

8. $|9| = -9$

Decide whether the given number is a solution of the equation.

9. $|3x + 1| = 12$; 3

10. $|6 - \frac{1}{2}x| = 14$; -40

11. $|\frac{1}{4}x - 3| = 1$; 8

12. $|\frac{2}{3}x + 2| = 10$; 6

Chapter 12 Review

Solve each of the following equations and inequalities. (DOK 2)

1. $19 - 8d = d - 17$

2. $7w - 8w = -4w - 30$

3. $6 + 16x = 16x - 12$

4. $6(b - 4) = 6b - 24$

5. $4x - 16 = 7x + 2$

6. $9w - 2 = -w - 22$

7. $\dfrac{-11c - 35}{4} = 4c - 2$

8. $5 + x - 3(x + 4) = -17$

9. $4(2x + 3) \geq 2x$

10. $7 - 3x \leq 6x - 2$

11. $\dfrac{5(n + 4)}{3} = n - 8$

12. $-y > 14$

13. $2(3x - 1) \geq 3x - 7$

14. $3(x + 2) = 3x - 10$

15. $-11|k| < -22$

16. $18 - 3|w| > -18$

17. $21 = -4 + |5x + 5|$

18. $|3x + 2| - 4 \geq -2$

(DOK 3)

19. Which line of the solution of the inequality $6x - 3 \geq 7 + 4x$ is incorrect?

    line 1 $6x - 3 \geq 7 + 4x$
    add 3 to both sides

    line 2 $6x - 3 \geq 4 + 4x$
    subtract $4x$ from both sides

    line 3 $2x \geq 4$

    line 4 $x \geq 2$

20. Mike and Jeff put their canoe in the river and paddled downstream at 9 miles per hour. They turned the canoe around and paddled upstream at 3 miles per hour to return to their car. How far did they go downstream if the whole trip took 6 hours?

Chapter 12   Solving Multi-Step Equations and Inequalities

Chapter 12 Test

1 Find the value of n. $19n - 57 = 76$

A 1
B 2
C 5
D 7

(DOK 2)

2 Solve: $3(x - 2) - 1 = 6(x + 5)$

A -4
B $-\frac{37}{3}$
C $\frac{23}{3}$
D 4

(DOK 2)

3 Simplify: $5(2x + 11) - 3 \times 5 = ?$

A $7x + 40$
B $7x + 20$
C $10x + 40$
D $10x - 4$

(DOK 2)

4 Solve: $39 + |10x - 8| > 41$

A $x = 1$
B $x > 1$ or $x < \frac{3}{5}$
C $x > \frac{3}{5}$
D $x > 1$ or $x < -1$

(DOK 2)

5 Solve: $4b - 8 < 56$

A $b > 16$
B $b < 12$
C $b < -12$
D $b < 16$

(DOK 2)

6 Which of the following is equivalent to $3(2x - 5) - 4(x - 3) = 7$?

A $x + 27 = 7$
B $2x - 3 = 7$
C $x - 27 = 7$
D $10x - 27 = 7$

(DOK 2)

7 Which of the following describes the solution set for $3(x - 2) + 1 - 3x = -5$?

A $x = -5$
B $x = -\frac{5}{3}$
C There are no solutions for x.
D All real numbers are solutions for x.

(DOK 2)

8 Solve for x. $14x + 84 = 154$

A 5
B 11
C 17
D 238

(DOK 2)

9 Which of the following is equivalent to $4 - 5x > 3(x - 4)$?

A $4 - 5x > 3x - 12$
B $4 - 5x > 3x - 1$
C $4 - 5x > 3x - 7$
D $4 - 5x > 3x + 12$

(DOK 2)

10 There is a new bike that Bianca has had her eye on for a few weeks. The bike costs $75. Her allowance is 10 dollars per week. If she saves 60% of her allowance each week, write an inequality that describes the minimum amount of weeks, y, that Bianca must save in order to buy that bike.

A $y > 75 - 0.6\,(10)$
B $y > 45\,(10)$
C $y > \dfrac{75}{0.6\,(10)}$
D $10 > \dfrac{75}{0.6y}$

(DOK 3)

Chapter 13
Solving Quadratic Equations

This chapter covers the following CCGPS standard(s):

Create Equations	A.CED.1
Complex Numbers	N.CN.7
Interpreting Functions	F.IF.8a
Solve Equations and Inequalities	A.REI.4a, A.REI.4b

13.1 Fundamental Theorem of Algebra (DOK 2)

The Fundamental Theorem of Algebra states that any polynomial of degree n has n "roots," otherwise known as "zeros." For example, the polynomial $x^4 + 7x - 9$ has three terms. The degree of the polynomial with one variable (x) is the largest exponent of that variable. In this case the degree is 4: The roots or zeros of a polynomial is where the polynomial is equal to zero (which can be determined from the factors). Since there are 4 factors of a polynomial with degree 4; there are 4 roots. However, it is important to keep in mind that the roots of a polynomial are not always real numbers. Sometimes, the roots are complex numbers, meaning they have a real part and an imaginary part. For example, the polynomial $x^2 - x + 1$ has two roots. However, the roots are complex: $0:5 - 0:866i$ and $0:5 + 0.866i$. So the roots may be real or complex numbers, and sometimes a combination of both. Remember that complex roots always come in conjugate pairs (meaning the sign in the middle changes, as they did in the previous example).

In the previous chapter, we factored polynomials such as $y^2 - 4y - 5$ into two factors:

$$y^2 - 4y - 5 = (y+1)(y-5)$$

In this chapter, we learn that any equation that can be put in the form $ax^2 + bx + c = 0$ is a quadratic equation if a, b, and c are real numbers and $a \neq 0$. $ax^2 + bx + c = 0$ is the standard form of a quadratic equation. To solve these equations, follow the steps below.

Example 1: Solve $y^2 - 4y - 5 = 0$

Step 1: Factor the left side of the equation.

$$y^2 - 4y - 5 = 0$$
$$(y+1)(y-5) = 0$$

Chapter 13 Solving Quadratic Equations

Step 2: If the product of these two factors equals zero, then the two factors individually must be equal to zero. Therefore, to solve, we set each factor equal to zero.

$$(y+1) = 0 \qquad\qquad (y-5) = 0$$
$$\underline{-1 \quad -1} \qquad\qquad \underline{+5 \quad +5}$$
$$y = -1 \qquad\qquad y = 5$$

The equation has two solutions: $y = -1$ and $y = 5$.

Check: To check, substitute each solution into the original equation.

When $y = -1$, the equation becomes:
$(-1)^2 - (4)(-1) - 5 = 0$
$1 + 4 - 5 = 0$
$0 = 0$

When $y = 5$, the equation becomes:
$5^2 - (4)(5) - 5 = 0$
$25 - 20 - 5 = 0$
$0 = 0$

Both solutions produce true statements.
The solution set for the equation is $\{-1, 5\}$.

Solve each of the following quadratic equations by factoring and setting each factor equal to zero. Check by substituting answers back in the original equation.

1. $x^2 + x - 6 = 0$
2. $y^2 - 2y - 8 = 0$
3. $a^2 + 2a - 15 = 0$
4. $y^2 - 5y + 4 = 0$
5. $b^2 - 9b + 14 = 0$
6. $x^2 - 3x - 4 = 0$
7. $y^2 + y - 20 = 0$
8. $d^2 + 6d + 8 = 0$
9. $y^2 - 7y + 12 = 0$
10. $x^2 - 3x - 28 = 0$
11. $a^2 - 5a + 6 = 0$
12. $b^2 + 3b - 10 = 0$
13. $a^2 + 7a - 8 = 0$
14. $c^2 + 3c + 2 = 0$
15. $x^2 - x - 42 = 0$
16. $a^2 + 5a - 6 = 0$
17. $b^2 + 7b + 12 = 0$
18. $y^2 + y - 12 = 0$
19. $a^2 - 3a - 10 = 0$
20. $d^2 + 10d + 16 = 0$
21. $x^2 - 4x - 12 = 0$

Quadratic equations that have a whole number and a variable in the first term are solved the same way as the previous page. Factor the trinomial, and set each factor equal to zero to find the solution set.

Example 2: Solve $2x^2 + 3x - 2 = 0$
$(2x - 1)(x + 2) = 0$
Set each factor equal to zero and solve:

13.2 Using the Quadratic Formula (DOK 2)

$$\begin{aligned} 2x - 1 &= 0 \\ +1 &+1 \\ \hline \frac{2x}{2} &= \frac{1}{2} \\ x &= \frac{1}{2} \end{aligned}$$

$$\begin{aligned} x + 2 &= 0 \\ -2 &-2 \\ \hline x &= -2 \end{aligned}$$

The solution set is $\left\{\frac{1}{2}, -2\right\}$.

Solve the following quadratic equations.

22. $3y^2 + 4y - 32 = 0$
23. $5c^2 - 2c - 16 = 0$
24. $7d^2 + 18d + 8 = 0$
25. $3a^2 - 10a - 8 = 0$
26. $11x^2 - 31x - 6 = 0$
27. $5b^2 + 17b + 6 = 0$
28. $3x^2 - 11x - 20 = 0$
29. $5a^2 + 47a - 30 = 0$
30. $2c^2 - 5c - 25 = 0$
31. $2y^2 + 11y - 21 = 0$
32. $5a^2 + 23a - 42 = 0$
33. $3d^2 + 11d - 20 = 0$
34. $3x^2 - 10x + 8 = 0$
35. $7b^2 + 23b - 20 = 0$
36. $9a^2 - 58a + 24 = 0$
37. $4c^2 - 25c - 21 = 0$
38. $8d^2 + 53d + 30 = 0$
39. $4y^2 + 37y - 30 = 0$
40. $8a^2 + 37a - 15 = 0$
41. $3x^2 - 41x + 26 = 0$
42. $8b^2 + 2b - 3 = 0$

13.2 Using the Quadratic Formula (DOK 2)

You may be asked to use the quadratic formula to solve an algebra problem known as a **quadratic equation**. The equation should be in the form $ax^2 + bx + c = 0$.

Example 3: Using the quadratic formula, find x in the following equation: $x^2 - 8x = -7$.

Step 1: Make sure the equation is set equal to 0.

$x^2 - 8x + 7 = -7 + 7$
$x^2 - 8x + 7 = 0$

The quadratic formula, $\dfrac{-b \pm \sqrt{b^2 - 4ac}}{2a}$, will be given to you on your formula sheet with your test.

Step 2: In the formula, a is the number x^2 is multiplied by, b is the number x is multiplied by and c is the last term of the equation. For the equation in the example, $x^2 - 8x + 7$, $a = 1$, $b = -8$, and $c = 7$. When we look at the formula we notice a $\pm$ sign. This means that there will be two solutions to the equation, one when we use the plus sign and one when we use the minus sign. Substituting the numbers from the problem into the formula, we have:

$$\frac{8 + \sqrt{8^2 - (4)(1)(7)}}{2(1)} = 7 \quad \text{or} \quad \frac{8 - \sqrt{8^2 - (4)(1)(7)}}{2(1)} = 1$$

The solutions are $\{7, 1\}$.

Copyright © American Book Company

Chapter 13   Solving Quadratic Equations

For each of the following equations, use the quadratic formula to find two solutions.

1. $x^2 + x - 4 = 2$
2. $2y^2 - 4y - 10 = 6$
3. $\frac{1}{2}a^2 + a - \frac{15}{2} = 0$
4. $y^2 - 5y + 7 = 3$
5. $b^2 - 11b + 2b + 20 = 6$
6. $7x^2 - 21x - 28 = 0$
7. $y^2 + y - 10 = 10$
8. $3d^2 + 18d + 22 = -2$
9. $y^2 - y + 15 = 6y + 3$
10. $4x^2 - 3x = 112 + 9x$
11. $2a^2 - 5a = a^2 - 6$
12. $3b - 6 = 4 - b^2$
13. $8 - a^2 - 7a = 0$
14. $4 = 2 - x^2 - 3x$
15. $\frac{x^2}{7} - 6 = \frac{x}{7}$
16. $7a^2 + a = 6a^2 + 6$
17. $b^2 + 14b + 5 = 7b - 7$
18. $15 = y^2 + 2y$
19. $a^2 = 3a + 10$
20. $d^2 + 14d + 18 = 2d - 2$
21. $\frac{x^2}{4} - x - 3 = 0$

13.3   Solving Quadratic Equations with Complex Roots (DOK 2)

Some quadratic equations do not have real number solutions. In this section, we will find complex solutions to each equation.

Example 4:   Find the roots of the quadratic equation $7x^2 - 8x + 3 = 0$.

Step 1:   The quadratic equation cannot be factored, so the quadratic formula must be used.

$$x = \frac{-b \pm \sqrt{b^2 - 4ac}}{2a} = \frac{-(-8) \pm \sqrt{(-8)^2 - 4(7)(3)}}{2(7)} = \frac{8 \pm \sqrt{-20}}{14}$$

Step 2:   As stated earlier, i is defined as $\sqrt{-1}$, so if $\frac{8 \pm \sqrt{-20}}{14}$ is rewritten as

$\frac{8 \pm \sqrt{20 \times (-1)}}{14}$, and it can be simplified to $\frac{8 \pm i\sqrt{20}}{14}$.

This can be simplified as follows:

$$\frac{8 \pm i\sqrt{20}}{14} = \frac{8 \pm i\sqrt{4 \times 5}}{14} = \frac{8 \pm 2i\sqrt{5}}{14} = \frac{4 \pm i\sqrt{5}}{7} = \frac{4}{7} \pm \frac{\sqrt{5}}{7}i.$$

Therefore, the roots of the equation are $x = \frac{4}{7} + \frac{\sqrt{5}}{7}i$ or $\frac{4}{7} - \frac{\sqrt{5}}{7}i$.

13.4 Solving the Difference of Two Squares (DOK 2)

Example 5: Find the roots of the quadratic equation $3x^2 - x + 5 = 0$.

Step 1: Plug the values for this equation into the quadratic formula: $a = 3$, $b = -1$, $c = 5$.

Step 2: Solve for x.
$$x = \frac{-(-1) \pm \sqrt{(-1)^2 - 4(3)(5)}}{2(3)} = \frac{1 \pm \sqrt{1-60}}{6} = \frac{1}{6} \pm \frac{\sqrt{-59}}{6}$$

Step 3: Apply the definition of i. $x = \frac{1}{6} \pm \frac{\sqrt{-59}}{6} = \frac{1}{6} \pm \frac{\sqrt{59}}{6}i$.

Find the roots of each of the following quadratic equations.

1. $9x^2 - 6x + 3 = 0$
2. $\frac{3}{4}x^2 - x + 12 = 0$
3. $-11x^2 + 10x - 3 = 0$
4. $14x^2 - 3x + \frac{1}{2} = 0$
5. $-x^2 - 3x - 16 = 0$
6. $8x^2 + 9x + 10 = 0$
7. $x(19x - 2) = -1$
8. $x(1 - 5x) = 5$
9. $-10x^2 + \frac{x}{2} - 7 = 0$
10. $8x^2 + 3x + 10 = 0$
11. $-6x^2 - 5x - 11 = 0$
12. $13x^2 + 12x + 7 = 0$
13. $4x^2 - 10x + 11 = 0$
14. $-5x^2 - x - 3 = 0$
15. $18x^2 - 5x + 1 = 0$

13.4 Solving the Difference of Two Squares (DOK 2)

To solve the difference of two squares, first factor. Then set each factor equal to zero.

Example 6: $25x^2 - 36 = 0$

Step 1: Factor the left hand side of the equation.
$25x^2 - 36 = 0$
$(5x + 6)(5x - 6) = 0$

Step 2: Set each factor equal to zero and solve.

$5x + 6 = 0$
$-6 = -6$
$\overline{5x = -6}$

$\frac{5x}{5} = -\frac{6}{5}$

$x = -\frac{6}{5}$

$5x - 6 = 0$
$+6 = +6$
$\overline{5x = 6}$

$\frac{5x}{5} = \frac{6}{5}$

$x = \frac{6}{5}$

Chapter 13 Solving Quadratic Equations

Find the solution sets for the following equations.

1. $25a^2 - 16 = 0$
2. $c^2 - 36 = 0$
3. $9x^2 - 64 = 0$
4. $100y^2 - 49 - 0$
5. $4b^2 - 81 = 0$
6. $d^2 - 25 = 0$
7. $9x^2 - 1 = 0$
8. $16a^2 - 9 = 0$
9. $36y^2 - 1 = 0$
10. $36y^2 - 25 = 0$
11. $d^2 - 16 = 0$
12. $64b^2 - 9 = 0$
13. $81a^2 - 4 = 0$
14. $64y^2 - 25 = 0$
15. $4c^2 - 49 = 0$
16. $x^2 - 81 = 0$
17. $49b^2 - 9 = 0$
18. $a^2 - 64 = 0$
19. $x^2 - 1 = 0$
20. $4y^2 - 9 = 0$
21. $t^2 - 100 = 0$
22. $16k^2 - 81 = 0$
23. $a^2 - 4 = 0$
24. $36b^2 - 16 = 0$

13.5 Solving Perfect Squares (DOK 2)

When the square root of a constant, variable, or polynomial results in a constant, variable, or polynomial without irrational numbers, the expression is a **perfect square**. Some examples are 49, x^2, and $(x-2)^2$.

Example 7: Solve the perfect square for x. $(x-5)^2 = 0$

    Step 1: Take the square root of both sides.
$$\sqrt{(x-5)^2} = \sqrt{0}$$
$$(x - 5) = 0$$

    Step 2: Solve the equation.
$$(x - 5) = 0$$
$$x - 5 + 5 = 0 + 5$$
$$x = 5$$

Example 8: Solve the perfect square for x. $(x-5)^2 = 64$

    Step 1: Take the square root of both sides.
$$\sqrt{(x-5)^2} = \sqrt{64}$$
$$(x - 5) = \pm 8$$
$$(x - 5) = 8 \text{ and } (x - 5) = -8$$

    Step 2: Solve the two equations.
$(x - 5) = 8$ and $(x - 5) = -8$
$x - 5 + 5 = 8 + 5$ and $x - 5 + 5 = -8 + 5$
$x = 13$ and $x = -3$

Solve the perfect square for x.

1. $(x-2)^2 = 0$
2. $(x+1)^2 = 0$
3. $(x+11)^2 = 0$
4. $(x-4)^2 = 0$
5. $(x-1)^2 = 0$
6. $(x+8)^2 = 0$
7. $(x+3)^2 = 4$
8. $(x-5)^2 = 16$
9. $(x-10)^2 = 100$
10. $(x+9)^2 = 9$
11. $(x-4.5)^2 = 25$
12. $(x+7)^2 = 36$
13. $(x+2)^2 = 49$
14. $(x-1)^2 = 4$
15. $(x+8.9)^2 = 49$
16. $(x-6)^2 = 81$
17. $(x-12)^2 = 121$
18. $(x+2.5)^2 = 64$

13.6 Completing the Square (DOK 2)

"Completing the Square" is another way of factoring a quadratic equation. To complete the square, convert the equation into a perfect square.

Example 9: Solve $x^2 - 10x + 9 = 0$ by completing the square.

Completing the square:

Step 1: The first step is to get the constant on the other side of the equation. Subtract 9 from both sides:
$x^2 - 10x + 9 - 9 = 0 - 9$
$x^2 - 10x = -9$

Step 2: Determine the coefficient of the x. The coefficient in this example is -10. Divide the coefficient by 2 and square the result.
$(-10 \div 2)^2 = (-5)^2 = 25$

Step 3: Add the resulting value, 25, to both sides:
$x^2 - 10x + 25 = -9 + 25$
$x^2 - 10x + 25 = 16$

Step 4: Now factor the $x^2 - 10x + 25$ into a perfect square:
$(x-5)^2 = 16$

Solving the perfect square:

Step 5: Take the square root of both sides.
$\sqrt{(x-5)^2} = \sqrt{16}$
$(x-5) = \pm 4$
$(x-5) = 4$ and $(x-5) = -4$

Step 6: Solve the two equations.
$(x-5) = 4$ and $(x-5) = -4$
$x - 5 + 5 = 4 + 5$ and $x - 5 + 5 = -4 + 5$
$x = 9$ and $x = 1$

Chapter 13 Solving Quadratic Equations

Solve for x by completing the square.

1. $x^2 + 2x - 3 = 0$
2. $x^2 - 8x + 7 = 0$
3. $x^2 + 6x - 7 = 0$
4. $x^2 - 16x - 36 = 0$
5. $x^2 - 14x + 49 = 0$

6. $x^2 - 4x = 0$
7. $x^2 + 12x + 27 = 0$
8. $x^2 + 2x - 24 = 0$
9. $x^2 + 12x - 85 = 0$
10. $x^2 - 8x + 15 = 0$

11. $x^2 - 16x + 60 = 0$
12. $x^2 - 8x - 48 = 0$
13. $x^2 + 24x + 44 = 0$
14. $x^2 + 6x + 5 = 0$
15. $x^2 - 11x + 5.25 = 0$

13.7 Real-World Quadratic Equations (DOK 3)

The most common real life situation that could be described by a quadratic equation is the motion of an object under the force of gravity. Two examples are a ball being kicked into the air or a rocket being shot into the air.

Example 10: A high school football player is practicing his field goal kicks. The equation below represents the height of the ball at a specific time.

$$s = -9t^2 + 45t$$

$t =$ amount of time in seconds

$s =$ height in feet

Question 1: Where will the ball be at 4 seconds?

Solution 1: Since there are only two variables, you will only need the value of one variable to solve the problem. Simply plug in the number 4 in place of the variable t and solve the equation as shown below.

$$s = -9(4)^2 + 45(4)$$

$$s = -9(16) + 180$$

$$s = -144 + 180$$

$$s = 36$$

At 4 seconds the ball will be 36 ft in the air.

13.7 Real-World Quadratic Equations (DOK 3)

Question 2: If the ball is 54 ft in the air, how much time has gone by?

Solution 2: This question is similar to the previous one, except that the given variable is different. This time you would be replacing s with 54 and then solve the equation.

$54 = -9t^2 + 45t$ Subtract 54 on both sides.

$0 = -9t^2 + 45t - 54$ Divide the entire equation by -9.

$0 = t^2 - 5t + 6$ Factor the equation.

$0 = (t-3)(t-2)$ Solve for t.

$t = 3 \quad t = 2$

For this question, we got 2 answers. The ball is 54 ft in the air when 2 and 3 seconds have gone by.

Example 11: John and Alex are kicking a soccer ball back and forth to each other. The equation below represents the height of the ball at a specific point in time. $s = -4t^2 + 24t$, where $t =$ amount of time in seconds and $s =$ height in feet

Question 1: How long does it take for the soccer ball to come back down to the ground?

Solution 1: Looking at this problem you can see that no value was given, although one was indirectly given. The question asks when will the ball come back down. This is just another way of asking, "When will the height of the ball be zero?" The value 0 will be used for s. Substitute 0 back in for s and solve.
$0 = -4t^2 + 24t$ Factor out the greatest common factor.
$0 = -4t(t-6)$ Set each factor to 0 and solve.
$t = 0 \quad t = 6$
It is clear that the ball is on the ground at 0 seconds, so the value of $t = 0$ is not the answer and the second value of t is used instead. It takes the ball 6 seconds to go up into the air and then come back down to the ground.

Chapter 13   Solving Quadratic Equations

Question 2:   What is the highest point the ball will go?

Solution 2:   This is asking what is the vertex of the equation. You will need to use the vertex formula. As a reminder, the quadratic equation is defined as $y = ax^2 + bx + c$, where $a \neq 0$. The quadratic equation can also be written as a function of x by substituting $f(x)$ for y, such as $f(x) = ax^2 + bx + c$. To find the point of the vertex of the graph, you must use the formula below.

$$\text{vertex} = \left(-\frac{b}{2a}, f\left(-\frac{b}{2a}\right)\right)$$

where $f\left(-\frac{b}{2a}\right)$ is the quadratic equation evaluated at the value $-\frac{b}{2a}$. To do this, plug $-\frac{b}{2a}$ in for x.

To use the vertex equation, put the original equation in quadratic form and find a and b.
$s = -4t^2 + 24t \Rightarrow -4t^2 + 24t + 0$.
Since a is the coefficient of t^2, $a = -4$. b is the coefficient of t, so $b = 24$.

Find the solution to $-\frac{b}{2a}$ by substituting the values of a and b from the equation into the expression.
$-\frac{b}{2a} = -\left(\frac{24}{2 \times -4}\right) = -\left(\frac{24}{-8}\right) = -(-3) = 3$

Find the solution to $f\left(-\frac{b}{2a}\right)$. We know that $-\frac{b}{2a} = 3$, so we need to find $f(3)$. To do this, we must substitute 3 into the quadratic equation for x.
$f(t) = -4t^2 + 24t$
$f(3) = -4(3)^2 + 24(3) = -4(9) + 72 = -36 + 72 = 36$

The vertex equals $(3, 36)$. This means at 3 seconds, the ball is 36 feet in the air. Therefore, the highest the soccer ball will go is 36 ft.

13.7 Real-World Quadratic Equations (DOK 3)

Solve the following quadratic problems.

1. Eric is at the top of a cliff that is 500 ft from the ocean's surface. He is waiting for his friend to climb up and meet him. As he waits he decides to start casually tossing pebbles off the side of the cliff. The equation that represents the height of his pebble tosses is $s = -t^2 + 5t + 500$, where $s =$ distance in feet and $t =$ time in seconds.

   (A) How long does it take the pebble to hit the water?

   (B) If fifteen seconds have gone by, what is the height of the pebble from the ocean?

   (C) What is the highest point the pebble will go?

2. Devin is practicing golf at the driving range. The equation that represents the height of his ball is $s = -0.5t^2 + 12t$, where $s =$ distance in feet and $t =$ time in seconds.

   (A) What is the highest his ball will ever go?

   (B) If the ball is 31.5 ft in the air, how many seconds have gone by?

   (C) How long will it take for the ball to hit the ground?

3. Jack throws a ball up in the air to see how high he can get it to go. The equation that represents the height of the ball is $s = -2t^2 + 20t$, where $s =$ distance in feet and $t =$ time in seconds.

   (A) How high will the ball be at 7 seconds?

   (B) If the ball is 48 feet in the air, how many seconds have gone by?

   (C) How long does it take for the ball to go up and come back down to the ground?

4. Kali is jumping on her super trampoline, getting as high as she possibly can. The equation to represent her height is $s = -5t^2 + 20t$, where $s =$ distance in feet and $t =$ time in seconds.

   (A) What is the highest Kali can jump on her trampoline?

   (B) How high will Kali be at 4 seconds?

   (C) If Kali is 18.75 ft in the air, how many seconds have gone by?

Chapter 13 Solving Quadratic Equations

Chapter 13 Review

Factor and solve each of the following quadratic equations. (DOK 2)

1. $16b^2 - 25 = 0$
2. $a^2 - a - 30 = 0$
3. $x^2 - x = 6$
4. $100x^2 - 49 = 0$
5. $81y^2 = 9$
6. $y^2 = 21 - 4y$
7. $y^2 - 7y + 8 = 16$
8. $6x^2 + x - 2 = 0$
9. $3y^2 + y - 2 = 0$
10. $b^2 + 2b - 8 = 0$
11. $4x^2 + 19x - 5 = 0$
12. $8x^2 = 6x + 2$
13. $2y^2 - 6y - 20 = 0$
14. $-6x^2 + 7x - 2 = 0$
15. $y^2 + 3y - 18 = 0$

Using the quadratic formula, find both solutions for the variable. (DOK 2)

16. $x^2 + 10x - 11 = 0$
17. $y^2 - 14y + 40 = 0$
18. $b^2 + 9b + 18 = 0$
19. $y^2 - 12y - 13 = 0$
20. $a^2 - 8a - 48 = 0$
21. $x^2 + 2x - 63 = 0$
22. $-3x^2 - 2x - 2 = 0$
23. $4x^2 + x + 5 = 0$
24. $2x^2 - 8x + 9 = 0$

Solve each of the following quadratic inequalities. (DOK 2)

25. $2x^2 + x - 15 \geq 0$
26. $-2x^2 + 21x + 11 < 0$
27. $x(3x + 1) \leq 24$
28. $-5x^2 - 9x - 12 > 0$
29. $x^2 + 8x + 30 \geq 0$
30. $6x^2 - 7x + 8 \leq 0$

Find the x-intercept(s) of the following quadratic equations. (DOK 2)

31. $y = 2x^2 - 8$
32. $y = -x^2 - 9$
33. $y = 5x^2 + 1$
34. $y = x^2 - 6$

Use the following information for questions 35–37. (DOK 3)

Branden is tossing a ball in the air. He knows the height of the ball is represented by $s = -12t^2 + 50t$, where s is the height and t is the time.

35. How high (in feet) will the ball be after 4 seconds?

36. How high (in feet) will the ball be after 1.5 seconds?

37. How long (in seconds) was the ball in the air before it came back down to the ground?

Chapter 13 Test

1 Solve: $4y^2 - 9y = -5$

  A $\{1, \frac{5}{4}\}$

  B $\{-\frac{3}{4}, -1\}$

  C $\{-1, \frac{4}{5}\}$

  D $\{\frac{5}{16}, 1\}$

(DOK 2)

2 Solve for y: $2y^2 + 13y + 15 = 0$

  A $\{\frac{3}{2}, \frac{5}{2}\}$

  B $\{\frac{2}{3}, \frac{2}{5}\}$

  C $\{-5, -\frac{3}{2}\}$

  D $\{5, -\frac{3}{2}\}$

(DOK 2)

3 Solve for x.

$x^2 - 3x - 18 = 0$

  A $\{-6, 3\}$
  B $\{6, -3\}$
  C $\{-9, 2\}$
  D $\{9, -2\}$

(DOK 2)

4 What are the values of x in the quadratic equation?

$x^2 + 2x - 15 = x - 3$

  A $\{-4, 3\}$
  B $\{-3, 4\}$
  C $\{-3, 5\}$
  D Cannot be determined

(DOK 2)

5 Solve the equation $(x+9)^2 = 49$

  A $x = -9, 9$
  B $x = -9, 7$
  C $x = -16, -2$
  D $x = -7, 7$

(DOK 2)

6 Solve the equation $c^2 + 8c - 9 = 0$ by completing the square.

  A $c = \{1, -9\}$
  B $c = \{-1, 9\}$
  C $c = \{3, 3\}$
  D $c = \{-3, -3\}$

(DOK 2)

7 Using the quadratic formula, solve the following equation:

$3x^2 = 9x$

  A $x = \{0, 1\}$
  B $x = \{3, 1\}$
  C $x = \{0, 3\}$
  D $x = \{3, -3\}$

(DOK 2)

8 Solve $6a^2 + 11a - 10 = 0$ using the quadratic formula.

  A $\{-\frac{2}{5}, \frac{3}{2}\}$

  B $\{\frac{2}{5}, \frac{2}{3}\}$

  C $\{-\frac{5}{2}, \frac{2}{3}\}$

  D $\{\frac{5}{2}, \frac{2}{3}\}$

(DOK 2)

Chapter 13   Solving Quadratic Equations

9 Which of the following quadratic inequalities has no real solution?

A $3x^2 - 17x - 6 < 0$
B $-4x^2 - 11x + 3 > 0$
C $5x^2 - 7x - 6 \leq 0$
D $-6x^2 - 3x - 8 \geq 0$

(DOK 2)

10 Which of the following graphs represents $y = 2x^2$?

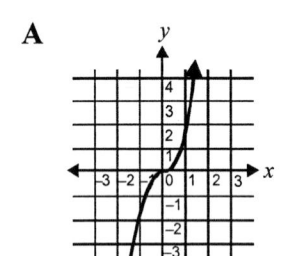

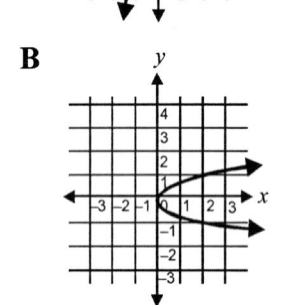

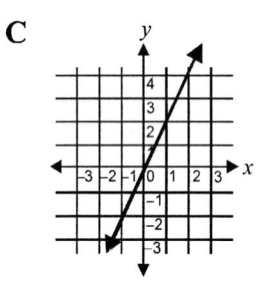

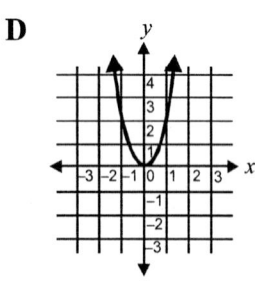

A

B

C

D

(DOK 2)

11 Which of the following solutions is correct for the quadratic inequality $6x^2 - 41x - 7 \leq 0$?

A $x \leq -6$ or $x \geq 7$
B $x \leq -\frac{1}{6}$ or $x \geq 7$
C $-6 \leq x \leq 7$
D $-\frac{1}{6} \leq x \leq 7$

(DOK 2)

12 Which of the following expressions is a root of the quadratic equation $-7x^2 - x - 3 = 0$?

A $-\dfrac{1}{14} - \dfrac{\sqrt{83}}{14}$

B $1 + \sqrt{83}i$

C $-\dfrac{1}{14} - \dfrac{\sqrt{83}}{14}i$

D $1 - \sqrt{83}i$

(DOK 2)

13 What are the x-intercept(s) of the equation $y = -2x^2 + 8$?

A $(-2, 0)$ and $(2, 0)$
B $(-4, 0)$ and $(4, 0)$
C $(8, 0)$
D There are no x-intercepts.

(DOK 2)

14 What are the x-intercept(s) of the equation $y = x^2 + 9$

A $(9, 0)$
B $(-3, 0)$ and $(3, 0)$
C $(-9, 0)$ and $(9, 0)$
D There are no x-intercepts.

(DOK 2)

Chapter 13 Test

15 Looking at the graph below, what is the solution to the inequality $y \leq x^2 - 10x + 19$?

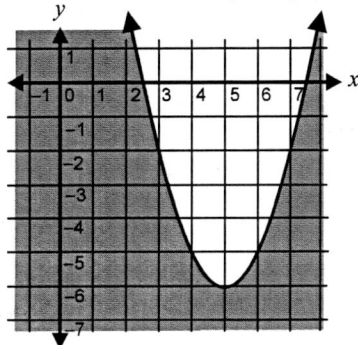

A $x = 5 \pm \sqrt{6}$
B $x < 5 - \sqrt{6}$ or $x > 5 + \sqrt{6}$
C $5 - \sqrt{6} \leq x \leq 5 + \sqrt{6}$
D $x \leq 5 - \sqrt{6}$ or $x \geq 5 + \sqrt{6}$

(DOK 3)

16 Eric and Jansen are throwing a baseball back and forth to each other. The height of the ball is represented by the equation $s = -8t^2 + 32t$. What is the highest point the ball will go?

A 16 feet
B 32 feet
C 64 feet
D 80 feet

(DOK 3)

Chapter 14
Systems of Equations and Inequalities

This chapter covers the following CCGPS standard(s):

Analytic Geometry | Create Equations | A.CED.2, A.CED.4

14.1 Solving Systems of Linear Equations by Substitution (DOK 2)

You can solve systems of equations by using the substitution method.

Example 1: Find the point of intersection of the following two equations:

Equation 1: $x - y = 3$

Equation 2: $2x + y = 9$

Step 1: Solve one of the equations for x or y. Let's choose to solve equation 1 for x.

Equation 1: $x - y = 3$

$x = y + 3$

Step 2: Substitute the value of x from equation 1 in place of x in equation 2.

Equation 2: $2x + y = 9$
$2(y + 3) + y = 9$
$2y + 6 + y = 9$
$3y + 6 = 9$
$3y = 3$
$y = 1$

14.1 Solving Systems of Linear Equations by Substitution (DOK 2)

Step 3: Substitute the solution for y back in equation 1 and solve for x.

Equation 1: $x - y = 3$
$x - 1 = 3$
$x = 4$

Step 4: The solution set is $(4, 1)$. Substitute $(4, 1)$ in both of the equations to check.

Equation 1: $x - y = 3$ Equation 2: $2x + 9 = 9$
$4 - 1 = 3$ $2(4) + 1 = 9$
$3 = 3$ $8 + 1 = 9$
                         $9 = 9$

The point $(4, 1)$ is common for both equations. This is the **point of intersection**.

For each of the following pairs of equations, find the point of intersection using the substitution method.

1. $x + 2y = 8$
   $2x - 3y = 2$

2. $x - y = -5$
   $x + y = 1$

3. $x - y = 4$
   $x + y = 2$

4. $x - y = -1$
   $x + y = 9$

5. $-x + y = 2$
   $x + y = 8$

6. $x + 4y = 10$
   $x + 5y = 10$

7. $2x + 3y = 2$
   $4x - 9y = -1$

8. $x + 3y = 5$
   $x - y = 1$

9. $-x = y - 1$
   $x = y - 1$

10. $x - 2y = 2$
    $2y + x = -2$

11. $5x + 2y = 1$
    $2x + 4y = 10$

12. $3x - y = 2$
    $5x + y = 6$

13. $2x + 3y = 3$
    $4x + 5y = 5$

14. $x - y = 1$
    $-x - y = 1$

15. $x = y + 3$
    $y = 3 - x$

Chapter 14 Systems of Equations and Inequalities

14.2 Graphing Systems of Linear Inequalities (DOK 2)

We solve systems of inequalities best graphically. Look at the following example.

Example 2: Sketch the solution set of the following system of inequalities:

$y > -2x - 1$ and $y \leq 3x$

Step 1: Graph both inequalities on a Cartesian plane.

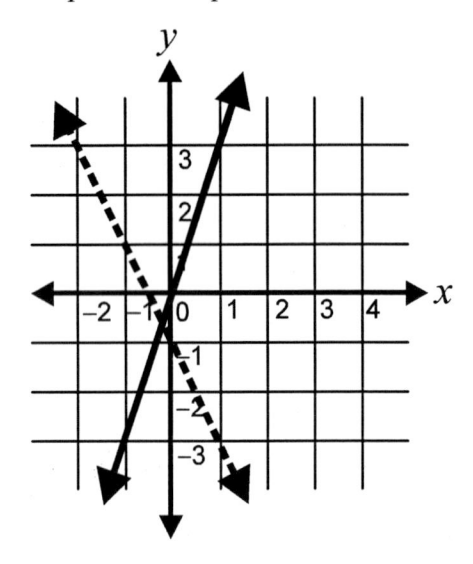

Step 2: Shade the portion of the graph that represents the solution set to each inequality.

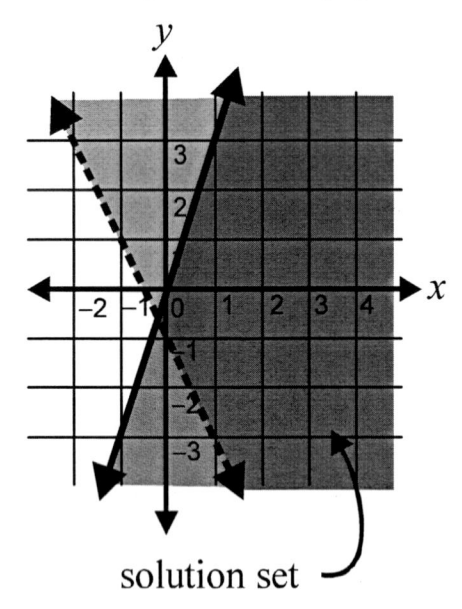

solution set

Step 3: Any shaded region that overlaps is the solution set of both inequalities.

14.2 Graphing Systems of Linear Inequalities (DOK 2)

Graph the following systems of inequalities on your own graph paper. Shade and identify the solution set for both inequalities.

1. $2x + 2y \geq -4$
   $3y < 2x + 6$

2. $7x + 7y \leq 21$
   $8x < 6y - 24$

3. $9x + 12y < 36$
   $34x - 17y > 34$

4. $-11x - 22y \geq 44$
   $-4x + 2y \leq 8$

5. $24x < 72 + 36y$
   $11x + 22y \leq -33$

6. $15x - 60 < 30y$
   $20x + 10y < 40$

7. $-12x + 24y > -24$
   $10x < -5y + 15$

8. $y \geq 2x + 2$
   $y < -x - 3$

9. $3x + 4y \geq 12$
   $y > -3x + 2$

10. $-3x \leq 6 + 2y$
    $y \geq -x - 2$

11. $2x - 2y \leq 4$
    $3x + 3y \leq -9$

12. $-x \geq -2y - 2$
    $-2x - 2y > 4$

Chapter 14 Systems of Equations and Inequalities

14.3 Graphing Systems of Three Linear Inequalities (DOK 3)

Example 3: Graph the solution set for the following inequalities:
$x - y \geq 0$
$y < 2x$
$5x + 6y \geq 1$

Step 1: Put the inequalities in slope-intercept form.

$$x - y \geq 0 \qquad y < 2x \qquad 5x + 6y \geq 1$$
$$-y \geq -x \qquad \qquad \qquad 6y \geq 1 - 5x$$
$$y \leq x \qquad \qquad \qquad \quad y \geq \tfrac{1}{6} - \tfrac{5}{6}x$$

Step 2: Graph the inequalities on a Cartesian plane.

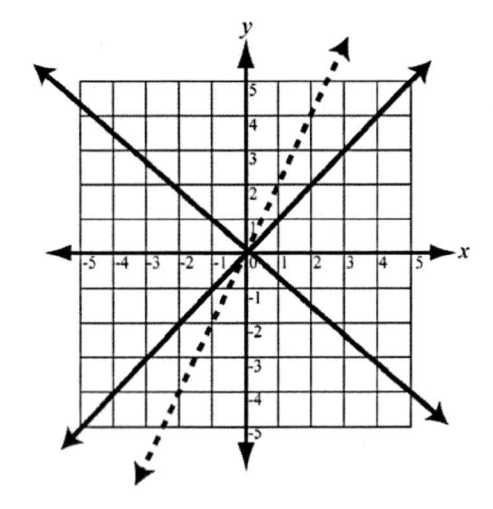

Step 3: Shade the portion of the graph that represents the solution set to each inequality.

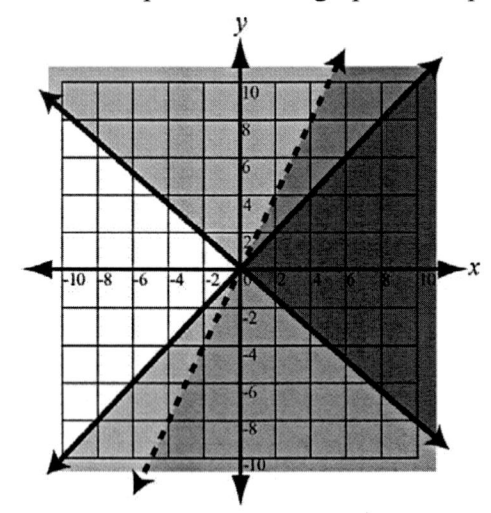

Step 4: The shaded region that overlaps all three shaded solution areas is the solution set of the three inequalities. This is represented by the darkest portion in the graph above.

14.3 Graphing Systems of Three Linear Inequalities (DOK 3)

Graph the following and shade the solution.

1. $y \geq 0$
   $x + y \geq 5$
   $3x + 4y \geq 18$

2. $x \geq -6$
   $y < 10$
   $x < y$

3. $x + 2y \geq 10$
   $y \geq -1$
   $x < 0$

4. $x - 3y > -6$
   $5x - 3y < 9$
   $x + 4y > -4$

5. $-x + 4y > 0$
   $x + y \leq 1$
   $x + 3y > -1$

6. $x + y \leq 4$
   $2x + y \geq -1$
   $x + y \geq -2$

7. $y + 7 > 3x$
   $3x + 3y < -12$
   $y > 11$

8. $13x + 5y \geq 2$
   $-x + 4y \leq -10$
   $x < -1$

9. $x + 2y < 12$
   $2x - y \geq 6$
   $x + 2y \geq 4$

14.4 Manipulating Formulas and Equations (DOK 2)

Sometimes you are given a formula such as $A = l \times w$ (A = area, l = length, and w = width) and you need to solve for w. For example: The area of a playground is 4500 square feet. The length is 600 feet. What is the width of the playground? Starting with $A = l \times w$, you need to solve for w. You need to have w on one side of the equation and all the other variables on the other.

$A = l \times w \Rightarrow \dfrac{A}{l} = \dfrac{l \times w}{l} \Rightarrow \dfrac{A}{l} = w \Rightarrow w = \dfrac{A}{l}$ You have solved $A = l \times w$ for w.

Solve each of the following formulas and equations for the given variable.

1. $C = 2\pi r$ for r
2. $I = PRT$ for R
3. $V = \pi r^2 h$ for h
4. $A = \frac{1}{2}bh$ for h
5. $d = 4a + 3c$ for c
6. $h = 6a + 9c^2$ for a
7. $y = 4xz$ for z
8. $5t = 9y + 22$ for y
9. $17 - 9m = n - 23$ for n
10. $7x + 4 = \frac{9y}{4}$ for y
11. $8 + 2a = 5b - 6$ for b
12. $A = s^2$ for s
13. $a^2 + b^2 = c^2$ for a
14. $I = PRT$ for P
15. $x = 4a + 7$ for a
16. $9 - 5y = 6x + 2$ for x
17. $D = rt$ for r
18. $A = lw$ for w
19. $a^2 + b^2 = c^2$ for b^2
20. $C = 2dr$ for r
21. $V = \pi r^2 h$ for r
22. $V = \frac{1}{3}Bh$ for B
23. $A = \pi r^2$ for r
24. $S = 4\pi r^2$ for r
25. $y = \frac{1}{4}x + 5$ for x
26. $x = -\frac{1}{5}y - 3$ for y
27. $a = \dfrac{b}{3}$ for b
28. $c = 3d + \frac{2}{5}$ for d
29. $g = \frac{2}{3}h - 2$ for h
30. $r = s^2$ for s
31. $F = \frac{9}{5}C + 32$ for C
32. $y = mx + b$ for m

Chapter 14 Review

Find the common solution for each of the following pairs of equations, using the substitution method. (DOK 2)

1. $x - y = 2$
   $x + 4y = -3$

2. $x + y = 1$
   $x + 3y = 1$

3. $-4y = -2x + 4$
   $x = -2y - 2$

4. $2x + 8y = 20$
   $5y = 12 - x$

5. $x = y - 3$
   $-x = y + 3$

6. $-2x + y = -3$
   $x - y = 9$

Graph the following systems of inequalities on your own graph paper. Shade the solution set for both inequalities. (DOK 3)

7. $x + 2y \geq 2$
   $2x - y \leq 4$

8. $20x + 10y \leq 40$
   $3x + 2y \geq 6$

9. $6x + 8y \leq -24$
   $-4x + 8y \geq 16$

10. $14x - 7y \geq -28$
    $3x + 4y \leq -12$

11. $2y \geq 6x + 6$
    $2x - 4y \geq -4$

12. $9x - 6y \geq 18$
    $3y \geq 6x - 12$

Find the point of intersection for each pair of equations by adding and/or subtracting the two equations. (DOK 2)

13. $2x + y = 4$
    $3x - y = 6$

14. $x + 2y = 3$
    $x + 5y = 0$

15. $x + y = 1$
    $y = x + 7$

16. $2x + 4y = 5$
    $3x + 8y = 9$

17. $2x - 2y = 7$
    $3x - 5y = \frac{5}{2}$

18. $x - 3y = -2$
    $y = -\frac{1}{3}x + 4$

Chapter 14 Systems of Equations and Inequalities

Chapter 14 Test

1 What is the intersection point of the graphs of the equations $x = y + 3$ and $y = 3 - x$?

A $(3, 0)$
B $(0, 3)$
C $(3, 3)$
D $(-3, 0)$

(DOK 2)

2 What is the intersection point of the graphs of the equations $2x + 3y = 2$ and $4x - 9y = -1$?

A $(2, 3)$
B $\left(\frac{1}{3}, \frac{1}{2}\right)$
C $\left(\frac{1}{2}, \frac{1}{3}\right)$
D $(3, 2)$

(DOK 2)

3 What is the solution of the graphs of the equations $-x = y - 1$ and $x = y - 1$?

A $(0, 1)$
B $(1, 0)$
C $(-2, -1)$
D $(2, 1)$

(DOK 2)

4 What is the solution of the graphs of the equations $2x - y = 2$ and $4x - 9y = -3$?

A $\left(\frac{1}{2}, -1\right)$
B $\left(1, \frac{3}{2}\right)$
C $\left(-\frac{3}{2}, 1\right)$
D $\left(\frac{3}{2}, 1\right)$

(DOK 2)

5 At what point do the graphs of the equations $x - 4y = 6$ and $-x - y = -1$ intersect?

A $(2, 1)$
B $(2, -1)$
C $(-2, 1)$
D $(-2, -1)$

(DOK 2)

6 Which inequalities are represented by the graph?

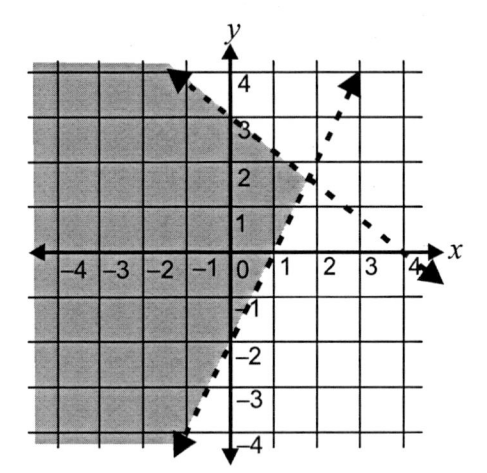

A $2x + 2y \geq -4$
   $3y < 2x + 6$

B $7x + 7y \leq 21$
   $8x < 6y - 24$

C $9x + 12y < 36$
   $34x - 17y < 34$

D $-11x - 22y \geq 44$
   $-4x + 2y \leq 8$

(DOK 3)

7 Which graph represents $2x - 2y \leq 4$ and $3x + 3y \leq -9$?

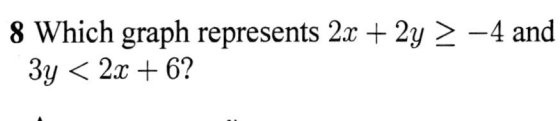

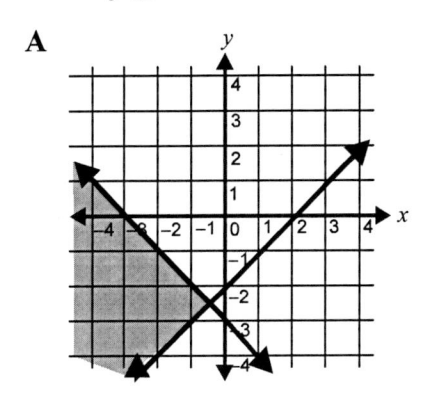

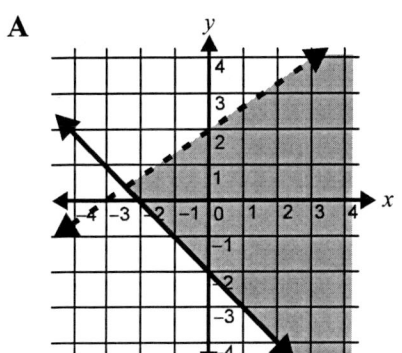

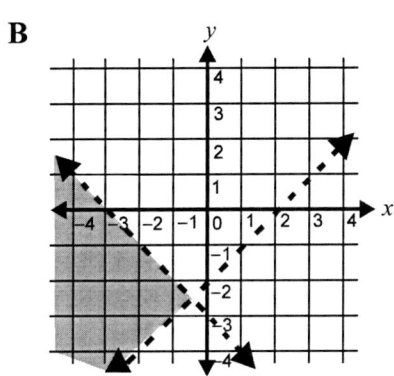

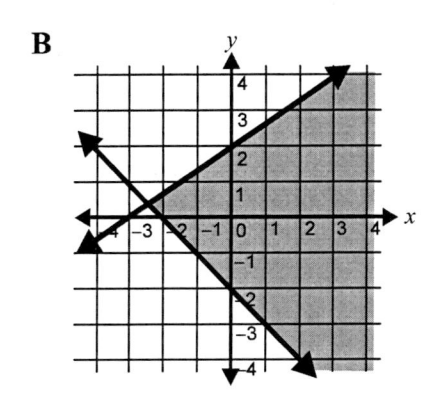

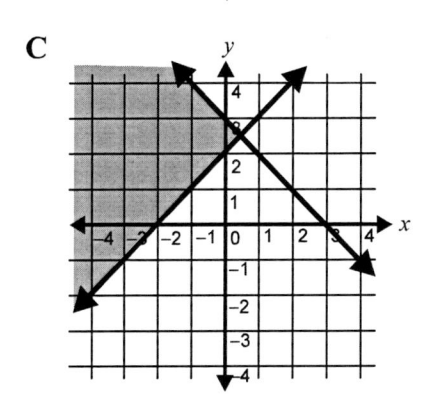

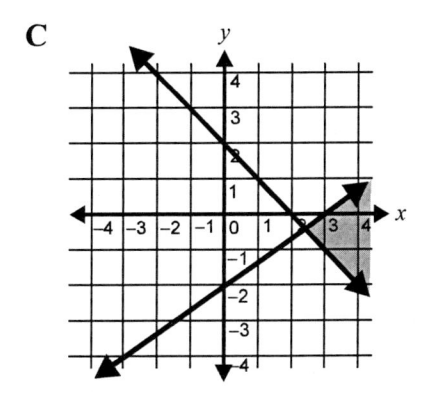

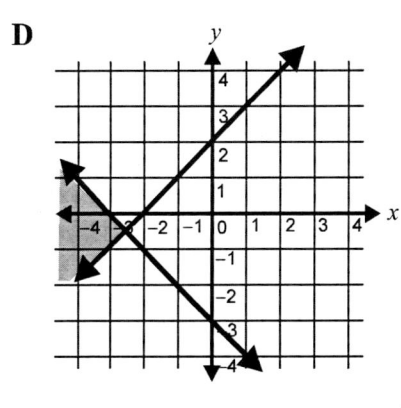

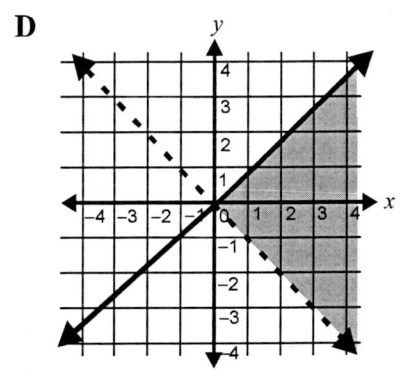

A

B

C

D

8 Which graph represents $2x + 2y \geq -4$ and $3y < 2x + 6$?

A

B

C

D

(DOK 2)

(DOK 2)

Chapter 14 Systems of Equations and Inequalities

9 Two numbers have a sum of 210 and a difference of 30. What are the two numbers?

A 140 and 170
B 170 and 40
C 150 and 60
D 120 and 90

(DOK 3)

10 Lucy is getting snacks for her party tonight. She wants to keep the price under thirty dollars. She can buy three 12 packs of soda and 2 bags of chips for $27.33 or she can buy two 12 packs of soda and three bags of chips for $25.27. How much is one bag of chips?

A $4.23
B $2.62
C $6.29
D $1.60

(DOK 3)

11 The sum of the digits of a two-digit number is seven. When the digits are reversed, the number is increased by 27. What is the number?

A 25
B 34
C 16
D 27

(DOK 3)

12 Which ordered pair is not the solution of the following systems of inequalities?

$$3x + 4y \geq -4$$
$$y - x > -6$$
$$x < 5$$

A (2, 4)
B (−1, 6)
C (0, 8)
D (8, 1)

(DOK 2)

13 Which inequalities are represented by the graph?

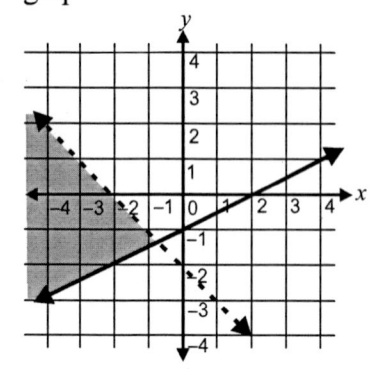

A $-x \geq -2y - 2$
   $-2x - 2y > 4$

B $2x - 2y \leq 4$
   $3x + 3y \leq -9$

C $-3x \leq 6 + 2y$
   $y \geq -x - 2$

D $3x + 4y \geq 12$
   $y > -3x + 2$

(DOK 3)

14 Match the graph below with the system of inequalities.

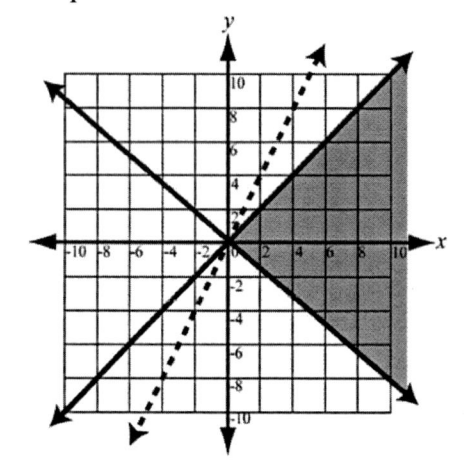

A $y \geq x$
   $y < 2x$
   $y \geq -\frac{5}{6}x + \frac{1}{6}$

B $y \leq x$
   $y < 2x$
   $y \geq -\frac{5}{6}x + \frac{1}{6}$

C $y \leq x$
   $y > 2x$
   $y \leq -\frac{5}{6}x + \frac{1}{6}$

D $y \leq x$
   $y > 2x$
   $y = -\frac{5}{6}x + \frac{1}{6}$

(DOK 2)

Chapter 15
Polynomial Functions

This chapter covers the following CCGPS standard(s):

Arithmetic with Polynomial and Rational Expressions	A.APR.1, A.REI.7
Creating Expressions	A.CED.1
Interpreting Functions	F.IF.4, F.IF.5, F.IF.7a, F.IF.8a
Building Functions	F.BF.3
Linear and Exponential Models	F.LE.3

15.1 Solving Polynomial Equations Analytically (DOK 3)

To solve a polynomial equation algebraically, it is necessary to first write the equation in standard form. The equation can then often be solved by factoring. There are other methods used. However, those will be discussed in later chapters

Example 1: Solve the equation $2x + x^2 = 15$ for x.

Step 1: Write the equation in standard form.
To solve the equation $2x + x^2 = 15$ for x, first it's necessary to write the equation in standard form, which means that the terms on the left side of the equation are in order from the highest degree to the lowest degree and the right side of the equation is 0. This is done as follows.
$2x + x^2 = 15$
$2x + x^2 - 15 = 15 - 15$
$2x + x^2 - 15 = 0$
$x^2 + 2x - 15 = 0$

Step 2: Factor the left side of the equation.
Since the equation is quadratic, with the middle term on the left side having a coefficient of 2 and the last term being -15, the left side can be factored as follows.
$x^2 + 2x - 15 = 0$
$(x + 5)(x - 3) = 0$

Chapter 15   Polynomial Functions

Step 3:   Solve for x.
Since $(x+5)(x-3) = 0$, either $x+5 = 0$ or $x-3 = 0$. This means that x can be solved for by setting each factor equal to 0.
$x + 5 = 0$
$x + 5 - 5 = 0 - 5$
$x = -5$
or
$x - 3 = 0$
$x - 3 + 3 = 0 + 3$
$x = 3$
Therefore, the solution to the equation $2x + x^2 = 15$ is $x = -5$ and 3.

Solve each of these equations for x.

1. $x^2 = 4x + 12$
2. $x^2 - 2x = 35$
3. $x + x^2 = 72$
4. $2x^2 + 29x = 15$
5. $12x + 4 = -9x^2$
6. $25x^2 = 144$

Example 2:   Solve the equation $x^3 + 27x = 40 - 12x^2$ for x.

Step 1:   Write the equation in standard form.
To solve the equation $x^3 + 27x = 40 - 12x^2$ for x, first it's necessary to write the equation in standard form, which means that the terms on the left side of the equation are in order from the highest degree to the lowest degree and the right side of the equation is 0. This is done as follows.
$x^3 + 27x = 40 - 12x^2$
$x^3 + 27x - 40 + 12x^2 = 40 - 12x^2 - 40 + 12x^2$
$x^3 + 27x - 40 + 12x^2 = 0$
$x^3 + 12x^2 + 27x - 40 = 0$

Step 2:   Use the Rational Root Theorem to find all of the possible rational roots of the equation.
Since the equation is now in standard form, all of its possible rational roots can be found by dividing all of the positive and negative factors of the last term on the left side by all of the positive and negative factors of the coefficient of the first term. Since the last term on the left side is 40, all of its positive and negative factors are $\pm 1, \pm 2, \pm 4, \pm 5, \pm 8, \pm 10, \pm 20,$ and ± 40. Also, since the coefficient of the first term is 1, all of its positive and negative factors are ± 1. This means that all of the possible rational roots of the equation are

$$x = \frac{\pm 1}{\pm 1}, \frac{\pm 2}{\pm 1}, \frac{\pm 4}{\pm 1}, \frac{\pm 5}{\pm 1}, \frac{\pm 8}{\pm 1}, \frac{\pm 10}{\pm 1}, \frac{\pm 20}{\pm 1}, \text{ and } \frac{\pm 40}{\pm 1}, \text{ or}$$

$x = \pm 1, \pm 2, \pm 4, \pm 5, \pm 8, \pm 10, \pm 20,$ and ± 40.

15.1 Solving Polynomial Equations Analytically (DOK 3)

Step 3: Use the Factor Theorem to find a rational root.

According to the Factor Theorem, if a root of the equation is plugged in for x, the output of the equation becomes 0. With this in mind, all of the possible rational roots of the equation can be tested. First, $x = 1$ and $x = -1$ can be tested as follows.

$$f(x) = x^3 + 12x^2 + 27x - 40$$

$$f(1) = (1)^3 + 12(1)^2 + 27(1) - 40 = 1 + 12 + 27 - 40 = 0$$

$$f(-1) = (-1)^3 + 12(-1)^2 + 27(-1) - 40 = -1 + 12 - 27 - 40 = -56$$

This means that $x = 1$ is a solution to the equation, but $x = -1$ is not. At this point, only one of the solutions to the equation has been found, but the Fundamental Theorem of Algebra states that the equation has 3 solutions, since the highest exponent on the left side of the equation is 3. While it's possible to keep using the Factor Theorem to test the other possible rational roots, a faster approach may be to produce a quadratic equation to find the two additional solutions.

Step 4: Use synthetic division to produce a quadratic equation.

Since $x = 1$ is a solution to the equation $x^3 + 12x^2 + 27x - 40 = 0$, a factor of $x^3 + 12x^2 + 27x - 40$ must be $x - 1$, so $x^3 + 12x^2 + 27x - 40$ can be divided by $x - 1$ to produce one side of the quadratic equation. The division is done synthetically as follows.

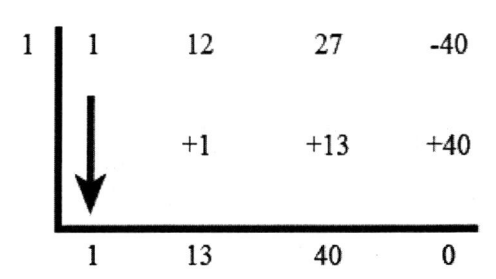

```
1 | 1    12    27    -40
  |      +1   +13   +40
  |_____
    1    13    40     0
```

Since $x^3 + 12x^2 + 27x - 40$ divided by $x - 1$ equals $x^2 + 13x + 40$, the quadratic equation that is produced is $x^2 + 13x + 40 = 0$.

Chapter 15 Polynomial Functions

Step 5: Factor the left side of the quadratic equation.

Since the middle term on the left side of the quadratic equation has a coefficient of 13 and the last term is 40, the left side can be factored as follows.

$x^2 + 13x + 40 = 0$

$(x + 5)(x + 8) = 0$

Step 6: Solve for x.

Since $(x + 5)(x + 8) = 0$, either $x + 5 = 0$ or $x + 8 = 0$. This means that x can be solved for as follows.

$x + 5 = 0$

$x + 5 - 5 = 0 - 5$

$x = -5$

or

$x + 8 = 0$

$x + 8 - 8 = 0 - 8$

$x = -8$

Therefore, the solution to the equation $x^3 + 27x = 40 - 12x^2$ is $x = -8, -5,$ and 1.

Solve each of these equations for x.

1. $x^3 + 9x^2 = -24 - 26x$

2. $x^3 + 24 = 6x^2 + 4x$

3. $x^3 + 2x^2 = 55x + 56$

4. $x^3 + 45 = 5x^2 + 41x$

5. $x^3 + 47x = 12x^2 + 60$

6. $11x^2 + 16x = 84 - x^3$

15.1 Solving Polynomial Equations Analytically (DOK 3)

Example 3: Solve the equation $x^4 + 2x^3 + 2x = -1$ for x. Include both real and complex solutions.

Step 1: Write the equation in standard form.

To solve the equation $x^4 + 2x^3 + 2x^2 + 2x = -1$ for x, first it's necessary to write the equation in standard form, which means that the terms on the left side of the equation are in order from the highest degree to the lowest degree and the right side of the equation is 0. This is done as follows.

$x^4 + 2x^3 + 2x = -1$

$x^4 + 2x^3 + 2x + 1 = -1 + 1$

$x^4 + 2x^3 + 2x + 1 = 0$

Step 2: Use the Rational Root Theorem to find all of the possible rational roots of the equation.

Since the equation is now in standard form, all of its possible rational roots can be found by dividing all of the positive and negative factors of the last term on the left side by all of the positive and negative factors of the coefficient of the first term. Since the last term on the left side is 1, all of its positive and negative factors are ± 1. Also, since the coefficient of the first term is 1, all of its positive and negative factors are ± 1. This means that all of the possible rational roots of the equation are $x = \frac{\pm 1}{\pm 1}$, or $x = \pm 1$.

Step 3: Use the Factor Theorem to find a rational root.

According to the Factor Theorem, if a root of the equation is plugged in for x, the left side of the equation becomes 0. With this in mind, $x = 1$ and $x = -1$ can be tested as follows.

$f(x) = x^4 + 2x^3 + 2x + 1$

$f(1) = (1)^4 + 2(1)^3 + 2(1) + 1 = 1 + 2 + 2 + 1 = 8$

$f(-1) = (-1)^4 + 2(-1)^3 + 2(-1) + 1 = 1 - 2 + 2 - 2 + 1 = 0$

This means that $x = -1$ is a solution to the equation, but $x = 1$ is not. At this point, only one of the solutions to the equation has been found, but the Fundamental Theorem of Algebra states that the equation has 4 solutions, since the highest exponent on the left side of the equation is 4. However, there are no additional possible rational roots to test, so the only option is to use polynomial division to produce a cubic equation.

Chapter 15 Polynomial Functions

Step 4: Use polynomial division to produce a cubic equation.

Since $x = -1$ is a solution to the equation $x^4 + 2x^3 + 2x^2 + 2x + 1 = 0$, a factor of $x^4 + 2x^3 + 2x^2 + 2x + 1$ must be $x + 1$, so $x^4 + 2x^3 + 2x^2 + 2x + 1$ can be divided by $x + 1$ to produce one side of the cubic equation. The division is done synthetically as follows.

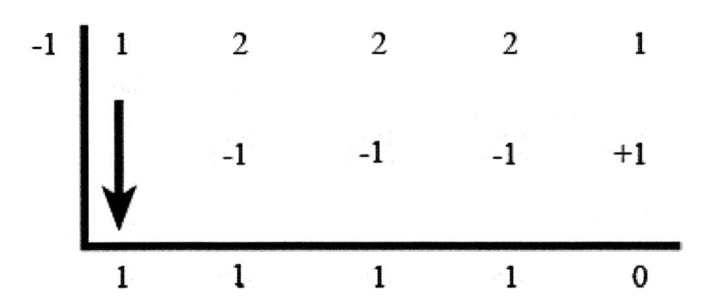

```
-1 |  1    2    2    2    1
   |      -1   -1   -1   +1
   |_____
      1    1    1    1    0
```

Since $x^4 + 2x^3 + 2x^2 + 2x + 1$ divided by $x + 1$ equals $x^3 + x^2 + x + 1$, the cubic equation that is produced is $x^3 + x^2 + x + 1 = 0$.

Step 5: Use the Factor Theorem to check if the rational root already found occurs more than once.

Even though it has already been determined that $x = -1$ is a solution to the equation $x^4 + 2x^3 + 2x^2 + 2x + 1 = 0$, it may be a solution more than once. To find out if this is the case, the root $x = -1$ can be tested again with the cubic equation $x^3 + x^2 + x + 1 = 0$ as follows.

$$f(x) = x^3 + x^2 + x + 1$$

$$f(-1) = (-1)^3 + (-1)^2 + (-1) + 1 = -1 + 1 + (-1) + 1 = 0$$

This means that $x = -1$ is a solution to the equation $x^4 + 2x^3 + 2x^2 + 2x + 1 = 0$ twice, or in other words, that it is a solution of multiplicity 2. However, again there are no additional possible rational roots to test, so the only option is to use polynomial division to produce a quadratic equation.

15.1 Solving Polynomial Equations Analytically (DOK 3)

Step 6: Use polynomial division to produce a quadratic equation.

Since $x = -1$ is a solution to the equation $x^3 + x^2 + x + 1 = 0$, a factor of $x^3 + x^2 + x + 1$ must be $x + 1$, so $x^3 + x^2 + x + 1$ can be divided by $x + 1$ to produce one side of the quadratic equation. The division is done synthetically as follows.

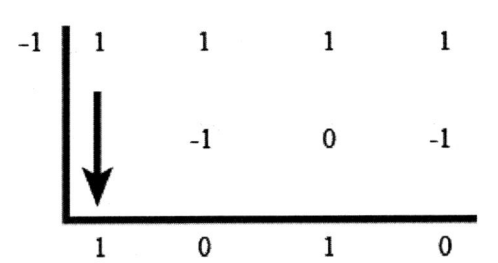

```
-1 | 1    1    1    1
   |     -1    0   -1
   |_____
     1    0    1    0
```

Since $x^3 + x^2 + x + 1$ divided by $x + 1$ equals $x^2 + 1$, the quadratic equation that is produced is $x^2 + 1 = 0$.

Step 7: Use the quadratic formula to find the remaining roots.

Now that there is a quadratic equation in the form $ax^2 + bx + c = 0$, the quadratic formula can be used to solve it. The quadratic formula is

$$x = \frac{-b \pm \sqrt{b^2 - 4ac}}{2a},$$ and in this case $a = 1$, $b = 0$, and $c = 1$. Thus, the

remaining roots can be found as follows.

$$x = \frac{-b \pm \sqrt{b^2 - 4ac}}{2a} = \frac{0 \pm \sqrt{0^2 - 4(1)(1)}}{2(1)} =$$

$$\frac{\pm\sqrt{-4}}{2} = \frac{\pm i\sqrt{4}}{2} = \frac{\pm 2i}{2} = \pm i$$

This means that the solution to the equation $x^4 + 2x^3 + 2x^2 + 2x = -1$ is $x = -1, -i,$ or i.

Solve each of these equations for x. Include both real and complex solutions.

7. $x^3 + x^2 - 8x = 12$

8. $x^3 - 5x^2 + 48 = 8x$

9. $x^4 + 5x^2 + 4 = 2x^3 + 8x$

10. $x^4 + 20x = 4x^3 + x^2 + 20$

11. $x^4 + 10x^3 + 33x^2 + 40x = -16$

12. $x^4 + 18x^2 + 81 = 6x^3 + 54x$

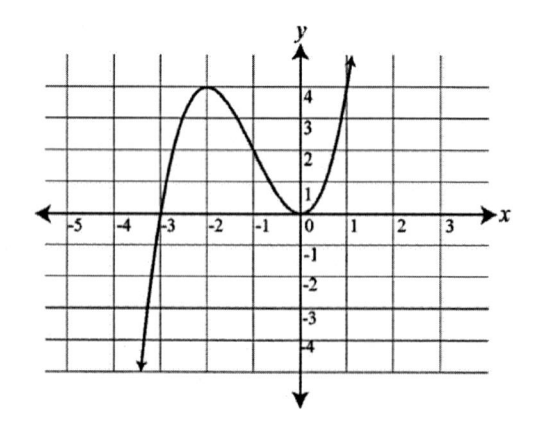

Chapter 15 Polynomial Functions

15.2 Solving Polynomial Equations Graphically (DOK 2)

There are two ways to solve a polynomial equation graphically.

1) Graph the left side of the equation, graph the right side of the equation on the same coordinate grid, and determine the x-coordinates of the points where the graphs intersect.

2) Write the polynomial equation in standard form, graph the left side of the equation, and determine the x-coordinates of the points where the graph crosses the x-axis.

If the first approach is used and the graphs do not intersect, or if the second approach is used and the graph does not cross the x-axis, then the polynomial equation has no real solution. When solving polynomial equations graphically, only real solutions, and not complex solutions, can be found.

Example 4: Solve the equation $x^3 + 3x^2 = x + 3$ graphically.

Step 1: Graph the left side of the equation.
First, the function $f(x) = x^3 + 3x^2$ should be graphed as follows.

> **Calculator:**
> On the TI-83/84 graphing calculator, this graph can be produced by performing the following steps.
>
> a. Press $\boxed{Y=}$.
>
> b. Enter X^3 + 3X^2 after $Y_1 =$.
>
> c. Press $\boxed{\text{WINDOW}}$.
>
> d. Enter the following values.
>
> Xmin = –6
> Xmax = 4
> Xscl = 1
> Ymin = –5
> Ymax = 5
> Yscl = 1
> Xres = 1
>
> e. Press $\boxed{\text{GRAPH}}$.

242

15.2 Solving Polynomial Equations Graphically (DOK 2)

Step 2: Graph the right side of the equation on the same coordinate grid.

Next, the function $g(x) = x + 3$ should be graphed on the same coordinate grid as follows.

Calculator:
On the TI-83/84 graphing calculator, this graph can be produced by performing the following steps.

a. Press $\boxed{Y=}$.

b. Enter X + 3 after $Y_2 =$.

c. Press $\boxed{\text{WINDOW}}$.

d. Enter the following values.

Xmin = –6
Xmax = 4
Xscl = 1
Ymin = –5
Ymax = 5
Yscl = 1
Xres = 1

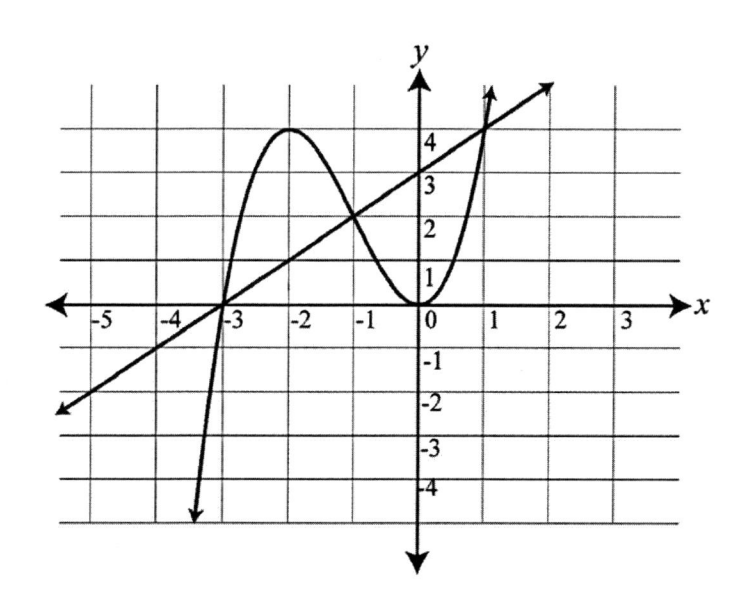

e. Press $\boxed{\text{GRAPH}}$.

Step 3: Determine the x-coordinates of the points where the graphs intersect.

Calculator:
On the TI-83/84 graphing calculator, the points of intersection of the graphs can be found by performing the following steps.

a. Press $\boxed{\text{2ND}}$ $\boxed{\text{TRACE}}$.

b. Press $\boxed{5}$.

c. Press $\boxed{\text{ENTER}}$ twice.

d. Move the cursor as close as possible to a point of intersection.

e. Press $\boxed{\text{ENTER}}$.

f. Repeat the steps above for the remaining points of intersection.

Since the graphs intersect at the points $(-3, 0)$, $(-1, 2)$, and $(1, 4)$, the solution to the equation $x^3 + 3x^2 = x + 3$ is $x = -3, -1$, and 1.

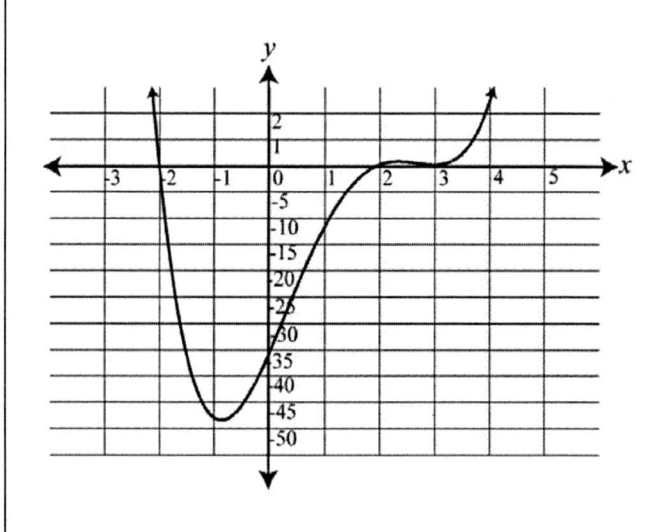

Chapter 15 Polynomial Functions

Example 5: Solve the equation $x^4 + 5x^2 + 24x = 6x^3 + 36$ graphically.

Step 1: Write the equation in standard form.

To write the equation $x^4 + 5x^2 + 24x = 6x^3 + 36$ in standard form, the terms on the left side of the equation should be in order from the highest degree to the lowest degree and the right side of the equation should be 0. The equation can be converted to standard form as follows.

$$x^4 + 5x^2 + 24x = 6x^3 + 36$$

$$x^4 + 5x^2 + 24x - 6x^3 - 36 = 6x^3 + 36 - 6x^3 - 36$$

$$x^4 + 5x^2 + 24x - 6x^3 - 36 = 0$$

$$x^4 - 6x^3 + 5x^2 + 24x - 36 = 0$$

Step 2: Graph the left side of the equation.

Now, the function $f(x) = x^4 - 6x^3 + 5x^2 + 24x - 36$ should be graphed as follows.

Calculator:
On the TI-83/84 graphing calculator, this graph can be produced by performing the following steps.

a. Press $\boxed{Y=}$.

b. Enter X^4 – 6X^3 + 5X^2 + 24X – 36 after $Y_1 =$.

c. Press $\boxed{\text{WINDOW}}$.

d. Enter the following values.

Xmin = –4
Xmax = 6
Xscl = 1
Ymin = –55
Ymax = 15
Yscl = 15
Xres = 1

e. Press $\boxed{\text{GRAPH}}$.

244   Copyright © American Book Company

15.2 Solving Polynomial Equations Graphically (DOK 2)

Step 3: Determine the x-coordinates of the points where the graph crosses the x-axis.

> **Calculator:**
> On the TI-83/84 graphing calculator, the points where the graph crosses the x-axis can be found by performing the following steps.
>
> a. Press $\boxed{\text{2ND}}$ $\boxed{\text{TRACE}}$.
>
> b. Press $\boxed{2}$.
>
> c. Move the cursor to the left of a point where the graph crosses the x-axis.
>
> d. Press $\boxed{\text{ENTER}}$.
>
> e. Move the cursor to the right of the point where the graph crosses the x-axis.
>
> f. Press $\boxed{\text{ENTER}}$ twice.
>
> g. Repeat the steps above for the remaining points where the graph crosses the x-axis.

Since the graph crosses the x-axis at the points $(-2, 0)$, $(2, 0)$, and $(3, 0)$, the solution to the equation $x^4 + 5x^2 + 24x = 6x^3 + 36$ is $x = -2, 2,$ and 3.

Solve each of the following equations graphically.

1. $x^3 + x^2 = 9x + 9$

2. $x^3 + 32 = 16x + 2x^2$

3. $x^3 + 11x^2 + 23x = 35$

4. $6x^2 + x^3 = -12x - 8$

5. $44x + x^3 = 48 + 12x^2$

6. $2x^3 + 15x^2 + 24x = 16$

7. $x^3 + 8x^2 = 15x + 54$

8. $x^3 = 21x + 20$

9. $x^4 + x^3 + 50 = 27x^2 + 25x$

10. $x^4 + 2x^3 = 35x^2 + 72x + 36$

11. $x^4 + 5x + 14 = 5x^3 + 15x^2$

12. $x^5 + 4x + 10x^2 = 5x^3 + 2x^4 + 8$

Chapter 15 Polynomial Functions

Example 6: How many distinct real solutions does the equation $5x^2 + x^4 = -6$ have?

Step 1: Write the equation in standard form.

To write the equation $5x^2 + x^4 = -6$ in standard form, the terms on the left side of the equation should be in order from the highest degree to the lowest degree and the right side of the equation should be 0. The equation can be converted to standard form as follows.

$$5x^2 + x^4 = -6 \to 5x^2 + x^4 + 6 = 0 \to x^4 + 5x^2 + 6 = 0$$

Step 2: Graph the left side of the equation.

Now, the function $f(x) = x^4 + 5x^2 + 6$ should be graphed as follows.

Calculator:
On the TI-83/84 graphing calculator, this graph can be produced by performing the following steps.

a. Press $\boxed{Y=}$.

b. Enter X^4 + 5X^2 + 6 after Y_1 = .

c. Press $\boxed{\text{WINDOW}}$.

d. Enter the following values.

Xmin = –5
Xmax = 5
Xscl = 1
Ymin = –10
Ymax = 40
Yscl = 5
Xres = 1

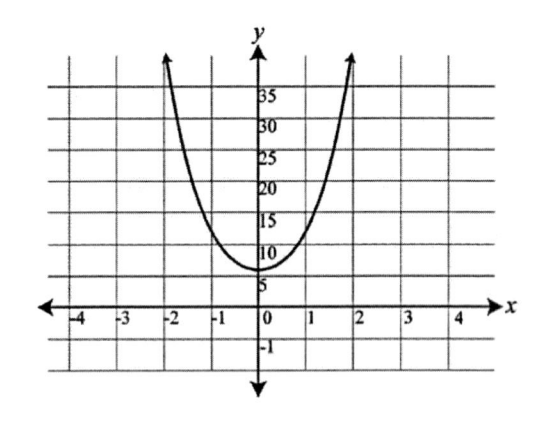

e. Press $\boxed{\text{GRAPH}}$.

Step 3: Count the number of times the graph crosses the x-axis.

The number of distinct real solutions the equation has is equal to the number of times its graph crosses the x-axis. Since the graph does not cross the x-axis, the equation has 0 distinct real solutions.

Determine the number of distinct real solutions to each of the following equations.

1. $6x^2 + x^4 = -5$
2. $x^3 + 4x = 28 + 7x^2$
3. $x^4 = 4x^2 + 45$
4. $3x^3 + x^5 + 2x = 9x^2 + 3x^4 + 6$
5. $x^6 + 27x^4 + 51x^2 = -25$
6. $x^6 + 8x^4 = -9x^2 + 18$

15.3 Solving Polynomial Inequalities Analytically (DOK 3)

To solve a polynomial inequality analytically, first replace the inequality sign with an equal sign to produce a polynomial equation. Then, write the polynomial equation in standard form, which means that the terms on the left side of the equation are in order from the highest degree to the lowest degree, and the right side of the equation is 0. The real roots of the equation can then be found by simply factoring, or if necessary, by using tools such as the Rational Root Theorem, the Factor Theorem, polynomial division, and the quadratic formula. Once the real roots are found, different regions of a number line can be tested to determine which regions satisfy the original inequality, and the solution to the inequality can be written in interval notation.

Example 7: Write the solution to the inequality $x^2 - x < 6$ in interval notation.

Step 1: Replace the inequality sign with an equals sign to produce a polynomial equation.

When $<$ is replaced with an equal sign, the inequality $x^2 - x < 6$ becomes $x^2 - x = 6$.

Step 2: Write the polynomial equation in standard form.

To write the equation $x^2 - x = 6$ in standard form, the terms on the left side of the equation should be in order from the highest degree to the lowest degree and the right side of the equation should be 0. The equation can be changed to standard form as follows.

$x^2 - x = 6$

$x^2 - x - 6 = 6 - 6$

$x^2 - x - 6 = 0$

Step 3: Factor the left side of the equation.

Since the equation is quadratic, with the middle term on the left side having a coefficient of -1 and the last term being -6, the left side can be factored as follows.

$x^2 - x - 6 = 0$

$(x + 2)(x - 3) = 0$

Chapter 15   Polynomial Functions

Step 4:   Find the real roots of the equation.

Since $(x+2)(x-3) = 0$, either $x + 2 = 0$ or $x - 3 = 0$. This means that the real roots of the equation can be found as follows.

$x + 2 = 0$

$x + 2 - 2 = 0 - 2$

$x = -2$

and

$x - 3 = 0$

$x - 3 + 3 = 0 + 3$

$x = 3$

Therefore, the real roots of the equation are $x = -2$ and 3.

Step 5:   Divide a number line into different regions.

Now that the real roots of the equation are known, they can be plotted on a number line, and the number line can be divided into three different regions as follows.

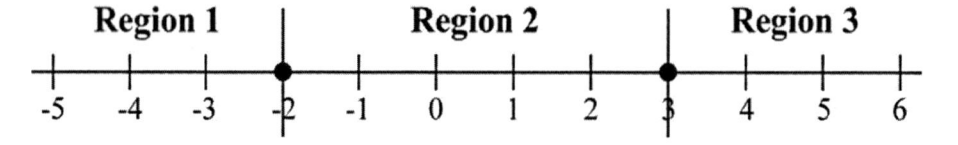

Region 1 | **Region 2** | **Region 3**

-5 -4 -3 -2 -1 0 1 2 3 4 5 6

Test each of the regions.

Any number from each of the regions can be tested with the original inequality to determine which regions satisfy the inequality. First, the number -4 from region 1 can be tested as follows.

$x^2 - x < 6$

$(-4)^2 - (-4) < 6$

$16 + 4 < 6$

$20 < 6$

Since the inequality $20 < 6$ is false, region 1 does not satisfy the inequality.

248   Copyright © American Book Company

15.3 Solving Polynomial Inequalities Analytically (DOK 3)

Step 7: Next, the number 1 from region 2 can be tested as follows.

$x^2 - x < 6$

$1^2 - 1 < 6$

$1 - 1 < 6$

$0 < 6$

Since the inequality $0 < 6$ is true, region 2 satisfies the inequality.

Step 8: Finally, the number 5 from region 3 can be tested as follows.

$x^2 - x < 6$

$(5)^2 - 5 < 6$

$25 - 5 < 6$

$20 < 6$

Since the inequality $20 < 6$ is false, region 3 does not satisfy the inequality.

Step 9: Write the solution to the inequality in interval notation.

Because the only region that satisfies the inequality is region 2, x must be greater than -2, but less than 3. Note that x is not equal to -2 or 3, since there is $<$ in the original inequality and not $\leq$. Therefore, the solution to the inequality in interval notation is $(-2, 3)$.

Write the solution to each of the following inequalities in interval notation.

1. $x^2 - 2x < 24$
2. $x^2 + 5 > -6x$
3. $x^2 + 3x \leq 70$
4. $18 + x^2 \geq 11x$
5. $x^2 < 5x + 24$
6. $4x^2 > 15x + 4$
7. $x^2 \leq x + 42$
8. $x^2 + 64 > 16x$
9. $9x^2 + 3x \geq 2$

Chapter 15   Polynomial Functions

Example 8: Write the solution to the inequality $x^3 + x^2 \geq 10x + 10$ in interval notation.

Step 1: Replace the inequality sign with an equal sign to produce a polynomial equation.

When $\geq$ is replaced with $=$, the inequality $x^3 + x^2 \geq 10x + 10$ becomes $x^3 + x^2 = 10x + 10$.

Step 2: Write the polynomial equation in standard form.

To write the equation $x^3 + x^2 = 10x + 10$ in standard form, the terms on the left side of the equation should be in order from the highest degree to the lowest degree and the right side of the equation should be 0. The equation can be changed to standard form as follows.

$x^3 + x^2 = 10x + 10$
$x^3 + x^2 - 10x - 10 = 10x + 10 - 10x - 10$
$x^3 + x^2 - 10x - 10 = 0$

Step 3: Use the Rational Root Theorem to find all of the possible rational roots of the equation.

Since the equation is now in standard form, all of its possible rational roots can be found by dividing all of the positive and negative factors of the last term on the left side by all of the positive and negative factors of the coefficient of the first term. Since the last term on the left side is 10, all of its positive and negative factors are ± 1, ± 2, ± 5, and ± 10. Also, since the coefficient of the first term is 1, all of its positive and negative factors are ± 1. This means that all of the possible rational roots of the equation are

$x = \dfrac{\pm 1}{\pm 1}, \dfrac{\pm 2}{\pm 1}, \dfrac{\pm 5}{\pm 1}$, and $\dfrac{\pm 10}{\pm 1}$, or $x = \pm 1, \pm 2, \pm 5$, and ± 10.

Step 4: Use the Factor Theorem to find a rational root.

According to the Factor Theorem, if a root of the equation is plugged in for x, the left side of the equation becomes 0. With this in mind, all of the possible rational roots of the equation can be tested. First, $x = 1$ and $x = -1$ can be tested as follows.

$f(x) = x^3 + x^2 - 10x - 10$
$f(1) = (1)^3 + (1)^2 - 10(1) - 10 = 1 + 1 - 10 - 10 = -18$
$f(-1) = (-1)^3 + (-1)^2 - 10(-1) - 10 = -1 + 1 + 10 - 1 = 0$

This means that $x = -1$ is a root of the equation, but $x = 1$ is not. At this point, only one of the roots of the equation has been found, but the Fundamental Theorem of Algebra states that the equation has 3 roots, since the highest exponent on the left side of the equation is 3. While it's possible to keep using the Factor Theorem to test the other possible rational roots, a faster approach may be to produce a quadratic equation to find the two additional roots.

15.3 Solving Polynomial Inequalities Analytically (DOK 3)

Step 5: Use polynomial division to produce a quadratic equation.

Since $x = -1$ is a root of the equation $x^3 + x^2 - 10x - 10 = 0$, a factor of $x^3 + x^2 - 10x - 10$ must be $x + 1$, so $x^3 + x^2 - 10x - 10$ can be divided by $x + 1$ to produce one side of the quadratic equation. The division is done synthetically as follows.

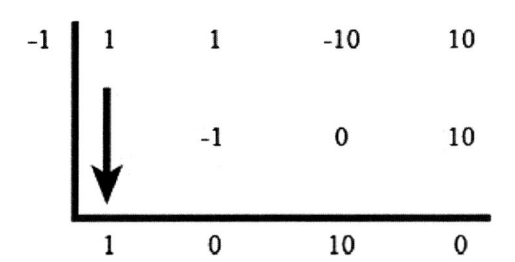

```
-1 |  1    1   -10    10
   |      -1    0     10
   |_____
      1    0    10    0
```

Since $x^3 + x^2 - 10x - 10$ divided by $x + 1$ equals $x^2 - 10$, the quadratic equation that is produced is $x^2 - 10 = 0$.

Step 6: Use the quadratic formula to find the remaining real roots.

Now that there is a quadratic equation in the form $ax^2 + bx + c = 0$ the quadratic formula can be used to solve it. The quadratic formula is $x = \dfrac{-b \pm \sqrt{b^2 - 4ac}}{2a}$, and in this case, $a = 1$, $b = 0$, and $c = -10$. Thus, remaining real roots can be found as follows.

$$x = \frac{-b \pm \sqrt{b^2 - 4ac}}{2a} = \frac{0 \pm \sqrt{0^2 - 4(1)(-10)}}{2(1)} = \frac{\pm\sqrt{40}}{2} = \frac{\pm 2\sqrt{10}}{2} = \pm\sqrt{10}$$

This means that three real roots of the equation $x^3 + x^2 - 10x - 10 = 0$ are $x = -1, -\sqrt{10}$, and $\sqrt{10}$.

Step 7: Divide a number line into different regions.

Now that the real roots of the equation are known, they can be plotted on a number line, and the number line can be divided into four different regions as follows.

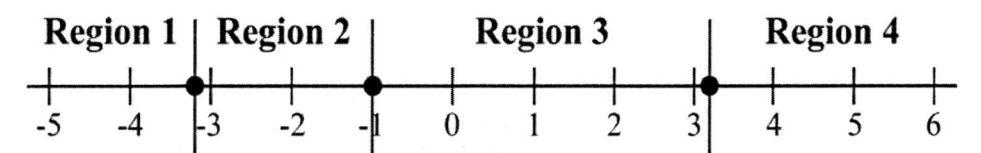

```
  Region 1 | Region 2 |    Region 3    | Region 4
  +--+--+--●--+--+--●--+--+--+--+--●--+--+--+--+
  -5 -4 -3 -2 -1  0  1  2  3  4  5  6
```

Copyright © American Book Company

251

Chapter 15   Polynomial Functions

Step 8: Test each of the regions.
Any number from each of the regions can be tested with the original inequality to determine which regions satisfy the inequality. First, the number -5 from region 1 can be tested as follows.
$(-5)^3 + (-5)^2 \geq 10(-5) + 10$
$-125 + 25 \geq -50 + 10$
Since the inequality $-100 \geq -40$ is false, region 1 does not satisfy the inequality.

Step 9: Next, the number -2 from region 2 can be tested as follows.
$(-2)^3 + (-2)^2 \geq 10(-2) + 10$
$-8 + 4 \geq -20 + 10$
$-4 \geq -10$
Since the inequality $-4 \geq -10$ is true, region 2 satisfies the inequality.

Step 10: Next, the number 1 from region 3 can be tested as follows.
$(1)^3 + (1)^2 \geq 10(1) + 10$
$1 + 1 \geq 10 + 10$
$2 \geq 20$
Since the inequality $2 \geq 20$ is false, region 3 does not satisfy the inequality.

Step 11: Finally, the number 5 from region 4 can be tested as follows.
$(5)^3 + (5)^2 \geq 10(5) + 10$
$125 + 25 \geq 50 + 10$
$150 \geq 60$
Since the inequality $150 \geq 60$ is false, region 4 satisfies the inequality.

Step 12: Write the solution to the inequality in interval notation.
Because the only regions that satisfy the inequality are region 2 and region 4, x must be greater than or equal to $-\sqrt{10}$ and less than or equal to -1, or it must be greater than or equal to $\sqrt{10}$. Therefore, the solution to the inequality in interval notation is $\left[-\sqrt{10}, -1\right] \cup \left[\sqrt{10}, \infty\right)$.

Write the solution to each of the following inequalities in interval notation.

1. $x^3 + 9x^2 + 23x < -15$
2. $14x + x^3 + 24 > 9x^2$
3. $x^3 + 5x^2 \leq 4x + 20$
4. $x^3 + 7x^2 \geq 63 + 9x$
5. $32 + x^3 < 2x^2 + 16x$
6. $2x^2 + x^3 > 30 + 15x$
7. $x^3 + 6x^2 \leq 8x + 48$
8. $25x + x^3 \geq 2x^2 + 50$
9. $x^3 + 16x + 3x^2 < -48$

15.4 Graph Transformation of $f(x) = ax^n$ (DOK 2)

The graph of the function $f(x) = ax^n$ can be transformed by vertically stretching or compressing it, reflecting it across the x-axis, horizontally stretching or compressing it, reflecting it across the y-axis, translating it right or left, and translating it up or down. Each of these transformations is described in the table below.

Transformation	Transformed Function	Conditions
Vertical Stretch or Compression	$f(x) = c \cdot ax^n$	c is a constant that has an absolute value greater than 1 for a vertical stretch and an absolute value less than 1 for a vertical compression
Reflection Across the x-axis	$f(x) = c \cdot ax^n$	c is a constant that is negative
Horizontal Stretch or Compression	$f(x) = a(d \cdot x)^n$	d is a constant that has an absolute value less than 1 for a horizontal stretch and an absolute value greater than 1 for a horizontal compression
Reflection Across the y-axis	$f(x) = a(d \cdot x)^n$	d is the constant that is negative
Translation Right or Left	$f(x) = a(x - h)^n$	h is a constant that is positive for a translation right and negative for a translation left
Translation Up or Down	$f(x) = ax^n + k$	k is a constant that is positive for a translation up and negative for a translation down

Example 9: The graph of the function $f(x) = 2x^2$ was transformed to produce the graph of the function $f(x) = -6x^2$ as shown below. Determine the transformations that were applied.

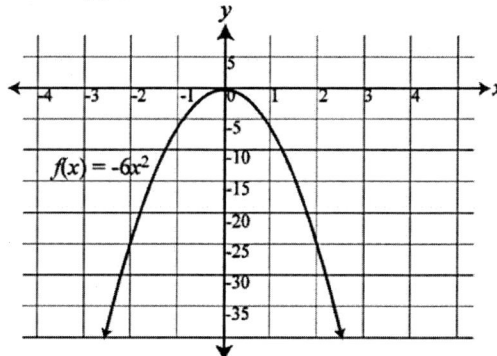

Step 1: Check for a vertical stretch or compression.
The function $f(x) = -6x^2$ is in the form $f(x) = c \cdot 2x^2$, with the value of c being -3. Since the absolute value of -3 is greater than 1, the graph of the function $f(x) = 2x^2$ has undergone a vertical stretch.

Step 2: Check for a reflection across the x-axis.
Also, since c is negative, the graph of the function $f(x) = 2x^2$ has undergone a reflection across the x-axis.

Chapter 15 Polynomial Functions

Example 10: The graph of the function $f(x) = 32x^3$ was transformed by a horizontal stretch, a horizontal compression, a reflection across the y-axis, or a combination of these transformations to produce the graph of the function $f(x) = 4x^3$ as shown below. Determine the transformations that were applied.

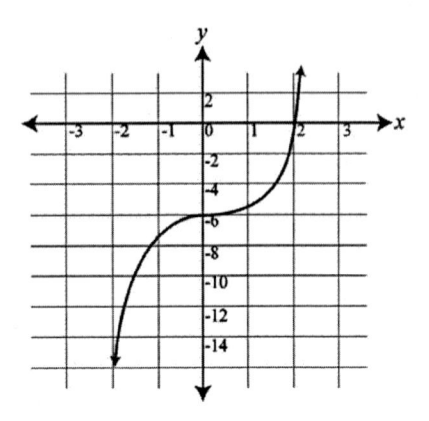

Step 1: Check for horizontal stretch or compression.

The function $f(x) = 4x^3$ is in the form $f(x) = 32(d \cdot x)^3$, with the value of d being $\frac{1}{2}$. Since the absolute value of $\frac{1}{2}$ is less than 1, the graph of the function $f(x) = 32x^3$ has undergone a horizontal stretch.

Step 2: Check for a reflection across the y-axis.

Also, since d is positive, the graph of the function $f(x) = 32x^3$ has not undergone a reflection across the y-axis.

Example 11: The graph of the function $f(x) = \dfrac{x^4}{12}$ was transformed by a translation right or left, a translation up or down, or a combination of these transformations to produce the graph of the function $f(x) = \dfrac{(x-2)^4}{12} + 5$ as shown below. Determine the transformations that were applied.

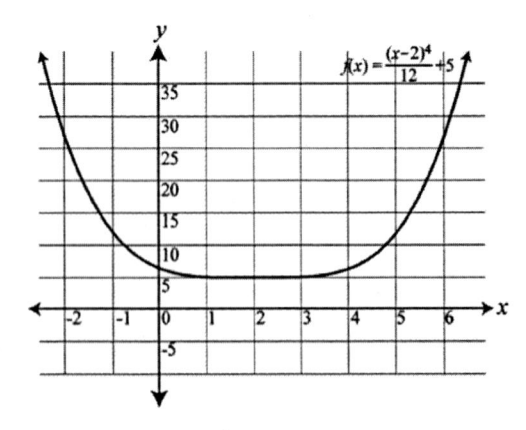

15.4 Graph Transformation of $f(x) = ax^n$ (DOK 2)

Step 1: Check for a translation right or left.

The function $f(x) = \dfrac{(x-2)^4}{12} + 5$ is in the form $f(x) = \dfrac{(x-h)^4}{12} + k$, with the value of h being 2. Since h is positive, the graph of the function $f(x) = \dfrac{x^4}{12}$ has undergone a translation right.

Step 2: Check for a translation up or down.

Also, the value of k is 5. Since k is positive, the graph of the function $f(x) = \dfrac{x^4}{12}$ has undergone a translation up.

The function $f(x) = 10x^3$ was transformed by a vertical stretch, a vertical compression, a reflection across the x-axis, or a combination of these to produce the graphs of each of the following functions. Determine the transformations that were applied to each.

1. $f(x) = -5x^3$
2. $f(x) = 40x^3$
3. $f(x) = -100x^3$
4. $f(x) = \dfrac{x^3}{10}$
5. $f(x) = \dfrac{25x^3}{2}$
6. $f(x) = -x^3$

The function $f(x) = x^5$ was transformed by a horizontal stretch, a horizontal compression, a reflection across the y-axis, or a combination of these to produce the graphs of each of the following functions. Determine the transformations that were applied to each.

7. $f(x) = -1024x^5$
8. $f(x) = \dfrac{7x^5}{3}$
9. $f(x) = 0.6x^5$
10. $f(x) = -1.9x^5$
11. $f(x) = -\dfrac{x^5}{243}$
12. $f(x) = 11x^5$

The function $f(x) = 3.5x^2$ was transformed by a translation right or left, a translation up or down, or a combination of these to produce the graphs of each of the following functions. Determine the transformations that were applied to each.

13. $f(x) = 3.5(x+7)^2 - 13$
14. $f(x) = 3.5(x-2.2)^2 + 8.9$
15. $f(x) = 3.5(x+10)^2$
16. $f(x) = 3.5x^2 + 475$
17. $f(x) = 3.5(x-46)^2 - 6$
18. $f(x) = 3.5(x+3)^2 + 20$

Chapter 15 Polynomial Functions

15.5 Multiplicity of Graphs of Polynomial Functions (DOK 3)

The graph of a polynomial function is affected by the multiplicity of its real zeros, its degree, and its lead coefficient. If the multiplicity of a real zero of a polynomial function is even – that is, if the real zero occurs an even number of times – the graph of the polynomial function touches the x-axis at this value of x, and if the multiplicity is odd, the graph passes through the x-axis at this value of x. The degree, or value of the highest exponent, and the lead coefficient of the function can affect the function's graph in different ways, depending on whether the degree is even or odd and whether the lead coefficient is positive or negative. The affects of the degree and lead coefficient are summarized in the table below.

	Even Degree	**Odd Degree**
Positive Lead Coefficient	• The graph has an absolute minimum. • The graph has no absolute maximum. • The domain of the function is $(-\infty, \infty)$, and the range is [absolute minimum, ∞]. • The end behavior of the graph is $f(x) \to \infty$ as $x \to \pm\infty$.	• The graph has no absolute maximum or minimum. • The domain of the function is $(-\infty, \infty)$, and the range is $(-\infty, \infty)$. • The end behavior of the graph is $f(x) \to -\infty$ as $x \to -\infty$ and $f(x) \to \infty$ as $x \to \infty$.
Negative Lead Coefficient	• The graph has an absolute maximum. • The graph has no absolute minimum. • The domain of the function is $(-\infty, \infty)$, and the range is $(-\infty,$ absolute maximum]. • The end behavior of the graph is $f(x) \to -\infty$ as $x \to \pm\infty$.	• The graph has no absolute maximum or minimum. • The domain of the function is $(-\infty, \infty)$, and the range is $(-\infty, \infty)$. • The end behavior of the graph is $f(x) \to \infty$ as $x \to -\infty$ and $f(x) \to -\infty$ as $x \to \infty$.

15.5 Multiplicity of Graphs of Polynomial Functions (DOK 3)

Example 12: For the polynomial function graphed below, determine the function's real zeros and whether the multiplicity of each zero is even or odd. Also determine whether the graph has an absolute maximum, whether the graph has an absolute minimum, whether the domain and range of the function are both $(-\infty, \infty)$, and the end behavior of the graph. Finally, determine whether the function's lead coefficient is positive or negative, and whether the function's degree is even or odd.

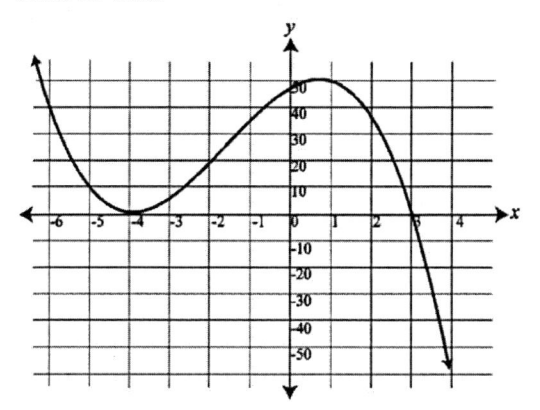

Step 1: Determine the function's real zeros.

Since the graph of the function touches the x-axis at $x = -4$ and passes through the x-axis at $x = 3$, the function's real zeros are $x = -4$ and $x = 3$.

Step 2: Determine whether the multiplicity of each zero is even or odd.

If the graph touches the x-axis, the multiplicity of the zero is even, and if it passes through the x-axis, the multiplicity of the zero is odd. This means that the multiplicity of the zero $x = -4$ is even and the multiplicity of the zero $x = 3$ is odd.

Step 3: Determine whether the graph has an absolute maximum.

Since the graph extends infinitely upward and to the left, the graph does not have an absolute maximum.

Step 4: Determine whether the graph has an absolute minimum.

Also, since the graph extends infinitely downward and to the right, the graph does not have an absolute minimum.

Step 5: Determine whether the domain and range of the function are both $(-\infty, \infty)$.

This means that the domain of the function is $(-\infty, \infty)$, and the range is $(-\infty, \infty)$.

Chapter 15 Polynomial Functions

Step 6: Determine the end behavior of the graph.

It also means that the end behavior of the graph is $f(x) \to \infty$ as $x \to -\infty$ and $f(x) \to -\infty$ as $x \to \infty$.

Step 7: Determine whether the function's lead coefficient is positive or negative.

The function's lead coefficient is negative.

Step 8: Determine whether the function's degree is even or odd.

The function's degree is odd.

For the polynomial function graphed below, determine each of the following.

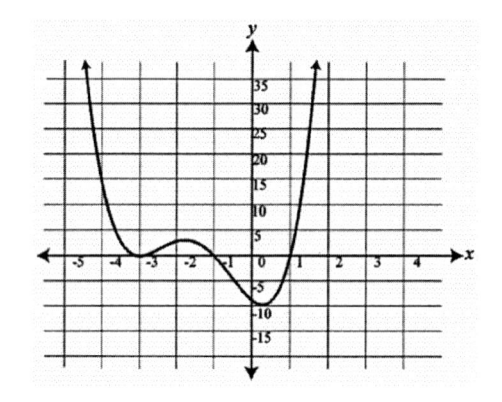

1. What are the function's real zeros?

2. Is the multiplicity of each zero even or odd?

3. Does the graph have an absolute maximum?

4. Does the graph have an absolute minimum?

5. Are the domain and range of the function both $(-\infty, \infty)$?

6. What is the end behavior?

7. Is the function's lead coefficient positive or negative?

8. Is the function's degree even or odd?

15.5 Multiplicity of Graphs of Polynomial Functions (DOK 3)

Example 13: For the polynomial function graphed below, determine the function's real zeros and whether the multiplicity of each zero is even or odd. Also determine whether the graph has an absolute maximum, whether the graph has an absolute minimum, whether the domain and range of the function are both $(-\infty, \infty)$, and the end behavior of the graph. Finally, determine whether the function's lead coefficient is positive or negative, and whether the function's degree is even or odd.

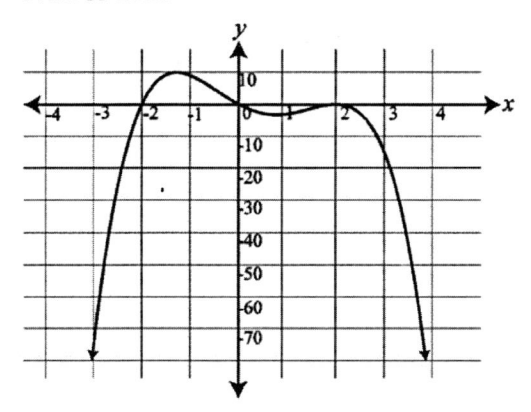

Step 1: Determine the function's real zeros.

Since the graph of the function passes through the x-axis at $x = -2$, $x = 0$, and $x = 2$.

Step 2: Determine whether the multiplicity of each zero is even or odd.

If the graph touches the x-axis, the multiplicity of the zero is even, and if it passes through the x-axis, the multiplicity of the zero is odd. This means that the multiplicity of the zero $x = 2$ is even and the multiplicities of the zeros $x = -2$ and $x = 0$ are odd.

Step 3: Determine whether the graph has an absolute maximum.

Since the graph does not extend infinitely upward, the graph has an absolute maximum.

Step 4: Determine whether the graph has an absolute minimum.

Also, since the graph extends infinitely downward and to the left and infinitely downward and to the right, the graph does not have an absolute minimum.

Step 5: Determine whether the domain and range of the function are both $(-\infty, \infty)$.

This means that the domain of the function is $(-\infty, \infty)$, but the range is not $(-\infty, \infty)$.

Copyright © American Book Company

Chapter 15 Polynomial Functions

Step 6: Determine the end behavior of the graph.

It also means that the end behavior of the graph is $f(x) \to -\infty$ as $x \to \pm\infty$.

Step 7: Determine whether the function's lead coefficient is positive or negative.

The function's lead coefficient is negative.

Step 8: Determine whether the function's degree is even or odd.

The function's degree is even.

For the polynomial function graphed below, determine each of the following.

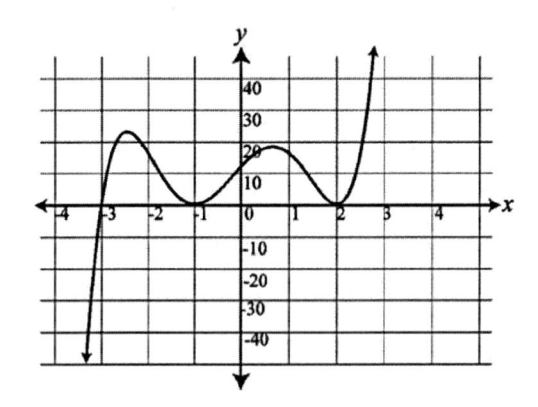

1. What are the function's real zeros?

2. Is the multiplicity of each zero even or odd?

3. Does the graph have an absolute maximum?

4. Does the graph have an absolute minimum?

5. Are the domain and range of the function both $(-\infty, \infty)$?

6. What is the end behavior?

7. Is the function's lead coefficient positive or negative?

8. Is the function's degree even or odd?

15.6 Real-World Polynomial Functions (DOK 3)

Example 14: While traveling through space from his planet to the next galaxy, Captain Coral uses the following formula to determine the distance in meters (d) between his initial location to his destination and the time (t) in seconds that it will take for him to reach his destination:
$d = 24t^4 + 54t^3 + 12t^2 + 3t$
How far is Captain Coral going to travel to his destination if it takes him only 25 seconds to get to his destination?

Step 1: Identify which numbers represent what variables.
$t = 25$

Step 2: Plug this variable into the equation and solve for distance (d) in meters.
$d = 24(25)^4 + 54(25)^3 + 12(25)^2 + 3(25)$

Step 3: $d = 937500 + 843750 + 7500 + 75$
$d = 10,226,325$ meters

Example 15: A ball is thrown directly upward from an initial height (s) of 100 feet with an initial velocity (v) of 30 feet per second (t). Use the following formula to solve for the height of the ball after 3 seconds:
$h = -16t^2 + vt + s$

Step 1: Identify which numbers represent what variables.
$v = 30 \quad s = 100 \quad t = 3$

Step 2: Plug these values into the equation and solve for the height (h) in feet.
$h = -16(3)^2 + 30(3) + 100$

Step 3: $h = -144 + 90 + 100$
$h = 46$ feet

Example 16: The sum of two integers is 13 and their product is 36. What is the value of each integer?

Step 1: Set up the sum and product of each integer as a system of equations.
$x + y = 13$
$xy = 36$

Step 2: Use the substitution method to combine the equations.
$y = 13 - x$
$xy = 36$
$x(13 - x) = 36$
$13x - x^2 = 36$

Chapter 15 Polynomial Functions

Step 3: Set the equation to 0 and factor.
$-x^2 + 13x - 36 = 0$
$-1(x^2 - 13x + 36) = 0$
$(x - 9)(x - 4) = 0$
$x = 9,\ x = 4$

Step 4: Therefore the value of the integers are 9 and 4. To check, plug in either 9 or 4 for the value of x and the other as the value of y into the system of equations to verify the sum and product.

For each of the following, solve for the missing variable.

1. If a painter uses 3 gallons of paint to cover a wall with a length of 120 feet and a height of 8 feet, how many gallons would he use on a surface area of 7,680 ft²?

2. The height of a cube is set at $(x + 1)$. Find the polynomial that represents the volume of the cube.

3. A rectangular swimming pool is three times as long as it is wide. A small concrete walkway surrounds the pool. The walkway is a constant 3 feet wide and has an area of 396 square feet. Find the dimensions of the pool.

4. A ball is thrown directly upward from an initial height (s) of 36 feet at an initial velocity, (v), of 45 feet per second. Use the following formula to solve for the height of the ball after 3 seconds: $h = -16t^2 + vt + s$.

5. Two positive integers have a sum of 9 and a product of 20. What is the value of each positive integer?

6. While on a mission in space, Captain Casey's crew uses the following formula to track an unidentifiable object's movement (in feet) in case it poses a threat to their ship, d being the distance travelled, and t being the time that has passed: $d = t4 + 9t^3 + t^2 - 81t + 70$. How far will the object travel if they track it for only 10 seconds?

Chapter 15 Review

Find real and complex roots of higher degree polynomial equations. (DOK 2)

1. $x^3 - 2x^2 - 13x - 10 = 0$

2. $x^3 + 6x^2 + 5x - 12 = 0$

3. $x^3 + 4x^2 - 5x - 20 = 0$

4. $x^3 - 3x^2 - 6x + 18 = 0$

Find all of the roots of each of the following equations. For each equation, one of the roots has been given. (DOK 2)

5. $x^3 + 2x^2 + 36x + 72 = 0$; $x = 6i$

6. $x^4 - 9x^3 + 24x^2 - 36x + 80 = 0$; $x = -2i$

Solve each of these equations for x. (DOK 2)

7. $x^2 + 8x = 48$

8. $x^3 + 6x^2 = 19x + 84$

Solve each of these equations for x. Include both real and complex solutions. (DOK 2)

9. $x^2 + 14x = -45$

10. $x^3 = 4x^2 + 29x + 24$

Solve each of these equations for x. Include both real and complex solutions. (DOK 3)

11. $x^4 + 6x^3 + 13x^2 + 24x = -36$

12. $x^4 + 56x = 4x^3 + 10x^2 + 56$

To solve the polynomial equation $x^3 - 6x^2 = x - 30$, both sides of the equation were graphed as shown below.

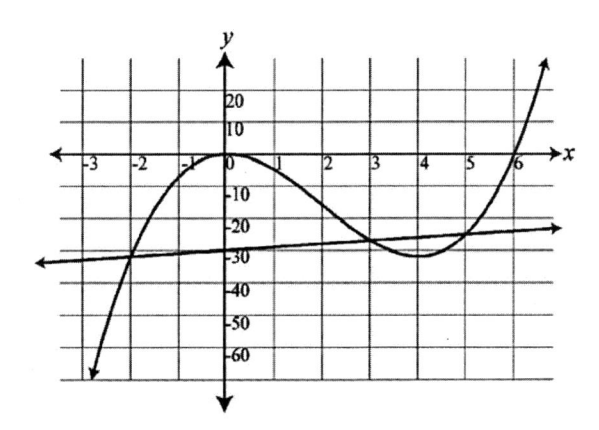

13. What two functions were graphed?

14. What is the solution to the polynomial equation $x^3 - 6x^2 = x - 30$?

Chapter 15   Polynomial Functions

To solve the polynomial equation $x^4 + x + 12 = x^3 + 13x^2$, the equation was written in standard form and then the left side of the equation was graphed as shown below. (DOK 3)

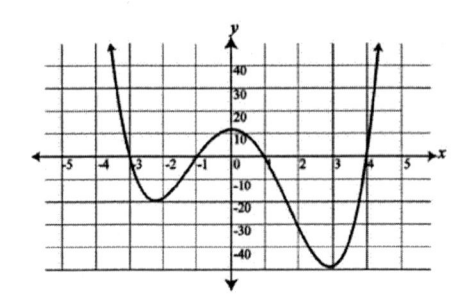

15. What function was graphed?

16. What is the solution to the polynomial equation $x^4 + x + 12 = x^3 + 13x^2$?

Determine the number of distinct real solutions to each of the following equations. (DOK 2)

17. $x^3 + 25x = 3x^2 + 75$

18. $x^6 + x^4 + 49x^2 = -49$

Determine each of the following for the polynomial function $f(x) = x^3 + 5x^2 - 8x - 12$. Round to the nearest tenth, when necessary. (DOK 2)

19. x-intercepts, y-intercept, and zeros

20. Relative and absolute maxima and minima

21. Domain and range

22. Intervals of increase and decrease

23. End behavior

The function $f(x) = 16x^2$ was transformed by a vertical stretch, a vertical compression, a reflection across the x-axis, a translation right or left, a translation up or down, or a combination of these to produce the graphs of each of the following functions. Determine the transformations that were applied to each. (DOK 3)

24. $f(x) = -8x^2$

25. $f(x) = 16(x-5)^2$

26. $f(x) = 48x^2 + 3$

The function $f(x) = 24x^3$ was transformed by a horizontal stretch, a horizontal compression, a reflection across the y-axis, a translation right or left, a translation up or down, or a combination of these to produce the graphs of each of the following functions. Determine the transformations that were applied to each. (DOK 3)

27. $f(x) = 3x^3$

28. $f(x) = 192x^3 - 7$

29. $f(x) = 24(x+2)^3$

Chapter 15 Review

For the polynomial function graphed below, answer problems 30–35. (DOK 2)

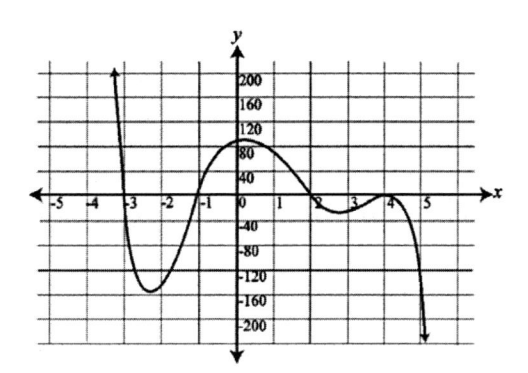

30. The function's real zeros and whether the multiplicity of each zero is even or odd

31. Whether the graph has an absolute maximum and/or an absolute minimum

32. Whether the domain and range of the function are both $(-\infty, \infty)$

33. The end behavior of the graph

34. Whether the function's lead coefficient is positive or negative

35. Whether the function's degree is even or odd

Read each problem and solve. (DOK 3)

36. A ball is dropped from the top of a building that is 1,250 feet tall. How long will it take the ball to reach the ground? Round to the nearest hundredth of a second. Use the formula $h = -16t^2 + t + s$.

37. If a painter uses 1 gallon of paint to paint walls in a room that are 8 feet tall and a total of 40 feet long, how many gallons would the painter use on a surface area of 2,400 ft^2?

38. The sum of two integers is 2. Their product is -15. What is the value of each integer?

39. A rectangular swimming pool is 3 times as long as it is wide. A small concrete walkway surrounds the pool at a constant 2-foot width and has an area of 288 ft^2. Find the dimensions of the pool.

Chapter 15 Test

1 According to the Rational Root Theorem, which of these is a possible rational root of the equation $2x^3 + 3x^2 - 23x - 12 = 0$?

A $x = \dfrac{1}{12}$

B $x = \dfrac{1}{6}$

C $x = \dfrac{1}{4}$

D $x = \dfrac{1}{2}$

(DOK 2)

2 To find the roots of the equation $x^3 - 8x^2 - 3x + 90 = 0$, all the possible rational roots were tested with the function $f(x) = x^3 - 8x^2 - 3x + 90$. Is $x = -3$ a root of the equation?

A No, because $f(-3) \neq 0$.
B No, because $f(-3) = 0$.
C Yes, because $f(-3) \neq 0$.
D Yes, because $f(-3) = 0$.

(DOK 2)

3 What are the roots of the equation $x^3 - 19x + 30 = 0$?

A $x = -5, -2,$ and 3
B $x = -5, 2,$ and 3
C $x = -2, 3,$ and 5
D $x = 2, 3,$ and 5

(DOK 2)

4. If $x^3 + 6x^2 - 19x - 24$ is divided by $x + 1$, what is the result?

A $x^2 - 5x - 24$
B $x^2 - 5x + 24$
C $x^2 + 5x - 24$
D $x^2 + 5x + 24$

(DOK 2)

5 What is the solution to the equation $x^3 + 12x^2 - 2x - 24 = 0$?

A $x = -12, -2i,$ and $2i$
B $x = -12, -\sqrt{2},$ and $\sqrt{2}$
C $x = 12, -2i,$ and $2i$
D $x = 12, -\sqrt{2},$ and $\sqrt{2}$

(DOK 2)

6 To solve the equation $x^3 - 2x^2 - 29x - 42 = 0$, the function $f(x) = x^3 - 2x^2 - 29x - 42$ was graphed. The graph of the function crosses the x-axis at which points?

A $(-3, 0), (-2, 0),$ and $(-7, 0)$
B $(-3, 0), (-2, 0),$ and $(7, 0)$
C $(0, -3), (0, -2),$ and $(0, -7)$
D $(0, -3), (0, -2),$ and $(0, 7)$

(DOK 2)

7 The graph of the function $f(x) = x^3 + 5x^2 + 2x - 8$ is shown below.

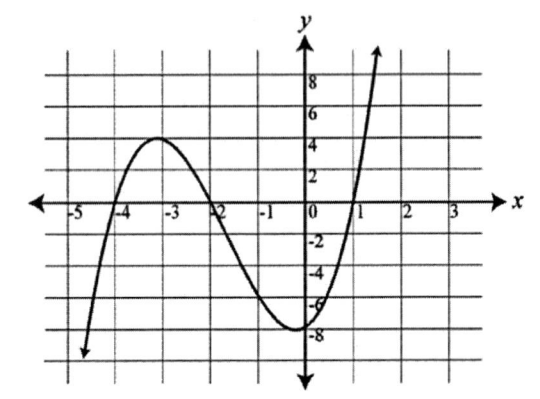

What is the solution to the equation $x^3 + 5x^2 + 2x - 8 = 0$?

A $x = -4, -2,$ and 1
B $x = -4, 2,$ and 1
C $x = -2, -1,$ and 4
D $x = -2, 1,$ and 4

(DOK 2)

Chapter 15 Test

8 How many distinct real solutions does the equation $x^5 - x^4 + 10x^3 - 10x^2 + 9x - 9 = 0$ have?

A 1
B 3
C 4
D 5

(DOK 2)

9 What is the y-intercept of the graph of the function $f(x) = x^3 - 12x^2 - x + 12$?

A $(12, 0)$
B $(0, -1)$
C $(0, 1)$
D $(0, 12)$

(DOK 2)

10 What is the domain of the function $f(x) = -x^2 - 50$?

A $(-\infty, -50]$
B $(-\infty, \infty)$
C $[-50, \infty)$
D $[0, \infty)$

(DOK 2)

11 If the graph of the function $f(x) = 10x^2$ were transformed by a horizontal stretch, the result could be the graph of which of these functions?

A $f(x) = 6x^2$

B $f(x) = 14x^2$

C $f(x) = 22x^2$

D $f(x) = 30x^2$

(DOK 3)

12 Suppose the only transformation applied to the graph of the function $f(x) = 6x^4$ was a reflection across y-axis. The result would be the graph of which of these functions?

A $f(x) = -6x^4 - 6$

B $f(x) = -6x^4$

C $f(x) = 6x^4$

D $f(x) = 6x^4 + 6$

(DOK 3)

13 Suppose the only transformation applied to the graph of the function $f(x) = 27x^5$ was a reflection across x-axis. The result would be the graph of which of these functions?

A $f(x) = -27x^5 - 27$

B $f(x) = -27x^5$

C $f(x) = 27x^5$

D $f(x) = 27x^5 + 27$

(DOK 3)

14 What are the intervals of increase of the function graphed below?

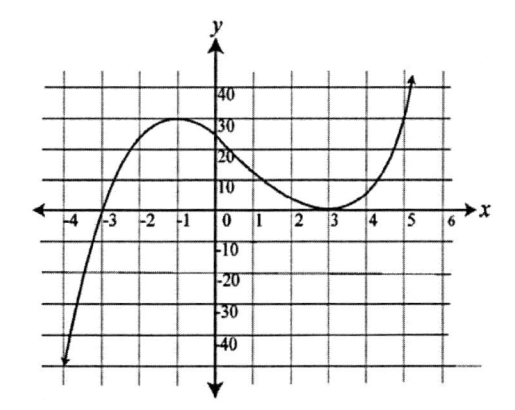

A $(-\infty, -3]$ and $[3, \infty)$
B $(-\infty, -1]$ and $[3, \infty)$
C $(-\infty, 1]$ and $[3, \infty)$
D $(-\infty, 3]$ and $[3, \infty)$

(DOK 2)

Copyright © American Book Company

Chapter 15   Polynomial Functions

15 The graph of the function $f(x) = 18x^7$ was translated 2 units to the left and 12 units up. The graph of which of these functions was produced?

A $f(x) = 18(x-2)^7 - 12$

B $f(x) = 18(x-2)^7 + 12$

C $f(x) = 18(x+2)^7 - 12$

D $f(x) = 18(x+2)^7 + 12$

(DOK 3)

16 Which of these functions has an absolute maximum?

A $f(x) = -26x^5$

B $f(x) = -7x^4$

C $f(x) = 9x^3$

D $f(x) = 11x^2$

(DOK 2)

17 A polynomial function has a zero of $x = 5$, which occurs two times, and a zero of $x = 8$, which occurs three times. Which of these statements is true?

A The multiplicity of $x = 5$ is odd, and the multiplicity of $x = 8$ is odd.

B The multiplicity of $x = 5$ is odd, and the multiplicity of $x = 8$ is even.

C The multiplicity of $x = 5$ is even, and the multiplicity of $x = 8$ is odd.

D The multiplicity of $x = 5$ is even, and the multiplicity of $x = 8$ is even.

(DOK 2)

18 A rectangular prism has a square base with length $(x+1)$. The height of the prism is $(x+4)$. Which of the following polynomials represents the volume of the prism?

A $x^2 + 2x + 1$

B $x^3 + 6x^2 + 9x + 4$

C $x^2 + 5x + 4$

D $x^3 - 9x^2 + 6x - 4$

(DOK 3)

19 Which of the functions has a range of all real numbers?

A $f(x) = -18x^6 - 7$

B $f(x) = -14x^4 + 3$

C $f(x) = 12x^3 + 5$

D $f(x) = 16x^2 - 1$

(DOK 2)

20 Which of these statements accurately describes the function graphed below?

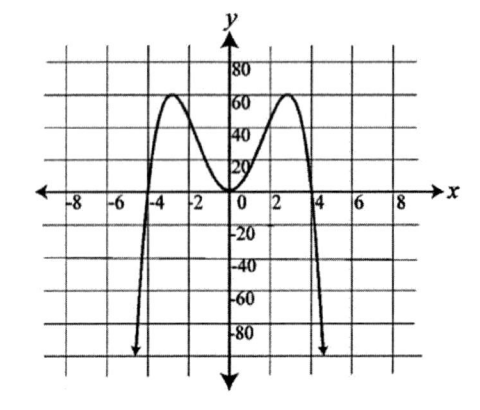

A It has an even degree and a negative lead coefficient.

B It has an even degree and a positive lead coefficient.

C It has an odd degree and a negative lead coefficient.

D It has an odd degree and a positive lead coefficient.

(DOK 2)

268   Copyright © American Book Company

Chapter 15 Test

21 A toy rocket is launched from the top of a building that is 100 feet tall and has an initial velocity of 300 feet per second. Using the formula: $h = -16r^2 + vt + s$, find the height of the rocket after 4 seconds.

A 36 feet
B 1,200 feet
C 1,044 feet
D 450 feet

(DOK 3)

22 The sum of two integers is -18 and their product is 45. What is the value of each integer?

A 3 and 15
B -3 and 15
C -3 and -15
D 3 and -15

(DOK 3)

23 Captain Krunchy's crew uses the following formula to track a shooting star in space where d is the distance travelled in meters, and t is the time passed in seconds: $d = t^4 - 2x^3 - 25r^2 - 26t + 120$. How far will the star have travelled if they tracked it for 30 seconds?

A 732,840 meter
B 100 meters
C 5,800 meters
D 3,999,110 meters

(DOK 3)

Chapter 16
Combining and Comparing Functions

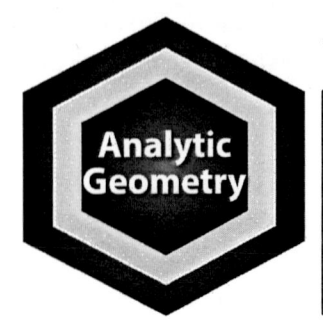

This chapter covers the following CCGPS standard(s):

Analytic Geometry

Reasoning with Equations	G.GPE.4
Interpreting Functions	F.IF.6, F.IF.9, S.ID.6
Building Functions	F.BF.1b, F.BF.3
Expressing Geometric Properties with Equations	G.GPE.2

16.1 Parabolas (DOK 3)

A **parabola** is curved line that can be expressed in standard form by the equation $y = ax^2 + bx + c$, where a, b, and c are constant real numbers. Each point on a parabola is equidistant from both a fixed straight line (the **directrix**) and a fixed point (the **focus**) that does not lie on the line.

Example 1: Find the distance between $(1, 5)$ and $(6, -1)$.

Substitute the values from the ordered pairs into the distance formula.

$$d = \sqrt{(y_2 - y_1)^2 + (x_2 - x_1)^2}$$
$$d = \sqrt{(-1 - 5)^2 + (6 - 1)^2}$$
$$d = \sqrt{(-6)^2 + (5)^2}$$
$$d = \sqrt{36 + 25}$$
$$d = \sqrt{61}$$

The distance between $(1, 5)$ and $(6, -1)$ is $\sqrt{61}$.

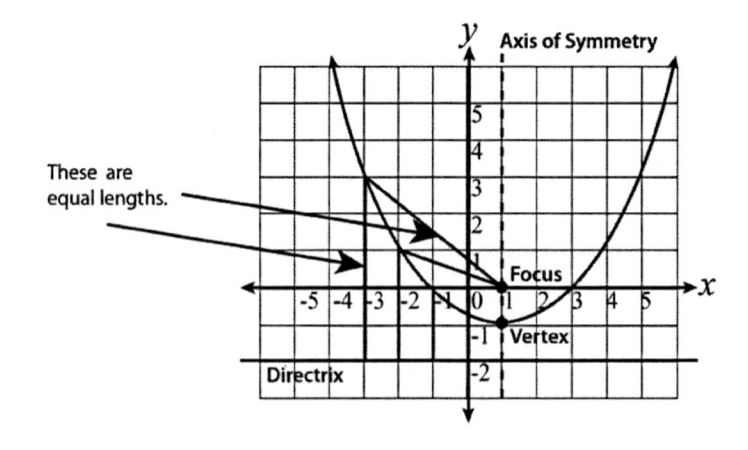

16.1 Parabolas (DOK 3)

We will only work with parabolas that open up towards the positive y-axis or down towards the negative y-axis. If the parabola opens up or down, then the directrix will be a horizontal line with an equation of the form, $y = c$.

Example 2: Find the equation of a parabola given a focus at $(1, 0)$ and directrix, $y = -2$.

Step 1: Let (x, y) be any point on our parabola. Use the distance formula to calculate the distance from the focus to the parabola and the distance from the parabola to the directrix.

Step 2: Find the distances:

$$(1, 0) \text{ to } (x, y) = \sqrt{(x-1)^2 + (y)^2}$$

$$(x, y) \text{ to the line } y = -2 = \sqrt{(x-x)^2 + (y-(-2))^2}$$
$$= \sqrt{(0)^2 + (y+2)^2}$$
$$= \sqrt{(y+2)^2}$$
$$= |y+2|$$

Step 3: The two distances found in step 2 are equal to each other. Set the two expressions equal to each other.

$$\sqrt{(x-1)^2 + (y)^2} = |y+2|$$

Step 4: Simplify the equation in step 3 to get y by itself.

$$\sqrt{(x-1)^2 + (y)^2} = |y+2|$$
$$\left(\sqrt{(x-1)^2 + (y)^2}\right)^2 = (|y+2|)^2$$
$$(x-1)^2 + (y)^2 = (y+2)^2$$
$$x^2 - 2x + 1 + y^2 = y^2 + 4y + 4$$
$$x^2 - 2x - 3 = 4y$$
$$\tfrac{1}{4}x^2 - \tfrac{1}{2}x - \tfrac{3}{4} = y$$

Answer: The equation of the parabola with focus of $(1, 0)$ and directrix of $y = -2$ is $y = \tfrac{1}{4}x^2 - \tfrac{1}{2}x - \tfrac{3}{4}$.

Determine the equation of the parabola in the form $y = ax^2 + bx + c$ given the focus and directrix.

1. Focus: $(11, 2)$, Directrix: $y = 0$

2. Focus: $(3, 4)$, Directrix: $y = 5$

3. Focus: $(2, 0)$, Directrix: $y = 0.5$

4. Focus: $(5, 0)$, Directrix: $y = 6$

5. Focus: $(2, 5)$, Directrix: $y = 3$

6. Focus: $(1, 2)$, Directrix: $y = -4$

16.2 Solutions of Equations (DOK 2)

We have already discussed in a previous section what order of operations are necessary to isolating a variable when solving simple equations. We have not, however, practiced with equations that involve two functions set equal to each other. The idea behind solving these more complex equations is to find for what values they are equal. In other words, to solve for the values at which their graphs intersect.

Example 3: Determine at which points $f(x) = x^2 + 2x + 1$ and $g(x) = x + 1$ intersect. There are multiple ways to solve this: graphically, algebraically, or by making a table of values.

Step 1: Set the functions equal to each other.
$x^2 + 2x + 1 = x + 1$

Step 2: Perform the order of operations necessary to isolate the variable.
$x^2 + 2x + 1 = x + 1$
$x^2 + x = 0$
$x(x + 1) = 0$

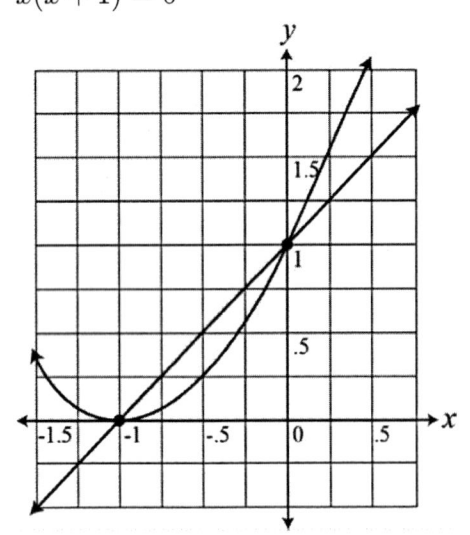

Step 3: Recognize that this statement is true for both values $x = 0$ and $x = -1$. Therefore, 0 and -1 are the solutions to the equation.
Making a table:

$f(x) = x^2 + 2x + 1$	
x	$f(x)$
-1	0
0	1
1	4
2	9
3	16

$g(x) = x + 1$	
x	$g(x)$
-1	0
0	1
1	2
2	3
3	4

The values of -1 and 0 are equal for both functions.

Answer: $x = -1, x = 0$

16.2 Solutions of Equations (DOK 2)

Determine where the graphs of the following functions intersect.

1. $f(x) = x^2 + 7x + 10$, $g(x) = 2x + 6$

2. $f(x) = 2^x$, $g(x) = e^{-x}$

3. $f(x) = |x|$, $g(x) = x^2 + 5x + 6$

4. $f(x) = \frac{1}{3}x + 2$, $g(x) = -x^2 + 2$

5. $f(x) = 3x + 4$, $g(x) = x^2 - 14$

6. $f(x) = 10^x - 99$, $g(x) = -x^2 + 5$

7. $f(x) = |x| - 10$, $g(x) = \frac{5}{4}x - 2$

8. $f(x) = 2^x$, $g(x) = (\frac{1}{2})^x$

9. $f(x) = x^2 + 8x + 15$, $g(x) = -x^2 - x + 8$

10. $f(x) = x^2 + 2x - 35$, $g(x) = \left|\dfrac{80 - x}{2}\right|$

11. $f(x) = \log(x) + 5$, $g(x) = 5^x$

12. $f(x) = 5x^2 - 44x + 120$, $g(x) = 11x - 30$

13. $f(x) = 3(x - 2)^2$, $g(x) = \log(x) + 7$

14. $f(x) = x + 6$, $g(x) = 5x^2$

15. $f(x) = (x + 4)^2$, $g(x) = \log(x^2) + 10$

Chapter 16   Combining and Comparing Functions

16.3   Finding Common Solutions of Functions (DOK 2)

Now that we know how to solve for the solutions of equations algebraically, we will now solve for the solutions graphically, or using tables. Keep in mind that the solution to an equation is where the two graphs intersect.

For what value(s) of x are the following functions equivalent?

1.

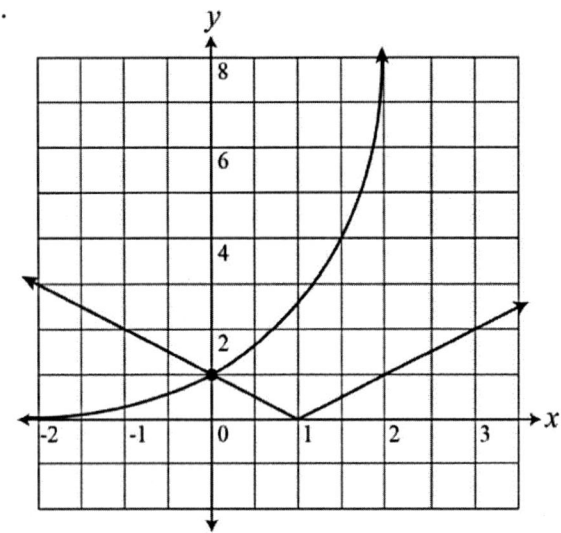

2.

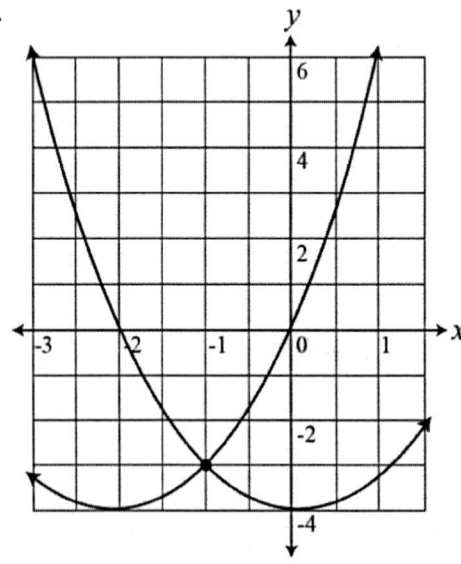

3.

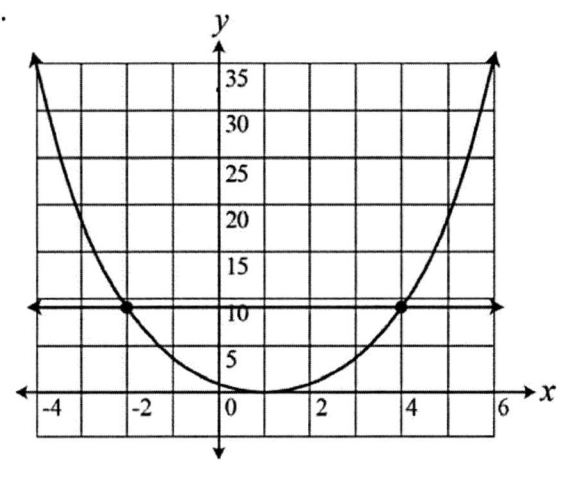

4.

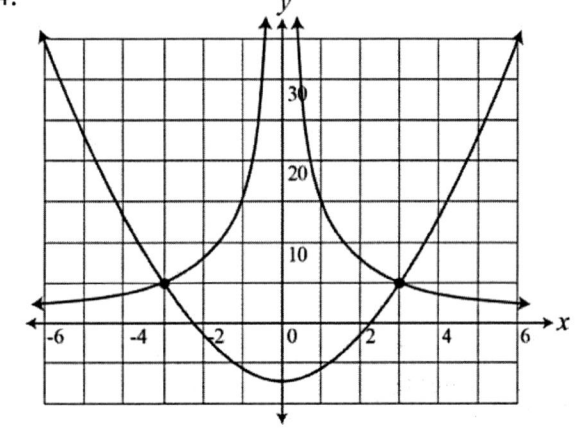

274   Copyright © American Book Company

16.4 Finding Rate of Change of a Function (DOK 2, 3)

5.

x	$f(x)$	$g(x)$
-2	4	2
-1	1	undefined
0	0	0
-1	1	$\frac{1}{2}$
2	4	$\frac{2}{3}$

6.

x	$f(x)$	$g(x)$
-2	-18	10
-1	-13	8
0	-8	6
1	-3	4
2	2	2

7.

x	$f(x)$	$g(x)$
-4	7	7
-3	0	undefined
-2	-5	-5
-1	-8	-2
0	-9	-1
1	-8	$-\frac{1}{2}$
2	-5	$-\frac{1}{5}$
3	0	0

8.

x	$f(x)$	$g(x)$
-4	-2	-2
-3	-2	0
-2	0	2
-1	4	4
0	10	6

16.4 Finding Rate of Change of a Function (DOK 2, 3)

To find the average rate of change of a function, $f(x)$ over a given interval $[a, b]$, we use the formula: $\frac{f(b) - f(a)}{b - a}$. We see that there is a similarity of this formula to the one we use to find the slope of a line: $\frac{y_2 - y_1}{x_2 - x_1}$. The difference is that a line has a constant rate of change, while other functions tend to increase or decrease over a certain interval. This is why the average rate of change may be different over two separate intervals of the same function.

Example 4: Find the average rate of change of the function $f(x) = x^2 + 3x + 2$ over the interval $[0, 5]$.

Step 1: Determine the value of $f(0)$ and $f(5)$.

$f(0) = 0^2 + 3(0) + 2 = 0 + 0 + 2 = 2$

$f(5) = 5^2 + 3(5) + 2 = 25 + 15 + 2 = 42$

Step 2: Plug the values into the formula and simplify:

$$\frac{f(b) - f(a)}{b - a} = \frac{42 - 2}{5 - 0} = \frac{40}{5} = 8$$

Step 3: The average rate of change of the function $f(x)$ is 8.

Chapter 16 Combining and Comparing Functions

Determine the average rate of change of the following functions over the given interval.

1. $f(x) = 2x^3 - 5x^2 + x - 7, [3, 8]$

2. $f(x) = 3x^3 - 6x^2 - 45x, [7, 12]$

3. $f(x) = 7 - 3x^2, [2, 5]$

4. $f(x) = 1.5^x, [-1, 0]$

5. $f(x) = 10x^2, [6, 10]$

6. $f(x) = 3x^2 - 2x + 1, [0, 8]$

7. $f(x) = x^3 - 3x + 4, [1, 7]$

8. $f(x) = 3x^3 - 5x^2 + 1, [-1, 4]$

9. $f(x) = x^4 - 5x, [-2, 3]$

10. $f(x) = x^2 + x - 2, [2, 9]$

11. On Monday, the price of gas was $3.34 per gallon. On Saturday, the price had risen to $4.12 per gallon. What is the average rate of change of the price of a gallon of gas from Monday to Saturday?

12. In 2003, the population of a town was 36,571 and rose to 72,636 by 2008. What was the average rate of change of the population from 2003 to 2008?

13. Greg delivers 24 ft² of tile for $320 and 55 ft² of tile for $630. What is the average rate of change of the cost as the number of square feet increases from 24 to 55?

14. An average 4-door sedan gets a fuel efficiency of 31 miles per gallon when driving at a speed of 60 mph. If the driver slows to a speed of 48 mph, the fuel efficiency increases to 37 miles per gallon. What is the average rate of change of the fuel efficiency as the speed drops from 60 mph to 48 mph?

15. Bill leaves on a road trip Sunday morning at 7:00 am and arrives at his destination at 2:00 p.m. When he began his trip, the car's odometer read 19,772 miles, and when he arrived it read 20,297 miles. What was his average speed for the trip?

16.5 Comparing Functions (DOK 3)

When two functions are each represented differently (algebraically, graphically, numerically in tables, or by verbal descriptions), it can be difficult to see their relationship with each other. This is why it is important to know how to identify functions and compare their properties in various ways.

Example 5: Which linear function below has a higher y-intercept?

$f(x) = -6x + 11$

x	$g(x)$
-2	-2
-1	3
0	8
1	13
2	18
3	23

Step 1: Identify the y-intercept. We know when looking at the algebraic representation of a linear function, that the value of the function when $x = 0$ is the y-intercept. Likewise when we look at a table, we look for the value of the function when $x = 0$, $f(x) = 11$, and $g(x) = 8$.

Step 2: Determine which y-intercept is larger. $11 > 8$ and so $f(x) > g(x)$.

Answer: $f(x)$ has a larger y-intercept.

1. Which function has the steeper slope? Explain your response.

   $f(x) = 3x - 8$

x	$g(x)$
-2	1
-1	3
0	5
1	7
2	9
3	11

2. Which function has the lower rate of change over the given interval? Explain your response.

   $f(x) = 3x^2 - 2x + 5, [3, 8]$

x	$g(x)$
3	16
4	27
5	42
6	61
7	84
8	111

3. Which function has the higher y-intercept? Explain your response.

   $f(x) = x^3 + 3x^2 + 8$

x	$g(x)$
-2	7
-1	6
0	3
1	-2
2	-9
3	-18

4. Which linear function has the smaller slope? Explain your response.

   $f(x) = x + 7.2$

   $g(x)$: is graphed below.

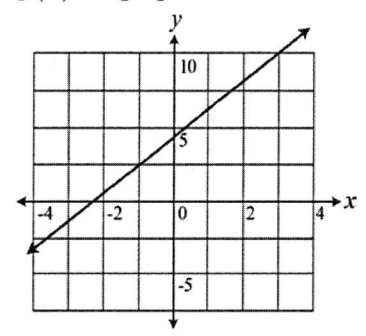

Chapter 16   Combining and Comparing Functions

5. Which function has the greater rate of change over the given interval? Explain your response.

$f(x) = \frac{x^2-9}{x+1}$, $[0, 5]$

$g(x)$ is graphed below.

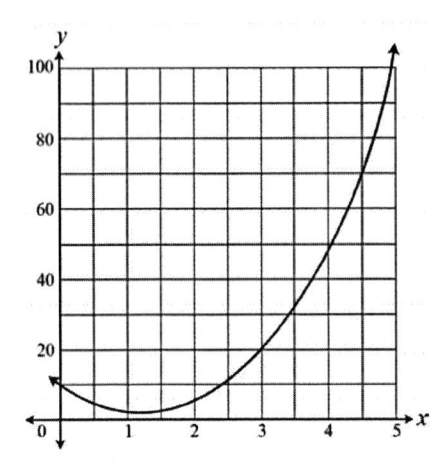

6. Which function has the greater maximum? Explain your response.

$f(x) = -x^2 + 17x + 7$
$g(x)$ is graphed below.

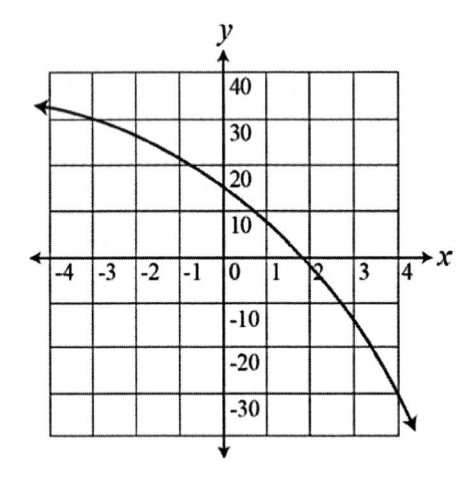

7. Which linear function has the steeper slope? Explain your response.
$g(x)$ is graphed below.

x	$f(x)$
-2	-4
-1	1
0	6
1	11
2	16
3	21

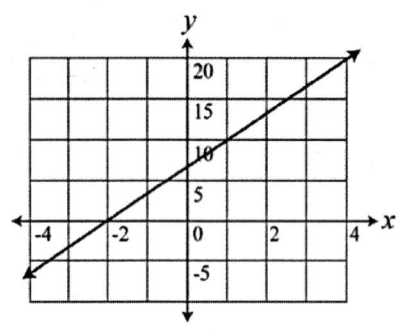

8. Which of the following functions has a smaller rate of change over the interval $[0, 3]$? Explain your response.
$g(x)$ is graphed below.

x	$f(x)$
0	0
1	1
2	4
3	9

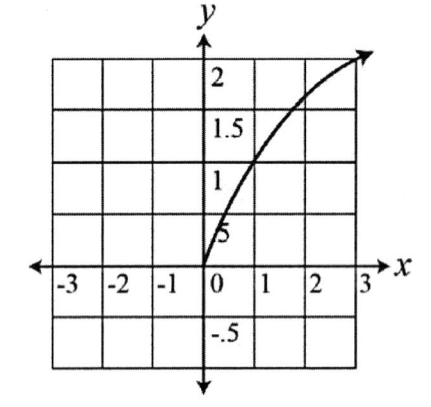

9. Which linear function has the greater y-intercept? Explain your response.
$g(x)$ is graphed below.

x	$f(x)$
-2	6
-1	7.5
0	9
1	10.5
2	12
3	13.5

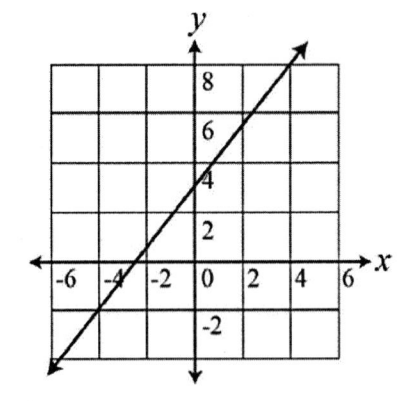

10. Which linear function has the greater slope?
    $f(x)$ is a function that represents the price of a pizza depending on the number of toppings. Each topping costs $0.50.
    $g(x) = 1.5x + 10.5$

11. Which function has the greater rate of change over the given interval?
    $f(x)$ is a function that represents your decision to invest money into a savings account. You initially deposit $500 and after 5 years, the account holds $800.
    $g(x) = -x^2 + 7x + 6$; $[0, 5]$

12. Which function has the lower minimum?
    $f(x)$ is a function that represents the dimensions of rectangular flags. The dimensions have a 2 : 1 ratio.
    $g(x) = x^2 + 9x + 18$

16.6 Comparing Real-Life Functions (DOK 3)

Match the following descriptions of the functions with their algebraic representation.

1. A cheese pizza costs $5.00 and every topping is an additional $1.50.

2. A savings account is opened with a deposit of $800. The interest rate is 2%, compounded annually.

3. The growth of bacteria quadruples every hour.

4. A toy rocket is launched from the ground with an initial velocity of 160 feet per second. Keep in mind that gravity has a velocity of -16 per second^2.

5. In a tournament, 64 teams compete for a trophy. The number of teams is divided by 2 at the end of each round.

6. The volume of a sphere is a function of the radius of the sphere.

7. An airplane appears to be descending at a steady rate of 112.5 feet per minute from 13,500 feet in the air.

8. A person who sews banners makes the banners with dimensions of a 2:1 ratio. The area of a banner is a function of the width of the banner.

(a) $a(x) - \$800(1.02^x)$

(b) $b(x) = \frac{4}{3}\pi x^3$

(c) $x(x) = -16x^2 + 160x$

(d) $d(x) = \$1.50 + \$5$

(e) $e(x) = 2x^2$

(f) $f(x) = 4e^x$

(g) $g(x) = 64(\frac{1}{2}^x)$

(h) $h(x) = 13,500 - 120x$

Chapter 16 Combining and Comparing Functions

16.7 Combining Functions (DOK 2)

There are 5 ways to combine functions: addition, subtraction, multiplication, division, and composition.

Example 6: Given $f(x) = x^2 + 6x + 9$ and $g(x) = 4x + 7$, evaluate $(f+g)(x)$, $(f-g)(x)$, $(fg)(x)$, (x), and $\left(\dfrac{f}{g}\right)(x)$, and $(f \circ g)(x)$.

Step 1: $(f + g) = x^2 + 6x + 9 + 4x + 7 = x^2 + 10x + 16$

Step 2: $(f - g) = x^2 + 6x + 9 - (4x + 7) = x^2 + 2x + 2$

Step 3: $(fg)(x) = (x^2 + 6x + 9)(4x + 7) = 4x^3 + 31x^2 + 78x + 63$

Step 4: $\left(\dfrac{f}{g}\right)(x) = \dfrac{4x^2 + 24x + 36}{4x + 12} = \dfrac{4(x^2 + 6x + 9)}{4(x + 3)} = \dfrac{\cancel{4(x+3)}(x+3)}{\cancel{4(x+3)}} = x + 3$

Step 5: $(f \circ g)(x) = (4x = 7)^2 + 6(4x + 7) + 9 = 16x^2 + 56x + 49 + 24x + 42 + 9 = 16x^2 + 80x + 100$

There are real life applications of these combinations.

Example 7: Boon works 40 hours a week at a furniture store. He makes $300 weekly salary, plus a 4% commission on sales over $2,500. Assume Boon sells enough this week to get the commission. Given the functions $f(x) = 0.04x$ and $g(x) = x - 2500$, what combination of these two functions represents Boon's commission?

Answer: The composition of these two functions represents Boon's commission. $f(g(x)) = 0.04(x - 2500)$.

Solve the problems below.

1. The number, n, of cellphones produced by some factory in one day after t hours, is given by $n = 900t - 9t^2$. If the cost C in dollars of producing, n, cellphones is $C(n) = 11{,}000 + 350n$, what is the cost C as a function of time t?

2. The price, p, of fish can be purchased at the market per pound, f, according to the equation $p = 2f + 18$. The amount of tax, t, due per pound of fish, f, can be calculated by $t = 0.03f + 1$. What is the total amount, (A) a customer would pay for f pounds of fish?

3. A rental car company charges $100 per day, represented by the function $d = 100$. There is an additional charge of $0.45 per mile travelled, represented by the function $r = 0.45m$. What is the function that represents the total cost, C, of renting a car for one day?

4. The price, p, of a camera and the quantity, x sold, are represented by the equation $p = -x + 12$ and the cost, C, of producing x units, is $C = \dfrac{x^2 + 1000}{15}$. Find the cost, C, as a function of price, p

5. Your investment, s, into your savings account is $800, such that $s = 800$. The savings account has an interest rate, r, of 9%, compounded annually, represented by the function, $r = 1.09t$. Find the total amount, B, in the account as a function of time, t.

6. A discount, d, of 20% on price, p, represented by the function, $d = 0.2p$, is offered on certain products in a store. Assuming someone purchases one of these items, and the full price, f, is represented by the function $f = p$, what is the function that represents the amount, A, the customer would pay for this item?

7. The surface area, S, of a spherical hot-air balloon is given by $S(r) = 4\pi r^2$ where r is the radius of the balloon. If the radius, r, increases with time, t, according to the formula, find the surface area, S, of the balloon as a function of time, t.

8. The total profit, t, made on all the tickets to attend a movie can be determined using the function $t = 8.25p - 1$, where p represents the number of people buying tickets. The total number of tickets, m, sold is represented by the function $m = p$. What is the function that represents the profit, R, made on each individual ticket?

16.8 Function Symmetry (DOK 2)

In Mathematics, functions can be defined as symmetrical with respect to the y-axis or the origin. To test equations for symmetry, it is helpful to remember the following:

$f(-x) = f(x)$ means the function is symmetrical with respect to the y-axis.
Being symmetrical with the y-axis means the function is **even**.

$f(-x) = -f(x)$ means the function is symmetrical with respect to the origin.
Being symmetrical with the origin means the function is **odd**.

$f(-x) \neq -f(x)$ or $f(x)$ means the function is not symmetrical.

Example 8: Test the following function for symmetry. $f(x) = x^4 + x^2 + 3$

Step 1: First, we need to substitute $-x$ in the function for x.
$f(-x) = (-x)^4 + (-x)^2 + 3$

Step 2: Carry out the operations.
$f(-x) = x^4 + x^2 + 3$

Step 3: Since $f(-x) = x^4 + x^2 + 3$ and $f(x) = x^4 + x^2 + 3$, then $f(-x) = f(x)$. This means $f(x) = x^4 + x^2 + 3$ is symmetrical with respect to the y-axis. The function $f(x) = x^4 + x^2 + 3$ is even.

Chapter 16   Combining and Comparing Functions

Example 9:   Test the following function for symmetry. $f(x) = x^5 + x^3 + x$

Step 1:   First, we need to substitute $-x$ in the function for x.
$f(-x) = (-x)^5 + (-x)^3 + (-x)$

Step 2:   Carry out the operations.
$f(-x) = -x^5 - x^3 - x$

Step 3:   Factor out a negative. $f(-x) = -x^5 - x^3 - x = -(x^5 + x^3 + x)$. Since $f(-x) = -(x^5 + x^3 + x)$ and $-f(x) = -(x^5 + x^3 + x)$, then $f(-x) = -f(x)$. This means $f(x) = x^5 + x^3 + x$ is symmetrical with respect to the origin. The function $f(x) = x^5 + x^3 + x$ is odd.

Example 10:   Test the following function for symmetry. $f(x) = x^7 + x^4 - x^2$

Step 1:   First, we need to substitute $-x$ in the function for x.
$f(-x) = (-x)^7 + (-x)^4 - (-x)^2$

Step 2:   Carry out the operations.
$f(-x) = -x^7 + x^4 - x^2$

Step 3:   In this case, we see that $f(-x) \neq -f(x)$ or $f(x)$, which means this function is neither even nor odd (no symmetry).

Determine whether the function is even, odd, or neither.

1. $f(x) = x^3 + x^2 + x + 1$

2. $f(x) = 2x^3 - 4x$

3. $f(x) = 7x^2 - 11$

4. $f(x) = 8x^2 + x^4$

5. $f(x) = 4x^3 + x$

6. $f(x) = 6x^3 + 2x^2 - x$

7. $f(x) = x^5 + x^3 + 11$

8. $f(x) = x^7 - x^4 + x^2 - 11$

16.9 Symmetry of Graphs of Functions (DOK 2)

Just like the symmetry of functions was determined algebraically, it can also be determined by looking at the graph. If you are unable to tell just by looking at the graph, you can check your answers using the following:

y-axis (even)   If (a, b) is on the graph, so is $(-a, b)$.
origin (odd)   If (a, b) is on the graph, so is $(-a, -b)$.
x-axis   If (a, b) is on the graph, so is $(a, -b)$.

Example 11: Determine if the graph of $y = x^2$ is symmetrical. If it is, is it odd, even, or neither?

Step 1: Graph the function.

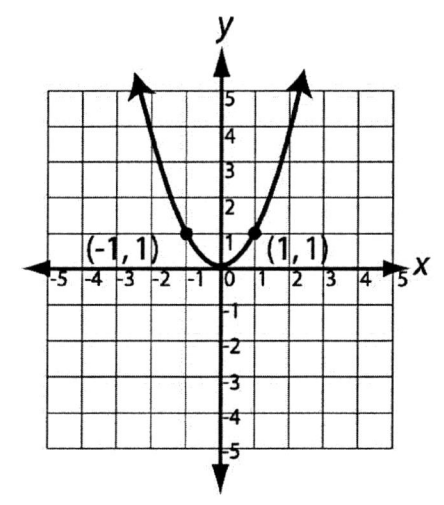

Looking at the graph we see that the graph is symmetrical about the y-axis. In other words, whatever the graph shows on one side of the y-axis, it is the same on the other side of the y-axis.

Step 2: We check using the points on the graph. We will use the point $(1, 1)$ as it is a solution to the equation.
According to the table above, if $(1, 1)$ is a solution to a graph that is symmetrical about the y-axis, then $(-1, 1)$ must also be a solution.
Is it? $y = (-1)^2 = 1$. Yes, it is.
We also know since the graph is symmetrical about the y-axis that the graph is even.

Chapter 16 Combining and Comparing Functions

Example 12: Determine if the graph of $x = y^2$ is symmetrical? If it is, is it odd, even, or neither?

Step 1: Graph the function.

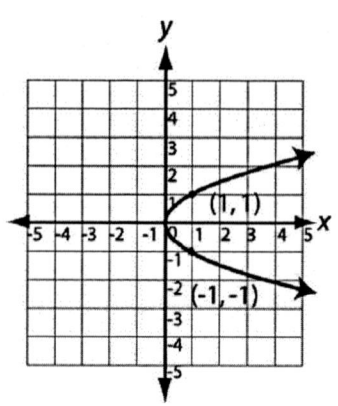

Looking at the graph we see that the graph is symmetrical about the x-axis. In other words, whatever the graph shows on one side of the x-axis, it is the same on the other side of the x-axis.

Step 2: We check using the points on the graph. We will use the point $(1, 1)$ as it is a solution to the equation. If $(1, 1)$ is a solution to a graph that is symmetrical about the x-axis, then $(1, -1)$ must also be a solution. Is it? Yes.
We also know since the graph is symmetrical about the x-axis that the graph is neither even or odd.

Example 13: Determine if the graph of $y = \dfrac{1}{x}$ is symmetrical? If it is, is it odd, even, or neither?

Step 1: Graph the function.

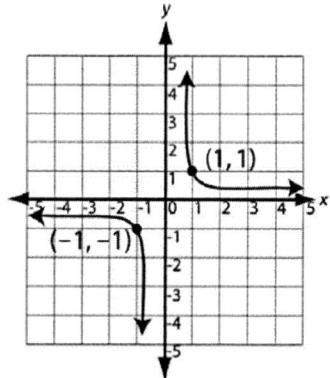

Looking at the graph we see that the graph is symmetrical about the origin. In other words, if you rotate one part of the graph 180° around the origin, it will be the exact same as the other half of the graph.

Step 2: We check using the points on the graph. We will use the point $(1, 1)$ as it is a solution to the equation. If $(1, 1)$ is a solution to a graph that is symmetrical about the origin, then $(-1, -1)$ must also be a solution. Is it? Yes.
We also know since the graph is symmetrical about the origin that the graph is odd.

16.9 Symmetry of Graphs of Functions (DOK 2)

Tell whether the graph is symmetrical with respect to the x-axis, y-axis, the origin, or neither. Also, tell whether the graph is odd, even, or neither.

1.

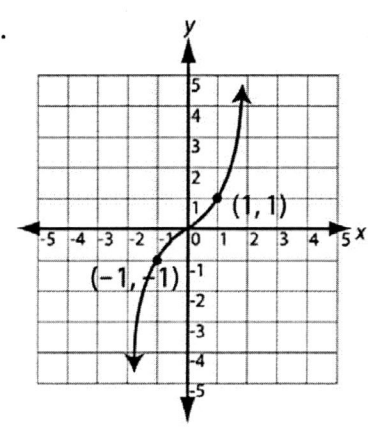

2.

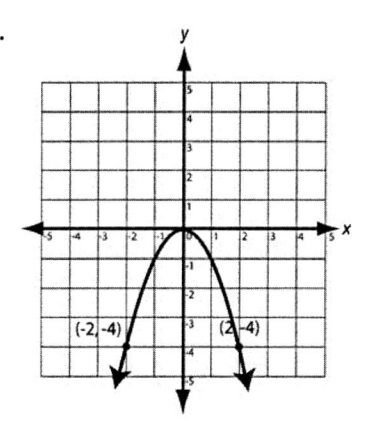

3.

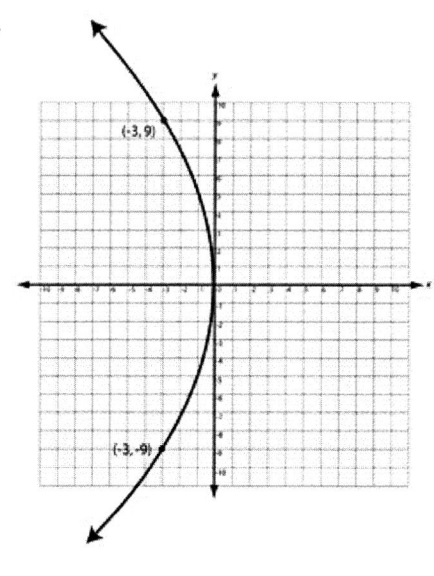

4.

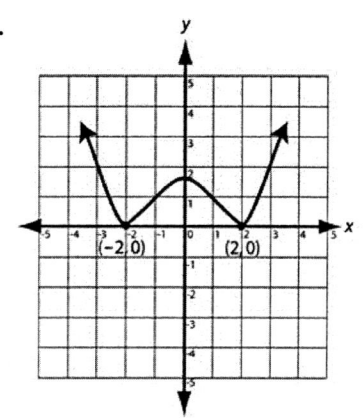

5.

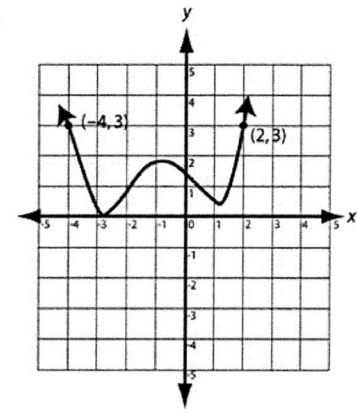

6.

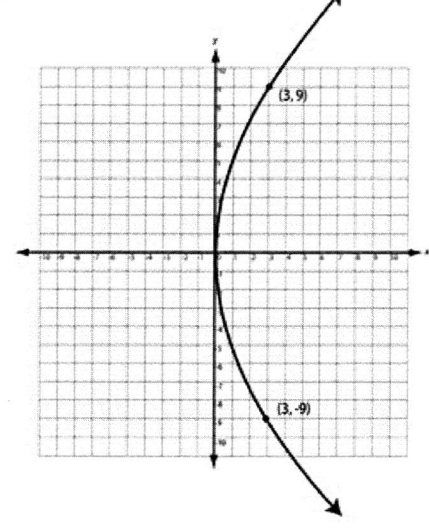

Chapter 16 Combining and Comparing Functions

16.10 Solutions of Equations (DOK 2)

An intersection point is a point where two functions meet. To find the intersection point, you must set the equations equal to each other, $f(x) = g(x)$.

Example 14: Find the intersection point(s) of $f(x) = x + 5$ and $g(x) = 2x + 6$.

Step 1: Find the value of x in the intersection point by setting the two equations equal to each other.

$$\begin{aligned} x + 5 &= 2x + 6 \quad &\text{Set the equations equal to each other.} \\ 5 &= x + 6 \quad &\text{Subtract } x \text{ from both sides and simplify.} \\ -1 &= x \quad &\text{Subtract 6 from both sides and simplify.} \end{aligned}$$

Step 2: We use the value of x to find the intersection point by substituting the x-value back into one of the original equations.

$$\begin{aligned} f(x) &= x + 5 \quad &\text{Choose an equation.} \\ f(-1) &= -1 + 5 \quad &\text{Substitute the } x\text{-value into the equation.} \\ f(-1) &= 4 \quad &\text{Simplify.} \end{aligned}$$

The intersection point is $(-1, 4)$.

Example 15: Find the intersection point(s) of $f(x) = x^2$ and $g(x) = -3x + 4$.

Step 1: Find the value of x in the intersection point by setting the two equations equal to each other.

$$\begin{aligned} x^2 &= -3x + 4 \quad &\text{Set the equations equal to each other.} \\ x^2 + 3x - 4 &= 0 \quad &\text{Move all terms to one side the equation.} \\ (x + 4)(x - 1) &= 0 \quad &\text{Factor.} \\ x = -4 \text{ or } x &= 1 \quad &\text{Solve for } x. \end{aligned}$$

Step 2: We use the values of x to find the intersection points by substituting the x-values back into one of the original equations.

$$\begin{aligned} f(-4) &= (-4)^2 \quad &\text{Substitute one } x\text{-value into an original equation.} \\ f(-4) &= 16 \quad &\text{Solve.} \end{aligned}$$

$$\begin{aligned} f(1) &= 1^2 \quad &\text{Substitute the other } x\text{-value to find the 2nd intersection point.} \\ f(1) &= 1 \quad &\text{Solve.} \end{aligned}$$

The intersection points are $(-4, 16)$ and $(1, 1)$.

Find the intersection points of the pairs of functions.

1. $f(x) = 2x + 3$ and $g(x) = -\frac{1}{2}x + 7$
2. $f(x) = 2x - 3$ and $g(x) = x + 4$
3. $f(x) = 5x - 1$ and $g(x) = 2x + 8$
4. $f(x) = x^2 + 6$ and $g(x) = -5x$
5. $f(x) = x^2 - x$ and $g(x) = 4x - 6$
6. $f(x) = x^2 - 3x$ and $g(x) = -4x + 6$
7. $f(x) = x^2 + 4x$ and $g(x) = 7x + 18$
8. $f(x) = 4x + 4$ and $g(x) = 16x + 4$

16.11 Modeling Data with Quadratic Functions (DOK 2)

When the change in a dependent variable changes at a constant rate in relation to the change in an independent variable, the relationship can be modeled with a quadratic function.

Example 16: A backyard swimming pool originally containing 128 m^3 of water developed a small leak, so the amount of water in the swimming pool decreased with time. The table below shows the amount of water that was in the swimming pool after certain numbers of hours had passed since the leak began.

Hours Since Leak Began	Amount of Water in Pool in m^3
0	128
1	126
2	120
3	110
4	96
5	78
6	56
7	30
8	0

Can the relationship between the number of hours after the leak began and the amount of water in the pool be modeled with a quadratic function? If so, plot the data and draw the parabola.

Step 1: Determine if the relationship can be modeled with a quadratic function.
From hour 1 to hour 2, the change in the amount of water in the pool was -6 m^3, from hour 2 to hour 3 it was -10 m^3, from hour 3 to hour 4 it was -14 m^3, and so on. The change in the amount of water in the pool is changing at a rate of -4 m^3 per hour. Since the change in the dependent variable decreases at a constant rate in relation to an increase in the independent variable, the relationship can be modeled with a quadratic function.

Step 2: Plot the data and draw the line.
The number of hours after the leak began is the independent variable (x), and the amount of water in the pool is the dependent variable (y).

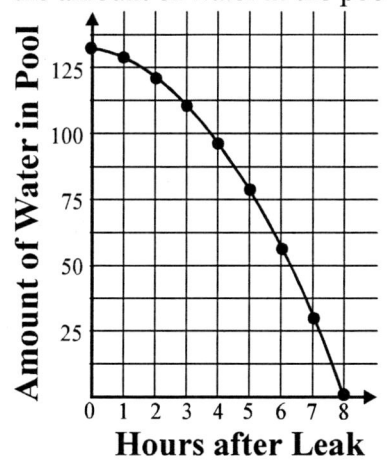

Copyright © American Book Company

287

Chapter 16 Combining and Comparing Functions

Example 17: A blog has been online for 5 days, and the change in the number of comments posted to the blog from one day to the next has been increasing at a constant rate. The relationship between the number of days the blog has been online and the total number of comments posted can be modeled with a quadratic function as shown below. What is the equation of the quadratic function?

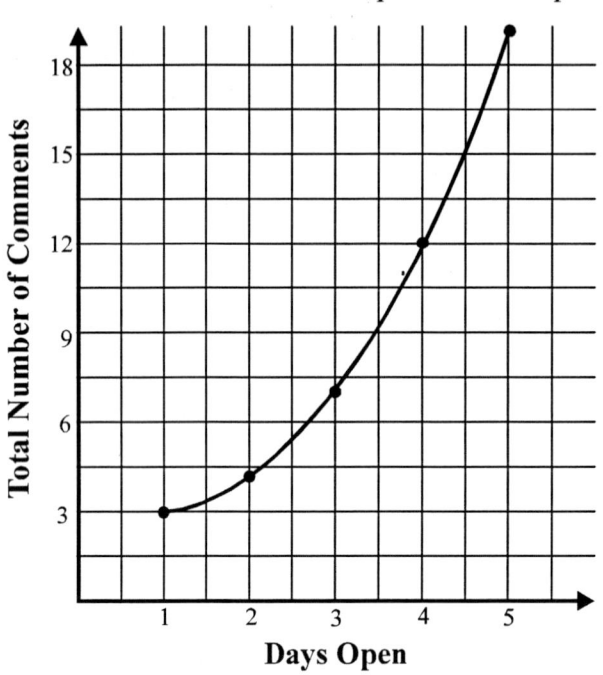

Total Number of Comments vs *Days Open*

Step 1: Find the vertex of the parabola.

Looking at the graph, we see the right side of a parabola with a vertex of $(1, 3)$. If this is the case, the equation of the parabola is $y = a(x - 1)^2 + 3$, where a is a constant.

Step 2: Substitute one point into the equation of the parabola and solve for a.

One point on the parabola is $(4, 12)$. Substitute and solve for a.

$12 = a(4 - 1)^2 + 3$

$12 = a(3)^2 + 3$

$12 = 9a + 3$

$12 - 3 = 9a + 3 - 3$

$9 = 9a$

$a = 1$

Since $a = 1$, the equation of the parabola is $y = 1(x - 1)^2 + 3$, or $y = (x - 1)^2 + 3$.
If all the other points on the parabola are tested, they all satisfy this equation, so the equation is, in fact, correct.

16.11 Modeling Data with Quadratic Functions (DOK 2)

Determine whether or not each of the following situations can be modeled with a quadratic function.

1. Every month the number of checks in a check book decreases by 4.

2. For every $50 decrease in the price of a computer, the change in the number of computers sold increases by 12.

3. The total number of donations made to a charitable organization is increasing at a rate of 28 per day.

4. A person's heart is beating at a rate of 70 beats per minute.

5. For every hour that passes, the change in temperature decreases by 0.2 degrees.

6. For every week that passes, the total amount of rainfall increases by 0.1 inches.

7. Every day the change in the number of visitors to a park over the previous day increases by 18.

8. For every student a college accepts, it spends $50 on facility maintenance.

9. For every point won in a debate, the change in a politician's approval rating increases by 0.5 percent.

What is the equation of the quadratic function that can be used to model the situation with each of the following pairs of data points? Assume the first data point is the vertex of the parabola.

10. (3 years, 5 km), (8 years, 55 km)

11. (1 mile, 6 min), (9 miles, 70 min)

12. (3 signs, 10 people), (4 signs, 9 people)

13. ($3.90, 44 units), ($4.30, 60 units)

14. (2 games, 7 points), (6 games, 55 points)

15. (5 cases, $70.00), (7 cases, $58.00)

Chapter 16 Combining and Comparing Functions

Chapter 16 Review

Determine where the graphs of the following functions intersect. (DOK 2)

1. $f(x) = \left|\dfrac{x^2}{4}\right|$, $g(x) = 3\log(x+7)$

2. $f(x) = -2x^2 + 4$, $g(x) = 2^x - 4$

3. $f(x) = |4x - 7|$, $g(x) = \dfrac{x^2 + 7x - 8}{(x-1)^2}$

4. $f(x) = \dfrac{2x}{3} + \dfrac{8}{3}$, $g(x) = 3x + 5$

For what value(s) of x are the following functions equivalent? (DOK 2)

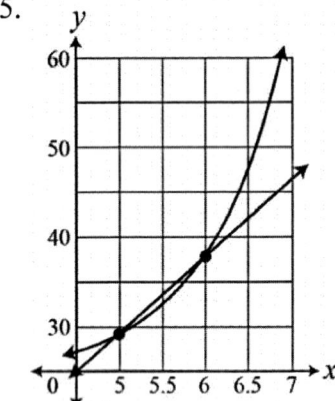

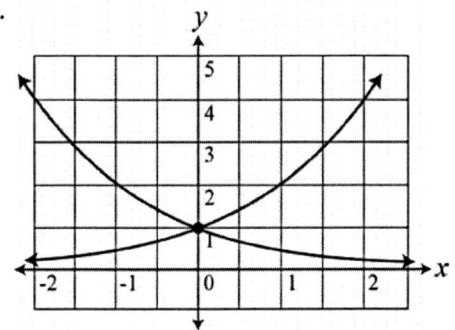

5.

7.

6.

x	$f(x)$	$g(x)$
0	9.9	-3
1	11.8	-1
2	11.8	1
3	9.9	3
4	8.1	5
5	7	7
6	6.4	9

8.

x	$f(x)$	$g(x)$
-1	$\frac{1}{2}$	1
0	1	0
1	2	1
2	4	4
3	8	9
4	16	16
5	32	25
6	64	36

9. What is the formula for finding the average rate of change of a function over a given interval?

10. Determine the average rate of change of the following function over the given interval:
$f(x) = x^6 - 9x^3 + 8x + 3$, $[4, 7]$

11. If a new sports car is valued at $21,000 and five years later it is valued at $15,000 then what is the average rate of change of its value during these five years?

12. After 6 seconds, a car had accelerated to 40 mph, and after 10 seconds the car had accelerated to 60 mph. What is the average rate of change of the speed of the car with respect to time?

Chapter 16 Review

13. Which function has a smaller slope?

   $f(x) = 3.2x + 9$ or

x	$g(x)$
-2	4.4
-1	6.7
0	9
1	11.3
2	13.6
3	15.9

14. Which function has the greater rate of change over the given interval?

   $f(x) = 7x - 9; [1, 3]$ or $g(x)$:

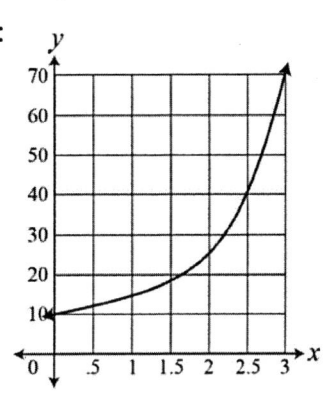

15. Which function has the lower y-intercept?

x	$f(x)$
-2	42
-1	30
0	20
1	12
2	6
3	2

   or $g(x)$:

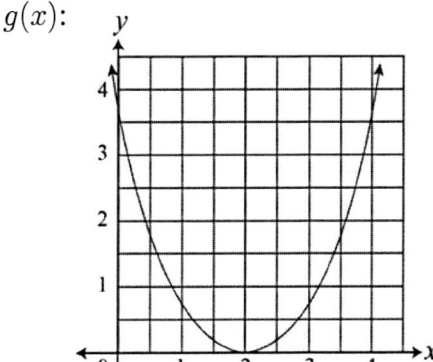

(DOK 3)

16. Which linear function has the steeper slope? $f(x)$ is a function that represents a hot air balloon descending at a constant rate of 2 feet per second. $g(x) = -x + 30$

17. A movie rental company charges $5 to rent a movie, represented by the function $r = 5$. There is a fee of $2.45 per day it is late, represented by the function $l = 2.45d$. What is the function that represents the total cost, C, of renting a movie if it is not returned on time?

18. Your investment, b, into your savings account is $700 such that $s = 700$. The savings account has an interest rate, r, of 12%, compounded annually, represented by the function $r = 1.12t$. Find the total amount, B, in the account as a function of time, t?

Chapter 16   Combining and Comparing Functions

19. The number, b, of bicycles produced by some factory in one day after, t, hours is given by $b = 40t - 13t2$. If the cost, C, in dollars of producing, b, bicycles is: $C(b) = 14000 + 75b$. What is the cost, C, as a function of time, t?

20. The price, p, of produce can be purchased at the market per pound, f, according to the equation $p = 1.2f + 4.7$. The amount of tax, t, due per pound of produce f can be calculated by $t = 0.03f$. What is the total amount, A, a customer would pay for f pounds of produce?

Solve the following problems about function symmetry. (DOK 1)

21. True or False. If a function $f(x)$, $f(-x) = -f(x)$ is true, then the function is odd.

22. True or False. If a function $f(x)$, $f(-x) = f(x)$ is true, then the function is even.

23. Fill in the blank. When all exponents in a function are ____, the function will be even.

24. Is the function $f(x) = x^6 + x^4 + 7$ even, odd, or neither?

25. Is the function $f(x) = x^5 + x^3 + 2$ even, odd, or neither?

26. Fill in the blank. When all exponents in a function are ____, the function will be odd.

27. Is the function $f(x) = x^4 - x + 2$ even, odd, or neither?

State whether the graphs are symmetrical with respect to the x-axis, y-axis, origin, or neither. (DOK 2)

28.

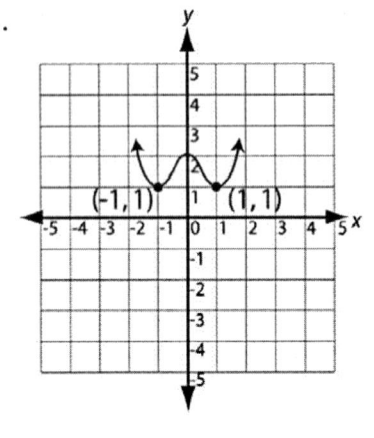

29.

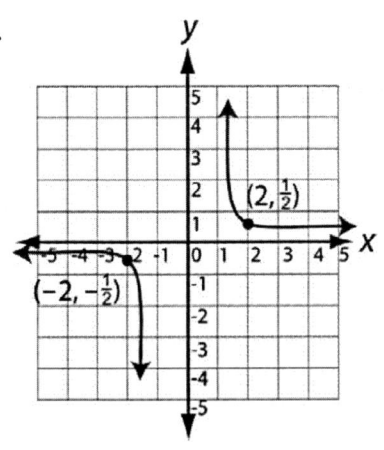

Determine the equation of the parabola in the form $y = ax^2 + bx + c$ given the focus and directrix. (DOK 3)

30. Focus: $(1, 2)$, Directrix: $y = 4$

31. Focus: $(7, 7)$, Directrix: $y = 5$

32. Focus: $(0, 1)$, Directrix: $y = 3$

33. Focus: $(-3, 4)$, Directrix: $y = -1$

Chapter 16 Test

1 Determine where the graphs of the following functions intersect:
$f(x) = x^2 - 9, g(x) = \dfrac{x-3}{x+3}$

A $x = -4, x = -2$
B $x = 4, x = 2$
C $x = -4, x = -2, x = 3$
D $x = -3, x = 2, x = 4$
(DOK 2)

2 Determine where the graphs of the following functions intersect:
$f(x) = log(2x + 6), g(x) = 6x$

A $x = -2, x = 0$
B $x = -2.494, x = 0.36$
C $x = 2.494, x = 0.36$
D $x = 0, x = 2$
(DOK 2)

3 Determine where the graphs of the following functions intersect:
$f(x) = |-2x^3 - 8x|, g(x) = \frac{1}{2}x + 10$

A $x = -1, x = 1$
B $x = -0.965, x = 1.036$
C $x = -1, x = 0$
D $x = -1.036, x = 0.965$
(DOK 2)

4 Determine where the graphs of the following functions intersect:
$f(x) = e^{2x}, g(x) = \dfrac{1}{x}$

A $x = -\dfrac{1}{e}, x = \dfrac{1}{e}$
B $x = -e, x = e$
C $x = -1, x = 1$
D $x = 0$
(DOK 2)

5 For what value(s) of x are the following functions equivalent?

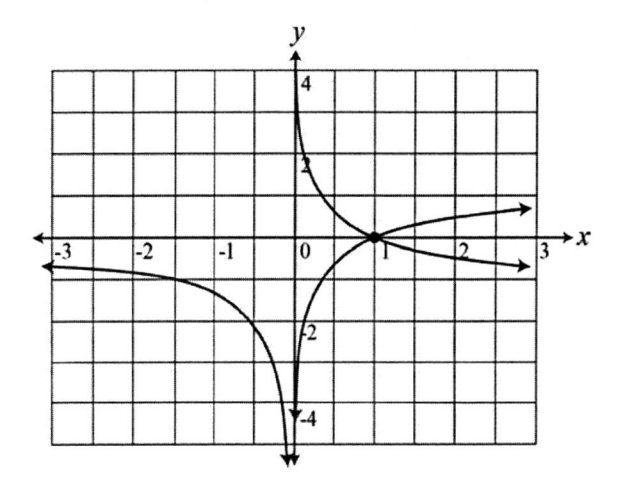

A $x = 1$
B $x = -1$
C $x = 0, x = -1$
D $x = 0, x = 1$
(DOK 2)

6 For what value(s) of x are the following functions equivalent?

x	$f(x)$	$g(x)$
-2	8	-8
-1	16	-1
0	undefined	-1
1	16	1
2	8	8
3	5.33	27
4	4	64

A $x = -2$
B $x = 21$
C $x = 8$
D $x = 4$
(DOK 2)

Chapter 16 Combining and Comparing Functions

7 For what value(s) of x are the following functions equivalent?

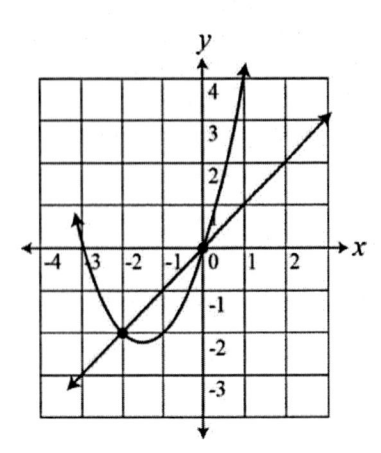

A $x = 0$

B $x = -2, x = 0$

C $x = 0, x = 2$

D $x = 0, x = 1$

(DOK 2)

8 For what value(s) of x are the following functions equivalent?

x	$f(x)$	$g(x)$
-2	undefned	-1.5
-1	undefned	-2
0	undefned	undefned
1	0	0
2	-0.301	-0.5
3	-0.4771	-0.6667
4	-0.6021	-0.75

A $x = $ undefned, $x = 0$

B $x = 0, x = 1$

C $x = 0$

D $x = 1$

(DOK 2)

9 Determine the average rate of change of the following function over the given interval:

$f(x) = x^2 + 4x - 7, [5, 8]$

A 14

B 17

C 38

D 89

(DOK 2)

10 The number of American cell phone users increased from 10 million in 1990 to 34 million in 1993. Find the average rate of change in number of users per year between 1990 and 1993.

A 24

B 24 million

C 8

D 8 million

(DOK 3)

11 The average local monthly electricity bill decreased from $210 in 2002 to $140 in 2007. Find the average rate of change of the monthly electricity bill between 2002 and 2007.

A $14

B $50

C −$14

D −$50

(DOK 3)

12 David goes camping in the mountains. When he pitches his tent at 6:00 p.m., the temperature is $82°F$. When he wakes up the next morning at 7:00 am, the temperature is $56°F$. What is the average rate of change in the temperature?

A $2°F$

B $-2°F$

C $26°F$

D $-26°F$

(DOK 3)

13 Which function has the greater rate of change over the given interval?

A $f(x) = 3x; [0, 3]$

B

x	$f(x)$
0	2
1	6
2	12
3	20

C $h(x) = 2^2$

D

x	$k(x)$
0	5
1	6
2	7
3	8

(DOK 3)

14 Which function has the smaller y-intercept?

A $f(x) = 8x + 2$

B $g(x)$:

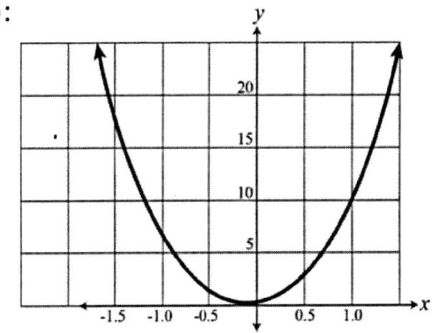

C $h(x) = \frac{1}{2}x + 5$

D

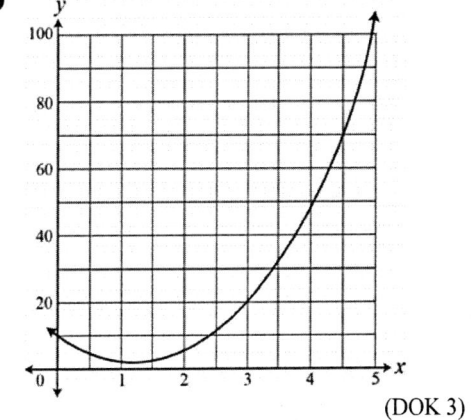

(DOK 3)

Chapter 16 Combining and Comparing Functions

15 Which function has the steeper slope?

A

x	$f(x)$
0	2
1	9
2	16
3	23
4	30
5	37

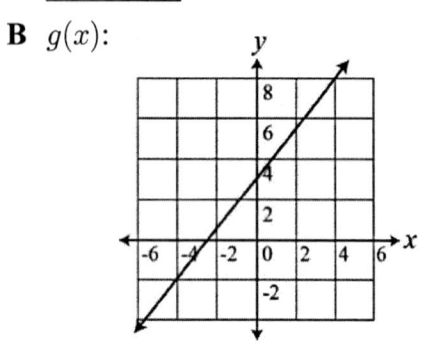

B $g(x)$:

(graph of a line passing through approximately (0,3) with positive slope on axes showing x from -6 to 6 and y from -2 to 8)

C

x	$f(x)$
0	1
1	4
2	7
3	10
4	13
5	16

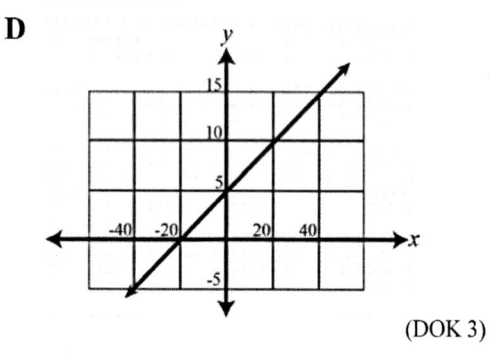

D

(graph of a line with positive slope on axes showing x from -40 to 40 and y from -5 to 15)

(DOK 3)

16 Which function has the smaller rate of change over the given interval?

A $f(x)$ is a function that represents the exponential growth of bacteria. There is initially one cell. The bacteria cells double every hour. After 5 hours there are 32 bacteria cells.

B

x	$g(x)$
0	12
1	21
2	36
3	57
4	84
5	117

C $h(x)$ is a function that represents a linear growth in a savings account. Every 7 days Mark gets paid for doing his chores. Initially he had $20 when he opened his bank account. Now after 3 weeks he has $41 in his savings account.

D

x	$k(x)$
0	−3
1	5
2	13
3	21
4	29
5	37

(DOK 3)

17 The price, p, of a telescope and the quantity, x, sold, obey the equation $p = x + 24$ and the cost, C, of producing x units is $C = \dfrac{x^2 + 2012}{17}$. Find the cost, C, as a function of price, p.

A $(C + P)(x) = \dfrac{x^2 + 17x + 2420}{17}$

B $(C - P)(x) = \dfrac{x^2 - 17x + 1604}{17}$

C $(C \times P)(x) = \dfrac{x^3 + 24x^2 - 2012x + 48288}{17}$

D $(C \circ P)(x) = \dfrac{x^2 + 48x + 2588}{17}$

(DOK 3)

18 A discount, d, of 45% on price, p, represented by the function: $d = 0.45p$, is offered on certain products in a store. Assuming someone purchases one of these items, and the full price, f, is represented by the function: $f = p$, what is the function that represents the amount, A, the customer would pay for this item?

A $(d + f)(p) = 1.45p$

B $(d - f)(p) = 0.55p$

C $(d/f)(p) = 0.45$

D $(d \circ f)(p) = 0.45p$

(DOK 3)

19 The total profit, t, made on all the tickets to attend a hockey game can be determined using the function: $t = 12.40p - 2$ where p represents the number of people buying tickets. The total number of tickets, m, sold is represented by the function: $m = p$. What is the function that represents the profit, R, made on each individual ticket?

A $R(p) = 11.40p - 2$

B $R(p) = 13.40p - 2$

C $R(p) = 12.40p^2 - 2p$

D $R(p) = \dfrac{12.40p - 2}{p}$

(DOK 3)

20 The surface area, S, of a spherical hot-air balloon is given by $S(r) = 4\pi r^2$, where, r, is the radius of the balloon. If the radius, r, increases with time, t, according to the formula, find the surface area, S, of the balloon as a function of time, t.

A $S(t) = 4\pi t^2$

B $S(t) = \frac{2}{3}\pi t^2$

C $S(t) = \frac{1}{9}\pi t^4$

D $S(t) = \dfrac{16\pi^2 t^4}{6}$

(DOK 3)

21 Is the following function even, odd, neither, or not enough information?

$f(x) = x^{10} + x^6 + x^2 + 4$

A even
B odd
C neither
D not enough information

(DOK 2)

Chapter 16   Combining and Comparing Functions

22 Is the following function even, odd, neither, or not enough information?

$f(x) = x^5 + x^4 + x^2 + x + 4$

A even
B odd
C neither
D not enough information

(DOK 2)

23 Is the following function even, odd, neither, or not enough information?

$f(x) = 2x^3 - 7x$

A even
B odd
C neither
D not enough information

(DOK 2)

24 Is the following function symmetrical?

$f(x) = 6x^2 - 13$

A yes, about the y-axis
B yes, about the x-axis
C yes, about the origin
D no

(DOK 2)

25 Is the following function symmetrical?

$f(x) = 7x^5 - 4x - 1$

A yes, about the y-axis
B yes, about the x-axis
C yes, about the origin
D no

(DOK 2)

26 Is the following graph symmetrical?

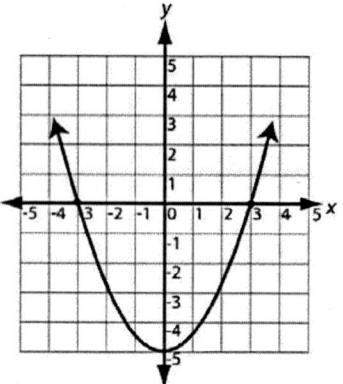

A yes, about the y-axis
B yes, about the x-axis
C yes, about the origin
D no

(DOK 2)

27 Which of the following equations is that of a parabola with a focus at $(2, 2)$ and a directrix $y = -2$?

A $y = \frac{1}{4}x^2 + x - 1$
B $y = \frac{1}{4}x^2 - x + 1$
C $y = \frac{1}{8}x^2 - \frac{1}{2}x + \frac{1}{2}$
D $y = \frac{1}{8}x^2 + \frac{1}{2}x - \frac{1}{2}$

(DOK 3)

28 Which of the following situations can best be modeled with a quadratic function?

A For every month that passes, the number of songs stored on an MP3 player increases by 22.
B The change in the number of students enrolled in a college over the previous semester is increasing by 58 each semester.
C The amount of yogurt consumed by citizens of a country is increasing by 25,000 ounces every month.
D The average number of minutes high school students sleep every night is decreasing by 3 minutes every 5 years.

(DOK 2)

Chapter 17
Sets

This chapter covers the following CCGPS standard(s):

	Content Standard
Conditional Probability and the Rules of Probability	S.CP.1

17.1 Set Notation (DOK 2)

A **set** contains an object or objects that are clearly identified. The objects in a set are called **elements** or **members**. A set can also be **empty**. The symbol ∅ is used to denote the empty set.

Members of a set can be described in two ways. One way is to give a **rule** of description that clearly defines the members. Another way is to make a **roster** of each member of the set, mentioning each member only <u>once</u> in any order.

$$\text{Rule} \qquad \qquad \text{Roster}$$
$$\{\text{the letters in the word "school"}\} = \{s, c, h, o, l\}$$

A **complement** of a set is everything that does not belong to the set. If there exists a set A, then the complement is denoted by A' or $\overline{A}$. Take the example above. Let $A = \{$the letters in the word "school"$\}$. The complement is the set that contains every other letter in the alphabet that is not in the word "school". $A' = \{a, b, d, e, f, g, i, j, k, m, n, p, q, r, t, u, v, w, x, y, z\}$.

The symbol $\in$ is used to show that an object is the member of a set. For example, $3 \in \{1, 2, 3, 4\}$.

The symbol $\notin$ means "is not a member of." For example, $8 \notin \{1, 2, 3, 4\}$.

The symbol $\neq$ means "is not equal to".

For example, **{the cities in Texas}** $\neq$ **{New York, Philadelphia, Nashville}**.

Read each of the statements below, and tell whether they are true or false.

1. {the days of the week} = {January, March, December}

2. {The first five letters of the alphabet} = {a, b, c, d, e}

3. $5 \notin \{$all odd numbers$\}$

4. Friday $\in$ {the days of the week}

5. {the last three letters of the alphabet} $\neq \{x, y, z\}$

6. $t \notin \{$the letters in the word "yellow"$\}$

7. {the letters in the word "funny"} $\in$ {the letters of the alphabet}

8. {The letters in the word "Alabama"} = {a, b, l, m}

9. {living unicorns} $\neq \emptyset$

10. {the letters in the word "horse"} $\neq$ {the letters in the word "shore"}

Copyright © American Book Company

Chapter 17 Sets

Identify each set by making a roster (see above). If a set has no members, use ∅.

11. {the letters in the word "hat" that are also in the word "thin"}

12. {the letters in the word "Mississippi"}

13. {the provinces of Canada that border the state of Texas}

14. {the letters in the word "kitchen" and also in the word "dinner"}

15. {the days of the week that have the letter "n"}

16. {the letters in the word "June" that are also in the word "April"}

17. {the letters in the word "instruments" that are not in the word "telescope"}

18. {the digits in the number "19,582" that are also in the number "56,871"}

17.2 Subsets (DOK 2)

If every member of set A is also a member of set B, then set A is a **subset** of set B.

The symbol $\subseteq$ means "is a subset of."

For example, every member of the set $\{1, 2, 3, 4\}$ is a member of the set $\{1, 2, 3, 4, 5, 6, 7\}$. Therefore, $\{1, 2, 3, 4\} \subseteq \{1, 2, 3, 4, 5, 6, 7\}$. The relationship is pictured in the following diagram:

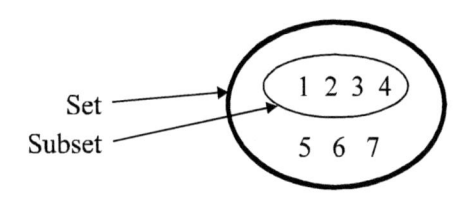

Set → 1 2 3 4
Subset → 5 6 7

The symbol $\not\subseteq$ means "is **not** a subset of." For example, not every member of the set {Ann, Sue} is a member of the set {Ann, John, Cindy}. Therefore, {Ann, Sue} $\not\subseteq$ {Ann, John, Cindy}.

Read each of the statements below, and tell whether they are true or false.

1. {a, b, c} $\not\subseteq$ {a, b, c, d}
2. {1, 3, 5, 7} $\subseteq$ {1, 2, 4, 5, 6, 7}
3. {a, e, i, o, u,} $\subseteq$ {the vowels of the alphabet}
4. {dogs} $\subseteq$ {poodles, bull dogs, collies}
5. {fruit} $\not\subseteq$ {apples, grapes, bananas}
6. {10, 20, 30, 40} $\subseteq$ {20, 30, 40, 50}
7. {English, Spanish, French} $\subseteq$ {languages}
8. {duck, swan, penguin} $\not\subseteq$ {mammals}
9. {1, 2, 5, 10} $\subseteq$ {whole numbers}
10. {Atlanta} $\not\subseteq$ {U.S. state capitals}

The set {Pam} has two subsets: {Pam} and ∅. The set {Emily, Brad} has 4 subsets: {Emily}, {Brad}, {Emily, Brad}, and ∅. The ∅ is a subset of any set.

For each of the following sets, list all of the possible subsets.

11. {a}   12. {1, 2}   13. {r, s, t}   14. {1, 3, 5, 7}   15. {Joe, Ed}

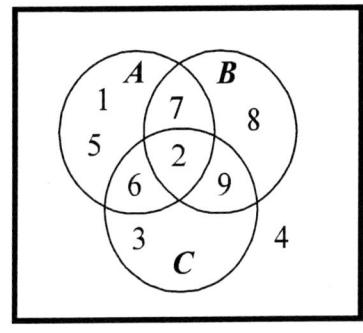

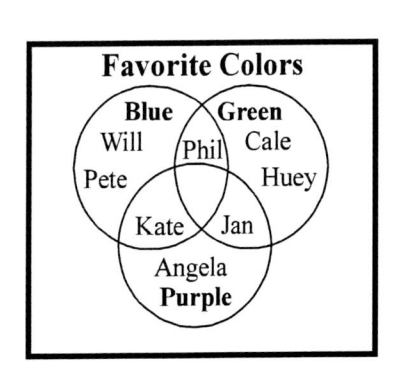

17.3 Intersection of Sets (DOK 2 & 3)

To find the **intersection** of two sets, you need to identify the members that the two sets share in common. The symbol for intersection is ∩. A **Venn diagram** shows how sets intersect.

Roster

$\{2, 4, 6, 8\} \cap \{4, 6, 8, 10\} = \{4, 6, 8\}$

The shaded area is the intersection of the two sets. It shows which numbers both sets have in common.

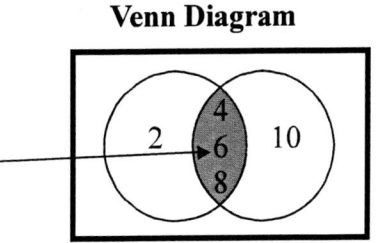

Venn Diagram

Find the intersection of the following sets.

1. {Ben, Jan, Dan, Tom} ∩ {Dan, Mike, Kate, Jan} =
2. {pink, purple, yellow} ∩ {purple, green, blue} =
3. {2, 4, 6, 8, 10} ∩ {1, 2, 3, 4, 5, 6, 7, 8, 9, 10} =
4. {a, e, i, o, u} ∩ {a, b, c, d, e, f, g, h, i, j, k} =
5. {pine, oak, walnut, maple} ∩ {maple, poplar} =
6. {100, 98, 95, 78, 62} ∩ {57, 82, 95, 98, 99} =
7. {orange, kiwi, coconut, pineapple} ∩ {pear, apple, orange} =

Look at the Venn diagram at the right to answer the questions below. Show your answers in roster form. Do the problem in parentheses first.

8. $A \cap B =$
9. $(A \cap B) \cap C =$
10. $A \cap C =$
11. $B \cap C =$
12. $(A \cap C) \cap B =$
13. blue ∩ green =
14. (purple ∩ blue) ∩ green =
15. blue ∩ purple =
16. (blue ∩ green) ∩ purple =
17. green ∩ purple =

Chapter 17 Sets

17.4 Union of Sets (DOK 2 & 3)

The **union** of two sets means to put the members of two sets together into one set without repeating any members. The symbol for union is ∪.

$$\{1, 2, 3, 4\} \cup \{3, 4, 5, 6\} = \{1, 2, 3, 4, 5, 6\}$$

The union of theses two sets is the shaded area in the Venn diagram below.

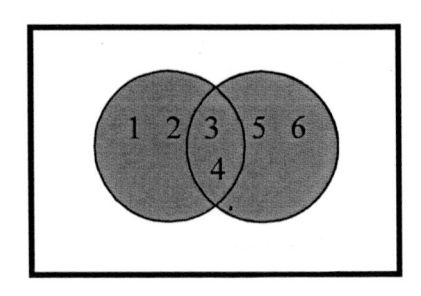

Find the union of the sets below.

1. {apples, pears, oranges} ∪ {pears, bananas, apples} =

2. {5, 10, 15, 20, 25} ∪ {10, 20, 30, 40} =

3. {Ted, Steve, Kevin, Michael} ∪ {Kevin, George, Kenny} =

4. {raisins, prunes, apricots} ∪ {peanuts, almonds, coconut} =

5. {sales, marketing, accounting} ∪ {receiving, shipping, sales} =

6. {beef, pork, chicken} ∪ {chicken, tuna, shark} =

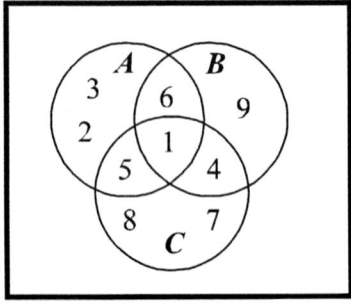

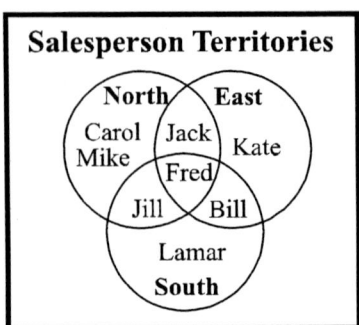

Refer to the following Venn diagrams to answer the questions below. Identify each of the following sets by roster.

7. $A \cup C =$

8. $C \cup B =$

9. $B \cup A =$

10. $A \cup B \cup C =$

11. North ∪ East =

12. North ∪ South =

13. East ∪ South =

14. North ∪ East ∪ South =

Salesperson Territories

17.5 Reading Venn Diagrams (DOK 3)

Venn diagrams are a visual way to see two or more variables. They show whether or not the variables intersect. A Venn diagram also shows the union of the events.

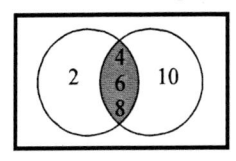

$\{2, 4, 6, 8\} \cap \{4, 6, 8, 10\} = \{4, 6, 8\}$
The shaded area is the intersection of the two sets.

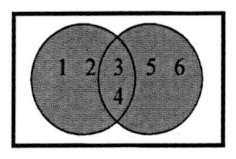

$\{1, 2, 3, 4\} \cup \{3, 4, 5, 6\} = \{1, 2, 3, 4, 5, 6\}$
The shaded area is the union of the two sets.

Example 1: Below is a Venn diagram of how many students play football and baseball. Find the intersection and the union of the two events below.

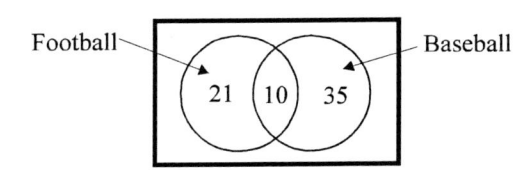

Step 1: First, you must figure out how many students play both sports by interpreting the diagram. Ten students play both sports, since they are counted on the football and baseball side. **The intersection is 10 students.**

Step 2: To find the union, you must add all the players together.
The union is $21 + 35 + 10 = 66$ **students**.

Look at the Venn diagram to answer the questions below. Note: 435 = the number of students that just have dogs.

1. Dogs ∩ Cats

2. Dogs ∪ Birds

3. Cats ∩ Birds

4. Dogs ∩ Cats ∩ Birds

5. Dogs ∪ Cats

6. Dogs ∪ Cats ∪ Birds

7. Dogs ∩ Birds

8. Cats ∪ Birds

9. Dogs ∪ (Cats ∪ Birds)

10. (Dogs ∩ Cats) ∩ Birds

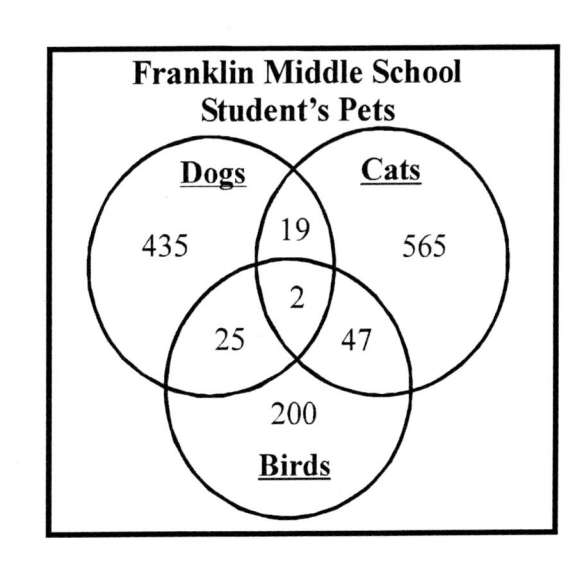

Chapter 17 Sets

Chapter 17 Review

Read each of the statements below, and tell whether they are true or false. (DOK 2)

1. {odd whole numbers} ⊆ {all whole numbers}
2. {yearly seasons} ≠ {spring, summer, fall, winter}
3. {Monday, Tuesday, Wednesday} ⊆ {days of the week}
4. United States of America ∉ {countries in North America}
5. {pink, black} ⊈ {primary colors}
6. {plants with red flowers} = ∅
7. {letters in the word "subsets"} = {b, u, s, e, t}
8. Milky Way ∉ {galaxies in the universe}
9. {Houston, Dallas } ⊆ {cities in Texas }
10. George Washington ∈ {former presidents of the United States}

Complete the following statements. (DOK 2)

11. {3, 6, 9, 12, 15} ∩ {0, 5, 10, 15, 20} =
12. {Felix, Mark, Kate} ∪ {Mark, Carol, Jack} =
13. {letters in "perfect"} ∩ {letters in "profit"} =
14. {Rome, London, Paris} ∩ {Italy, England, France} =
15. {black, white, gray} ∪ {red, white, blue} =
16. {1, 2, 3, 4, 5, 6} ∪ {2, 4, 6, 8, 10, 12} =

Refer to the Venn diagram to complete the following statements. Answers should be in roster form. (DOK 3)

17. basketball ∪ football ∪ baseball =
18. basketball ∩ football =
19. (football ∩ basketball) ∩ baseball =
20. baseball ∪ basketball =
21. football ∪ baseball =
22. baseball ∩ football =
23. basketball ∩ baseball =
24. football ∪ basketball =

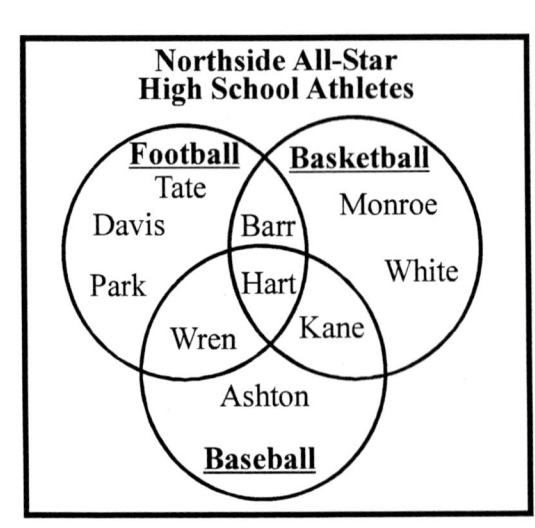

Northside All-Star High School Athletes

Football: Tate, Davis, Park
Football ∩ Basketball: Barr
Basketball: Monroe, White
Football ∩ Baseball: Wren
Basketball ∩ Baseball: Kane
Football ∩ Basketball ∩ Baseball: Hart
Baseball: Ashton

Chapter 17 Test

Use the following Venn diagram to answer questions 1 and 2.

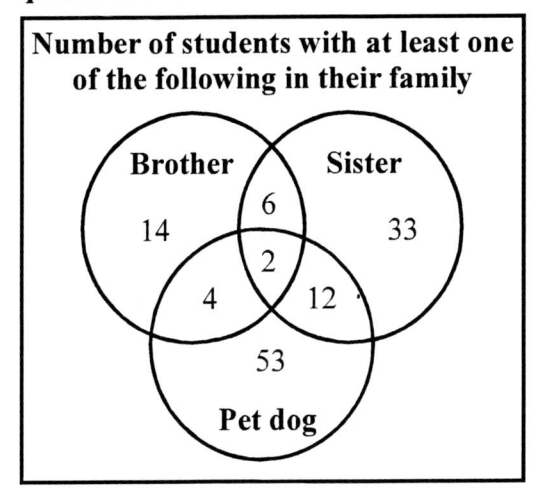

Number of students with at least one of the following in their family

Brother: 14
Sister: 33
Pet dog: 53
Brother ∩ Sister: 6
Brother ∩ Pet dog: 4
Sister ∩ Pet dog: 12
Brother ∩ Sister ∩ Pet dog: 2

1 In the Venn diagram above, how many members are in the following set? {sister ∪ pet dog}

- **A** 12
- **B** 14
- **C** 100
- **D** 110 (DOK 3)

2 In the Venn diagram above, how many members are in the following set? {brother} ∩ {sister}

- **A** 6
- **B** 8
- **C** 53
- **D** 55 (DOK 3)

3 Which of the following sets equals ∅?

- **A** {all whole numbers}
- **B** {all letters in the alphabet}
- **C** {all fish with feathers}
- **D** {all flowering plants} (DOK 2)

4

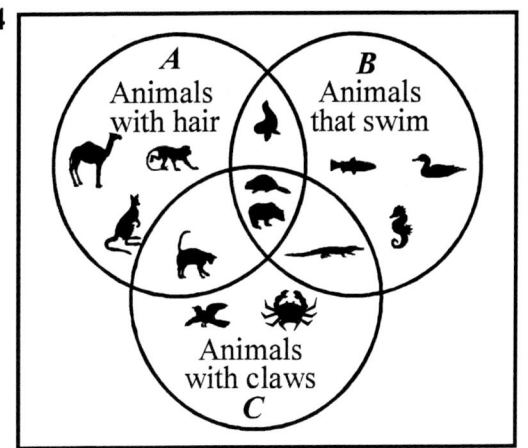

A: Animals with hair
B: Animals that swim
C: Animals with claws

According to the above diagram, which one of the following statements is false?

- **A** $A \cap B = \{\text{🐇}, \text{🐻}, \text{🦫}\}$ 
- **B** 🐊 $\in B$
- **C** $\{\text{🦘}, \text{🐻}, \text{🐪}\} \not\subseteq A$
- **D** $B \cup C \neq \emptyset$ (DOK 3)

5 {Karen, John, Sue} ∩ {John, Perry, Kay} =

- **A** ∅
- **B** {John}
- **C** {Karen, Sue, Perry, Kay}
- **D** {Karen, John, Sue, Perry, Kay} (DOK 2)

6 Friday _____ {the days of the week}

- **A** =
- **B** ≠
- **C** ∈
- **D** ∉ (DOK 2)

7 {Atlanta} _____ {state capitals}

- **A** ⊆
- **B** ⊈
- **C** ∩
- **D** ∉ (DOK 2)

Copyright © American Book Company

Chapter 17   Sets

8 $\{1, 3, 5, 7, 9\} \cap \{1, 2, 3, 4, 5, 6, 7, 8, 9, 10\} =$

A $\{3\}$
B $\{1, 3\}$
C $\{1, 3, 5, 7, 9\}$
D $\{1, 2, 3, 4, 5, 6, 7, 8, 9, 10\}$ (DOK 2)

9 $\{\text{pink, purple, yellow}\} \cap \{\text{yellow, orange}\} =$

A {pink, purple, orange}
B {pink, purple, yellow, orange}
C {yellow}
D {orange}   (DOK 2)

10 {vowels} ∩ {letters of the alphabet} =

A a, e, i, o, u
B all letters of the alphabet
C a, e, i, o, u, y
D all consonants   (DOK 2)

11 Which is the symbol for union?

A ∈
B ⊂
C ∩
D ∪   (DOK 1)

12 $\{5, 10, 15, 20, 25\} \cup \{10, 20, 30, 40\} =$

A $\{5, 10, 15, 20, 25, 30, 40\}$
B $\{10, 20\}$
C $\{5, 15, 25\}$
D $\{10, 20, 30, 40\}$ (DOK 2)

13 {eggs, bacon} ∪ {hot dog, bacon, pork} =

A {eggs, bacon, hot dog, pork}
B {eggs, hot dog, pork}
C {meat}
D {bacon}   (DOK 2)

14 What is the complement of A?
$A = \{a, b, c, d, e, f, g, m, n, o, q, r, s, v, w, y, z\}$?

A {a, b, c, d, e, f, g, m, n, o, q, r, s, v, w, y, z}
B {h, i, j, k, l, p, t, u, x}
C {i, x}
D {a, e, i, o, u}   (DOK 2)

Chapter 18
Probability

This chapter covers the following CCGPS standard(s):

	Content Standards
Conditional Probability and the Rules of Probability	S.CP.1, S.CP.3, S.CP.5, S.CP.6, S.CP.7
Using Probability to Make Decisions	S.CP.4 S.CP.2

18.1 Probability Terms (DOK 2)

Probability - the branch of mathematics that calculates the chance something will or will not happen.

Independent Events - the outcome of one event does not influence the outcome of the second event.

Dependent Events - the outcome of one event does influence the outcome of the second event.

Mutually Exclusive Events - two events that have no outcomes in common. These events cannot occur at the same time.

Conditional Probability - probability that a second event will happen, given that the first event has already occurred.

Equally Likely Outcomes - all outcomes of the event have the same chance of occurring.

Expected Value - the mean of a random variable.

Population - an entire group or collection about which we wish to draw conclusions.

Census - a count of the entire population

Sample - units selected to study from the population.

Sample Space - the set of all possible outcomes.

$P(A)$ - notation used to mean the probability of outcome 'A' occurring.

$P(A^c)$ **(Complement)** - notation used to mean the probability that outcome 'A' does not occur.

$P(A^c) = 1 - P(A)$

Chapter 18 Probability

18.2 Probability (DOK 2)

Probability is the chance something will happen. Probability is most often expressed as a fraction, a decimal, a percent, or can also be written out in words.

Example 1: Billy has 3 red marbles, 5 white marbles, and 4 blue marbles on the floor. His cat comes along and bats one marble under the chair. What is the **probability** it is a red marble?

Step 1: The number of red marbles will be the top number of the fraction. ⟶ 3

Step 2: The total number of marbles is the bottom number of the fraction. ⟶ 12

The answer may be expressed in lowest terms. $\frac{3}{12} = \frac{1}{4}$.

Expressed as a decimal, $\frac{1}{4} = 0.25$, as a percent, $\frac{1}{4} = 25\%$, and written out in words, $\frac{1}{4}$ is one out of four.

Example 2: Determine the probability that the pointer will stop on a shaded wedge or the number 1.

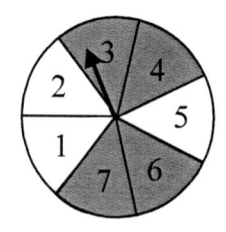

Step 1: Count the number of possible wedges that the spinner can stop on to satisfy the above problem. There are 5 wedges that satisfy it (4 shaded wedges and one number 1). The top number of the fraction is 5.

Step 2: Count the total number of wedges, 7. The bottom number of the fraction is 7. The answer is $\frac{5}{7}$ or **five out of seven.**

Example 3: Refer to the spinner above. If the pointer stops on the number 7, what is the probability that it will **not** stop on 7 the next time?

Step 1: Ignore the information that the pointer stopped on 7 the previous spin. The probability of the next spin does not depend on the outcome of the previous spin. Simply find the probability that the spinner will **not** stop on 7. Remember, if P is the probability of an event occurring, $1 - P$ is the probability of an event **not** occurring (it is the complement). In this example, the probability of the spinner landing on 7 is $\frac{1}{7}$.

Step 2: The probability that the spinner will not stop on 7 is $1 - \frac{1}{7}$ which equals $\frac{6}{7}$. The answer is $\frac{6}{7}$ or **six out of seven.**

18.2 Probability (DOK 2)

Find the probability of the following problems. Express the answer as a percent.

1. A computer chooses a random number between 1 and 50. What is the probability that you will guess the same number that the computer chose in 1 try?

2. There are 24 candy-coated chocolate pieces in a bag. Eight have defects in the coating that can be seen only with close inspection. What is the probability of pulling out a defective piece without looking?

3. Seven sisters have to choose which day each will wash the dishes. They put equal-sized pieces of paper in a hat, each labeled with a day of the week. What is the probability that the first sister who draws will choose a weekend day?

4. For his garden, Clay has a mixture of 12 white corn seeds, 24 yellow corn seeds, and 16 bicolor corn seeds. If he reaches for a seed without looking, what is the probability that Clay will plant a bicolor corn seed first?

5. Mom just got a new department store credit card in the mail. What is the probability that the last digit is an odd number?

6. Alex has a paper bag of cookies that holds 8 chocolate chip, 4 peanut butter, 6 butterscotch chip, and 12 ginger. Without looking, his friend John reaches in the bag for a cookie. What is the probability that the cookie is peanut butter?

7. An umpire at a little league baseball game has 14 balls in his pockets. Five of the balls are brand A, 6 are brand B, and 3 are brand C. What is the probability that the next ball he throws to the pitcher is a brand C ball?

8. What is the probability that the spinner's arrow will land on an even number?

9. The spinner in the problem above stopped on a shaded wedge on the first spin and stopped on the number 2 on the second spin. What is the probability that it will not stop on a shaded wedge or on the 2 on the third spin?

10. A company is offering 1 grand prize, 3 second place prizes, and 25 third place prizes based on a random drawing of contest entries. If your entry is one of the 500 total entries, what is the probability you will win a third place prize?

11. In the contest problem above, what is the probability that you will win the grand prize or a second place prize?

12. A box of a dozen doughnuts has 3 lemon cream-filled, 5 chocolate cream-filled, and 4 vanilla cream-filled. If the doughnuts look identical, what is the probability of picking a lemon cream-filled?

Chapter 18 Probability

18.3 More Probability (DOK 2)

Example 4: You have a cube with one number, 1, 2, 3, 4, 5, or 6, painted on each face. What is the probability that if you throw the cube 3 times, it will land with the number 2 face up each time?

If you roll the cube once, you have a 1 in 6 chance of getting the number 2. If you roll the cube a second time, you again have a 1 in 6 chance of getting the number 2. If you roll the cube a third time, you again have a 1 in 6 chance of getting the number 2. The probability of rolling the number 2 three times in a row is:

$$\frac{1}{6} \times \frac{1}{6} \times \frac{1}{6} = \frac{1}{216}$$

Find the probability that each of the following events will occur.

There are 10 balls in a box, each with a different digit on it: 0, 1, 2, 3, 4, 5, 6, 7, 8, or 9. A ball is chosen at random and then put back in the box.

1. What is the probability that if you pick out a ball 3 times, you will get number 7 each time?

2. What is the probability you will pick a ball with 5, then 9, and then 3?

3. What is the probability that if you pick out a ball 4 times, you will always get an odd number?

4. A couple has 4 children ages 9, 6, 4, and 1. What is the probability that they are all girls?

There are 26 letters in the alphabet, allowing a different letter to be on each of 26 cards. The cards are shuffled. After each card is chosen at random, it is put back in the stack of cards, and the cards are shuffled again.

5. What is the probability that when you pick 3 cards, you would draw first a "y", then an "e", and then an "s"?

6. What is the probability that you would draw 4 cards and get the letter "z" each time?

7. What is the probability that you draw twice and get a letter in the word "random" both times?

8. If you flip a coin 3 times, what is the probability you will get heads every time?

9. Marie is clueless about 4 of her multiple-choice answers. The possible answers are A, B, C, D, E, or F. What is the probability that she will guess all four answers correctly?

18.4 Independent and Dependent Events (DOK 3)

In mathematics, the outcome of an event may or may not influence the outcome of a second event. If the outcome of one event does not influence the outcome of the second event, these events are **independent**. However, if one event has an influence on the second event, the events are **dependent**. When someone needs to determine the probability of two events occurring, he or she will need to use an equation. These equations will change depending on whether the events are independent or dependent in relation to each other. When finding the probability of two **independent** events, multiply the probability of each favorable outcome together. Independent events use the **multiplication rule**. The multiplication rule for independent events is $P(A \text{ and } B) = P(A) P(B)$.

Example 5: One bag of marbles contains 1 white, 1 yellow, 2 blue, and 3 orange marbles. A second bag of marbles contains 2 white, 3 yellow, 1 blue, and 2 orange marbles. What is the probability of drawing a blue marble from each bag?

Solution: Probability of favorable outcomes

Bag 1: $P(A) = \dfrac{2}{7}$

Bag 2: $P(B) = \dfrac{1}{8}$

Probability of a blue marble from each bag = $P(A \text{ and } B)$.

$P(A \text{ and } B) = P(A) P(B) = \dfrac{2}{7} \times \dfrac{1}{8} = \dfrac{2}{56} = \dfrac{1}{28}$

In order to find the probability of two **dependent** events, you will need to use a different set of rules. For the first event, you must divide the number of favorable outcomes by the number of possible outcomes. For the second event, you must subtract one from the number of favorable outcomes **only if** the favorable outcome is the **same**. However, you must subtract one from the number of total possible outcomes. Finally, you must multiply the probability for event one by the probability for event two.

Example 6: One bag of marbles contains 3 red, 4 green, 7 black, and 2 yellow marbles. What is the probability of drawing a green marble, removing it from the bag, and then drawing another green marble without looking?

	Favorable Outcomes	Total Possible Outcomes
Draw 1	4	16
Draw 2	3	15
Draw 1 × Draw 2	12	240

Answer: $\dfrac{12}{240}$ or $\dfrac{1}{20}$

Example 7: Using the same bag of marbles, what is the probability of drawing a red marble, removing it, and then drawing a black marble?

	Favorable Outcomes	Total Possible Outcomes
Draw 1	3	16
Draw 2	7	15
Draw 1 × Draw 2	21	240

Answer: $\dfrac{21}{240}$ or $\dfrac{7}{80}$

Chapter 18   Probability

Find the probability of the following problems. Express the answer as a fraction.

1. Prithi has two boxes. Box 1 contains 3 red, 2 silver, 4 gold, and 2 blue combs. She also has a second box containing 1 black and 1 clear brush. What is the probability that Prithi selects a red comb from box 1 and a black brush from box 2?

2. Steve Marduke has two spinners in front of him. The first one is numbered 1–6, and the second is numbered 1–3. If Steve spins each spinner once, what is the probability that the first spinner will show an odd number and the second spinner will show a "1"?

3. Carrie McCallister flips a coin twice and gets heads both times. What is the probability that Carrie will get tails the third time she flips the coin?

4. Artie Drake spins a spinner which is evenly divided into 11 sections numbered 1–11. On the first spin, Artie's pointer lands on "8". What is the probability that the spinner lands on an even number the second time he spins the spinner?

5. Leanne Davis plays a game with a street entertainer. In this game, a ball is placed under one of three coconut halves. The vendor shifts the coconut halves so quickly that Leanne can no longer tell which coconut half contains the ball. She selects one and misses. The entertainer then shifts all three around once more and asks Leanne to pick again. What is the probability that Leanne will select the coconut half containing the ball?

6. What is the probability that Jane Robelot reaches into a bag containing 1 daffodil and 2 gladiola bulbs and pulls out a daffodil bulb, and then reaches into a second bag containing 6 tulip, 3 lily, and 2 gladiola bulbs and pulls out a lily bulb?

7. Terrell casts his line into a pond containing 7 catfish, 8 bream, 3 trout, and 6 northern pike. He immediately catches a bream. What are the chances that Terrell will catch a second bream the next time he casts his line?

8. Gloria Quintero enters a contest in which the person who draws his or her initials out of a box containing all 26 letters of the alphabet wins the grand prize. Gloria reaches in, draws a "G", keeps it, then draws another letter. What is the probability that Gloria will next draw a "Q"?

9. Vince Macaluso is pulling two socks out of a washing machine in the dark. The washing machine contains three tan, one white, and two black socks. If Vince reaches in and pulls out the socks one at a time, what is the probability that he will pull out two tan socks on his first two tries?

10. John Salome has a bag containing 2 yellow plums, 2 red plums, and 3 purple plums. What is the probability that he reaches in without looking and pulls out a yellow plum and eats it, then reaches in again without looking and pulls out a red plum to eat?

18.5 Mutually Exclusive Events (DOK 3)

Events are said to be **mutually exclusive** if they don't occur at the same time (no common outcome). This means the probability of event 1 (A) and event 2 (B) both occurring is zero, $P(A$ and $B) = 0$. Mutually exclusive events use the **addition rule**.

Let A and B be events. Let $P(A) =$ the probability of event A occurring and $P(B) =$ the probability of event B occurring.

The addition rule for mutually exclusive events is $P(A$ or $B) = P(A) + P(B)$.

The rule for sets that are **not** mutually exclusive is $P(A$ or $B) = P(A) + P(B) - P(A \cap B)$.

Example 8: A pair of diced is rolled. What is the probability that the sum of the dice rolled is either a 7 or a 2?

Step 1: Find the number of outcomes for rolling a 7.
There are six outcomes for rolling a sum of seven:
$(1,6), (6,1), (2,5), (5,2), (3,4), (4,3)$
$$P(7) = \frac{\text{\# of outcomes for rolling a sum of seven}}{\text{total number of outcomes}} = \frac{6}{36} = \frac{1}{6}$$

Step 2: Find the number of outcomes for rolling a sum of 2.
There is one outcome for rolling a two:
$(1,1)$
$$P(2) = \frac{\text{\# of outcomes for rolling a sum of two}}{\text{total number of outcomes}} = \frac{1}{36}$$

Step 3: Since the sum of the dice cannot be seven and two (it must be one or the other), then the events are mutually exclusive. Use the formula $P(A$ or $B) = P(A) + P(B)$ for mutually exclusive events to find the probability of either rolling a sum of 7 or a sum of 2.
Let $A =$ rolling a sum of 7 and $B =$ rolling a sum of 2.
$P(7$ or $2) = P(7) + P(2) = \frac{1}{6} + \frac{1}{36} = \frac{7}{36}$
The probability of rolling the sum of either a 7 or a 2 is $\frac{7}{36}$.

Find the probability of each event.

1. A pair of dice is rolled. What is the probability that the sum of the dice rolled is either an 11 or a 5?

2. A pair of dice is rolled. What is the probability that the sum of the dice rolled is either a 9 or a 3?

3. A pair of dice is rolled. What is the probability that the sum of the dice rolled is either a 6 or a 12?

4. Russell has a bag of 100 marbles. Fifty are red, and fifty are black. What is the probability of
   (A) picking 2 red marbles if the first one is NOT replaced?
   (B) picking two black and two red marbles without replacement?

Chapter 18 Probability

18.6 Conditional Probability (DOK 3)

Conditional probability is defined as the probability that a second event will happen, given that the first event has already occurred. When two events are dependent, the conditional probability of A given B is

$$P(A \text{ given } B) = P(A|B) = \frac{P(A \text{ and } B)}{P(B)} = \frac{P(A \cap B)}{P(B)}.$$

Example 9: A pair of dice is rolled. What is the probability that the sum of two dice will be greater than 7, A, if the first die rolled is a 4, B?

Step 1: Find A and B.
$A = $ total of two dice > 7
$B = 4$

Step 2: Find the number of outcomes for the sum of the dice. There are three outcomes, since the first die must be four and the sum of the dice must be greater than 7.
$A \text{ and } B = A \cap B = (4,4), (4,5), (4,6)$

Step 3: Find $P(A \text{ and } B)$ and $P(B)$.
$$P(A \text{ and } B) = \frac{\text{\# of outcomes for } A \text{ and } B}{\text{total \# of outcomes}} = \frac{3}{36} = \frac{1}{12}$$
$$P(B) = \frac{\text{\# of outcomes for } B}{\text{total \# of outcomes}} = \frac{1}{6}$$
$$P(A|B) = \frac{P(A \text{ and } B)}{P(B)} = \frac{\frac{1}{12}}{\frac{1}{6}} = \frac{1}{2}$$

The probability of rolling the sum of two dice that is greater than 7, given that the first dice must be 4 is $\frac{1}{2}$.

Example 10: The probability that UGA and GT both win a football game in the same weekend is 37%. The probability that just UGA wins is 68%. What is the probability that GT will win given that UGA has already won?

$$P(A|B) = \frac{P(A \text{ and } B)}{P(B)} = \frac{37\%}{68\%} = 54\%$$

Find the conditional probability.

1. Kelly took two math tests. Her teacher said 30% of the class passed both tests, but 52% passed the first test. What is the probability that Kelly, who passed the first test, also passed the second test?

2. In California, 88% of teenagers have a cell phone, and 76% have a cell phone and a MP3 player. What is the probability that a teenager has a MP3 player given that he or she also has a cell phone?

18.6 Conditional Probability (DOK 3)

3. Tonya is selling 2 kinds of cookies, chocolate chip and sugar. After going to 100 houses, she determines that 41% of people bought chocolate chip cookies, and 38% bought chocolate chip and sugar cookies. What percentage of people bought sugar cookies given that they already purchased chocolate chip cookies?

4. The Braves have a double-header on Saturday. The probability that they will win both games is 31%. The probability that they will win just the first game is 71%. What is the probability they will win the second game given that they have already won the first game?

5. During a storm, the probability of the power and the cable going out is 31%. The probability that just the cable will go out is 54%. What is the probability that the power will go out given that the cable is already out?

6. The Nutcracker is being performed at the Fox Theater. The probability that the show will sell out on Friday and Saturday night is 61%. The probability that it will sell out Friday night is 63%. What is the probability that the show will sell out Saturday night if it sold out Friday night?

7. During a race, the probability that a race car driver will have two flat tires is 13%. The probability that he will have one flat tire is 21%. Given that he has already had one flat tire, what is the probability that he will have another?

8. The probability of thunder and lightning during a summer storm is 19%. The probability of only having thunder is 27%. What is the probability that lightning will occur given that there is already thunder?

9. Matthew and Lexi decided to go out to dinner. The probability that they both order a hamburger is 7%. The probability that one of them orders a hamburger is 24%. What is the probability that Matthew will order a hamburger if Lexi has already ordered one?

10. A pair of dice is rolled. What is the probability that the sum of two dice will be greater than 9 if the first die rolled is a 6?

11. A pair of dice is rolled. What is the probability that the sum of two dice will be less than 5 if the second die rolled is a 2?

Chapter 18 Probability

18.7 The Multiplication Rule (DOK 2)

The general multiplication rule is a method for finding the probability that two events will happen. If the two events are independent, like flipping a coin twice, then use the formula $P(A \text{ and } B) = P(A) \times P(B)$. If the two events are dependent, use the formula $P(A \text{ and } B) = P(A) \times P(B|A)$.

Example 11: What is $P(A \text{ and } B)$ if $P(A) = 0.20$, $P(B) = 0.70$, and $P(B|A) = 0.40$?

Answer: Plug the numbers into the formula and solve.

$P(A \text{ and } B) = P(A) \times P(B|A)$

$P(A \text{ and } B) = (0.20) \times (0.40)$

$P(A \text{ and } B) = 0.08$

Answer the following probability problems.

1. The probability of Sally getting an A on her Chemistry exam is 0.76, and the probability of her getting an A on her Calculus exam is 0.72. The probability of her getting an A on her Calculus exam given that she got an A on her Chemistry exam is 0.65. What is the probability of Sally getting an A on her Chemistry exam and on her Calculus exam?

2. If $P(\text{11th Grade}) = 0.45$, $P(\text{Pie}) = 0.32$, and $P(\text{Pie}\,|\,\text{11th Grade}) = 0.28$, what is $P(\text{11th Grade and Pie})$?

3. If $P(\text{Woman}) = 0.15$, $P(\text{Wears Blue}) = 0.78$, and $P(\text{Wears Blue}\,|\,\text{Woman}) = 0.59$, what is $P(\text{Woman and Wears Blue})$?

4. If $P(\text{Man}) = 0.35$, $P(\text{Plays Video Games}) = 0.48$, and $P(\text{Plays Video Games}\,|\,\text{Man}) = 0.62$, what is $P(\text{Man and Plays Video Games})$?

5. If $P(\text{6th Grade}) = 0.23$, $P(\text{Loves the Color Red}) = 0.52$, and $P(\text{Loves the Color Red}\,|\,\text{6th Grade}) = 0.87$, what is $P(\text{6th Grade and Loves the Color Red})$?

18.8 Two-Way Frequency Tables (DOK 3)

A **two-way table** is a useful way to organize data that can be categorized by two variables.

The conditional probability of an event B is the probability that the event will occur given the knowledge that an event A has already occurred. This probability is written $P(B|A)$, notation for the probability of B given A.

$$P(B \text{ given } A) = P(B|A) = \frac{P(A \text{ and } B)}{P(A)}$$

Example 12: A poll was taken at Mipsy High School that asked students in each grade what their favorite subject was. The poll generated the following frequency (two-way) table.

Subject

Grade	Math	English	Science	Social Studies	Total
9th	520	232	105	20	877
10th	115	308	200	56	679
11th	205	105	68	225	603
12th	15	420	10	3	448
Total	855	1065	383	304	2607

What is the probability that a student likes Math given that they are in the 11th grade?

$$P(\text{Math}|\text{11th grade}) = \frac{P(\text{Math and 11th Grade})}{P(\text{11th Grade})}$$

$$P(\text{Math}|\text{11th grade}) = \frac{\frac{205}{2607}}{\frac{603}{2607}} = \frac{205}{603} = 0.33997 \sim 34\%$$

What is the probability that a student is in the 11th grade given that they like Math?

$$P(\text{11th grade}|\text{Math}) = \frac{P(\text{11th Grade and Math})}{P(\text{Math})}$$

$$P(\text{11th grade}|\text{Math}) = \frac{\frac{205}{2607}}{\frac{855}{2607}} = \frac{205}{855} = 0.23977 \sim 24\%$$

Chapter 18   Probability

Practice your knowledge.

1. You roll two dice. The first die shows a ONE and the other die rolls under the table, and you cannot see it. What is the probability that both die show ONE?

2. You draw two cards from a standard deck of 52 cards that has four suits with 13 cards each. What is the probability of drawing two cards of the same suit in a row? The cards are not replaced in the deck.

3. A new bag of golf tees contains 10 red tees, 10 orange tees, 10 green tees, and 10 blue tees. You empty the tees into your golf bag. What is the probability of grabbing out two tees of the same color for you and your partner?

4. In a library box, there are 8 novels, 8 biographies, and 8 war history books. If Jack selects two books at random, what is the probability of selecting two different kinds of books?

5. What is the probability that the sum of two die will be greater than 8, given that the first die is 6?

6. What is the probability of drawing two aces from a standard deck of 52 cards that has 4 aces, given that the first card is an ace? The cards are not returned to the deck.

7. A new MasterCard has been issued to 2,000 customers. Of these customers, 1,500 hold a Visa card, 500 hold an American Express card, and 40 hold a Visa card and an American Express card. Find the probability that a customer chosen at random holds a Visa card, given that the customer holds an American Express card.

8.

		Sport		
Grade		Snowboarding	Skiing	Ice Skating	Total
	6th	68	41	46	155
	7th	84	56	70	210
	8th	59	74	47	180
	Total	211	171	163	545

(A) What is the probability of selecting a student whose favorite sport is skiing?

(B) What is the probability of selecting a 6th grade student?

(C) If the student selected is a 7th grade student, what is the probability that the student prefers ice-skating?

(D) If the student selected prefers snowboarding, what is the probability that the student is a 6th grade student?

Chapter 18 Review

(DOK 2)

1. There are 50 students in the school orchestra in the following sections:

string section	woodwind	percussion	brass
25	15	5	5

   One student will be chosen at random to present the orchestra director with an award. What is the probability the student will be from the woodwind section?

2. Fluffy's cat treat box contains 6 chicken-flavored treats, 5 beef-flavored treats, and 7 fish-flavored treats. If Fluffy's owner reaches in the box without looking, and chooses one treat, what is the probability that Fluffy will get a chicken-flavored treat?

3. The spinner in figure A stopped on the number 5 on the first spin. What is the probability that it will not stop on 5 on the second spin?

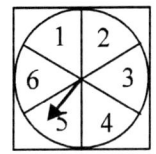

   Fig. A   Fig. B

4. Sherri turns the spinner in figure B above 3 times. What is the probability that the pointer lands on a shaded number all three times?

5. Three cakes are sliced into 20 pieces each. Each cake contains 1 gold ring. What is the probability that one person who eats one piece of cake from each of the 3 cakes will find the 3 gold rings?

6. Brianna tosses a coin 4 times. What is the probability she gets tails all four times?

7. Tempest has a bag with 4 red marbles, 3 blue marbles, and 2 yellow marbles. Does adding 4 purple marbles increase or decrease her chances that the first marble she draws at random will be red?

8. Simone has lived in Silver Spring for 250 days. In that time, her power has been out 3 days.

   (A) If the power outages happen at random, what is the probability that the power will be out tomorrow?

   (B) If the probability remains the same, how many days will she be without power for the next 10 years (3,652 days)? Round to the nearest whole number.

9. There are 20 balls in a box. 12 are green, 8 are black. If two balls are chosen without replacement, what is the probability that both are green balls?

Chapter 18 Probability

(DOK 3)

10. The probability that a student is at Ross University is 16%. The probability that a student plays football and is at Ross University is 9%. What is the probability that a student plays football given that they are already at Ross University?

11. At Texas State University, the probability that a student is taking Spanish is 72%. The probability that a student is taking Spanish and self defense is 28%. What is the probability that a student is taking self defense given that they are already registered to take Spanish?

(DOK 2)

12. Branden has a box of 50 cookies, 4 of which are peanut butter. On the way home, he randomly picks two cookies to eat as his snack. What is the probability he chose

    (A) 2 peanut butter cookies (no replacement)?

    (B) 2 peanut butter cookies if he puts the first one back?

    (C) 1 peanut butter cookie and 1 other cookie (no replacement)?

13. At the county fair, there are 5 unlabeled cakes. 3 are on table A. Two of the 3 were made by Ann, and the third was made by Betty. Two cakes (one of Ann's and one of Betty's) are on table B. The judge at the fair moves one cake from table A to table B. Jack then randomly chooses a cake from table B. What is the probability that he chose a cake made by Betty?

(DOK 3)

14. Drees works in a factory that makes combs. The following chart shows the number of defective combs per batch (1 batch is 500 combs):

Batch Number	1	2	3	4
Number of Defective Combs	34	23	33	26

If the factory produces 50 batches a month, how many defective combs should Drees expect the factory to produce?

Read the following, and answer questions 15–19. (DOK 2)

There are 9 slips of paper in a hat, each with a number from 1 to 9. The numbers correspond to a group of students who must answer a question when the number for that group is drawn. Each time a number is drawn, the number is put back in the hat.

15. What is the probability that the number 6 will be drawn twice in a row?

16. What is the probability that the first 5 numbers drawn will be odd numbers?

17. What is the probability that the second, third, and fourth numbers drawn will be even numbers?

Chapter 18 Review

18. What is the probability that the first five times a number is drawn it will be the number 5?

19. What is the probability that the first five numbers drawn will be 1, 2, 3, 4, 5 in that order?

Solve the following word problems. For questions 20–22, write whether the problem is "dependent" or "independent". (DOK 3)

20. Felix reaches into a 10-piece puzzle and pulls out one piece at random. This piece has two places where it could connect to other pieces. What is the probability that he will select another piece which fits the first one if he selects the next piece at random?

21. Barbara wants a piece of chocolate candy. She reaches into a bag which contains 8 peppermint, 5 butterscotch, 7 toffee, 3 mint, and 6 chocolate pieces and pulls out a toffee piece. She puts it back into the bag and then reaches in and pulls out another piece of candy. What is the probability that Barbara pulls out a chocolate piece on the second try?

22. Christen goes to a pet shop to purchase a guppy. The clerk reaches with a net into the tank containing 5 goldfish, 6 guppies, 4 miniature catfish, and 3 minnows and accidently pulls up a goldfish. He places the goldfish back in the water. The fish are swimming so fast, it is difficult to tell which fish the clerk will catch. What is the probability that he will catch a guppy on his second try?

23. A game is played in which 2 contestants take turns picking a letter and seeing if it appears in an unknown phrase. The object of the game is to guess what the phrase is. If a correct letter is guessed, the player gets another turn. However, if an incorrect letter is guessed, the game moves to the next player. In this game, men play against the women. Men can pick only consonants, and women can pick only vowels. Is this game fair? Why or why not?

Chapter 18 Test

1 There are 10 boys and 12 girls in a class. If one student is selected at random from the class, what is the probability it will be a girl?

A $\dfrac{1}{2}$

B $\dfrac{1}{22}$

C $\dfrac{6}{11}$

D $\dfrac{6}{5}$

(DOK 2)

2 David just got a new credit card in the mail. What is the probability the second digit of the credit card number is a 3?

A $\dfrac{1}{3}$

B $\dfrac{1}{10}$

C $\dfrac{2}{3}$

D $\dfrac{1}{5}$

(DOK 2)

3 Brenda has 18 fish in an aquarium. The fish are the following colors: 5 orange, 7 blue, 2 black, and 4 green. Brenda's trouble-making cat has grabbed a fish. What is the probability the cat grabbed a green fish?

A $\dfrac{2}{9}$

B $\dfrac{1}{18}$

C $\dfrac{2}{7}$

D $\dfrac{1}{4}$

(DOK 2)

4 In problem number 3, what is the probability that the cat did **not** grab an orange fish?

A $\dfrac{1}{3}$

B $\dfrac{13}{18}$

C $\dfrac{3}{4}$

D $\dfrac{13}{5}$

(DOK 2)

5 You have a cube with each face numbered 1, 2, 3, 4, 5, or 6. If you roll the cube 4 times, what is the probability that you will get the number 5 each time?

A $\dfrac{4}{1296}$

B $\dfrac{4}{5}$

C $\dfrac{1}{256}$

D $\dfrac{1}{1296}$

(DOK 2)

6 Katie spun a spinner 15 times and recorded her results in a table below. The spinner was divided into 6 sections numbered 1–6. The results of the spins are shown below.

1	3	6	5	1
6	2	4	3	4
2	5	1	4	5

Based on the results, how many times would 4 be expected to appear in 45 spins?

A 9
B 12
C 15
D 21

(DOK 2)

Chapter 18 Test

7 Carrie bought a large basket of 60 apples. When she got them home, she found 4 of the apples were green. If she buys 200 more apples, about how many green apples should she expect?

A 8
B 13
C 20
D 24   (DOK 2)

8 In a game using two numbered cubes, what is the probability of **not** rolling the same number on both cubes three times in a row?

A $\frac{125}{256}$

B $\frac{125}{216}$

C $\frac{27}{64}$

D $\frac{16}{27}$ (DOK 2)

9 Ben has a bag of 40 red triangles and 40 blue triangles. Without replacement, what is the probability of Ben choosing 2 red triangles?

A 0.20
B 0.25
C 0.30
D 0.35   (DOK 2)

10 Kyle has 4 different kinds of Halloween candy in a bag. He has 5 chocolate rolls, 6 lollipops, 4 chocolate bars, and 7 peanut butter candies. What kind of candy did Kyle pick out of the bag if the probability of picking that kind is $\frac{2}{11}$?

A chocolate roll
B lollipop
C chocolate bar
D peanut butter   (DOK 2)

11 There are 7 colored erasers in a bag. At least one is purple. At least two are green. At least three are orange. What color must the last eraser be if the probability of pulling that color from the bag is $\frac{4}{7}$?

A purple
B green
C orange
D brown   (DOK 2)

12 The table below shows the actual sum of the rolling of two cubes numbered 1 through 6. The two cubes were rolled 100 times.

Sum	Frequency
2	5
3	5
4	9
5	10
6	15
7	14
8	15
9	11
10	9
11	4
12	3

Using the information in the table, predict how many times a score of "7" would occur in 150 tries.

A 21
B 28
C 35
D 43   (DOK 3)

13 In Florida, 94% of people have a TV. 71% have a TV and a DVD player. What percent of people that have a TV also have a DVD player?

A 66%
B 70%
C 83%
D 76%   (DOK 3)

Copyright © American Book Company

Chapter 18   Probability

14 A standard deck of playing cards has 4 suits with each suit containing 13 cards: numbered cards 1 (Ace) though 10 and 3 face cards, a Jack, Queen, and King. This adds to a total of 52 cards in a standard deck: 40 numbered cards and 12 face cards. Samantha draws two cards out of the deck. What is the probability of her drawing one face card and one seven, with replacement?

A $\dfrac{4}{221}$

B $\dfrac{3}{26}$

C $\dfrac{3}{169}$

D $\dfrac{12}{13}$ (DOK 2)

15 A bag has 55 green marbles and 45 red marbles inside. If two green marbles are pulled from the bag at the same time, player one gets a point. If two red marbles are pulled from the bag at the same time, player two gets a point. If one red and one green marble are pulled at the same time, both players get a point. Is this a fair game?

A No. There are more red marbles than green marbles in the bag, giving player two an unfair advantage.

B Yes, this is a fair game.

C No. There are more green marbles than red marbles in the bag, giving player two an unfair advantage.

D No. There are more green marbles than red marbles in the bag, giving player one an unfair advantage.

(DOK 3)

16 Karen has two tests this week. The probability that she'll pass both is 62%. The probability she'll pass one is 74%. What's the probability she'll pass one if she already passed the other?

A 80%
B 87%
C 89%
D 84%   (DOK 3)

17 At Little Mario's Restaurant, 53% of customers typically order an appetizer with their meal, and 21% of customers order both an appetizer and a dessert with their meal. What is the probability that a customer will order a dessert, given that they have already ordered an appetizer?

A 40%
B 11%
C 32%
D 252%   (DOK 3)

18 Sarah has a bag of 100 marbles. 40 are blue, 40 are green, and 20 are purple. What is the probability of Sarah picking a purple marble on the first try?

A $\dfrac{1}{5}$

B $\dfrac{1}{4}$

C $\dfrac{1}{100}$

D $\dfrac{8}{25}$ (DOK 2)

A survey given at a high school asked students in each grade what sport they participated in. The two-way frequency table shown below has the results. Use the table for questions 19 and 20.

Sport

Grade	Football	Soccer	Rugby	Baseball	Total
9th Grade	35	200	102	140	477
10th Grade	74	133	93	122	422
11th Grade	110	72	81	131	394
12th Grade	182	42	84	93	401
Total	401	447	360	486	1694

19 What is the percentage of 12th grade students surveyed?

A 45%

B 24%

C 19%

D 11%

(DOK 2)

20 What is probability that a student plays soccer given that they are in 9th grade?

A 29%

B 32%

C 42%

D 87%

(DOK 2)

Index

30-60-90 Triangles, 87
45-45-90 Triangles, 87

AA Theorem, 47
AAS Congruence Theorem, 17
Acknowledgements, ii
Acute Angles Theorem, 17
Addition of Polynomials, 183
Addition Property of Equality, 14
Addition Rule, 313
Adjacent Congruent Angles Theorem, 16
Algebra
   multi-step problems, 195
   two step problems, 194
Alternate Exterior Angles
   defined, 13
   theorem, 16
Alternate Exterior Angles Theorem, 16
Alternate Interior Angles
   defined, 13
   theorem, 16
Alternate Interior Angles Theorem, 16
Angle Addition Postulate, 15
Angle Bisector
   constructing, 32
Angle-Side Theorem, 18
Angles
   alternate exterior, 13
   alternate interior, 13
   circumscribed, 102
   complementary, 13, 95
   congruent, 13
   consecutive exterior, 13
   consecutive interior, 13
   copying, 30
   corresponding, 13, 15
   inscribed, 102
   supplementary, 13, 15
   vertical, 13

Arc Length
   defined, 115
   measure of, 113
Arccosine, 90
Arcs
   defined, 102
   major, 102, 113
   measure of, 113
   minor, 102, 113
Arcsine, 90
Arctangent, 90
Area
   circle, 105
   sectors, 119
ASA, 15

Base Angles Theorem, 17
Binomials, 182
Bisector, 32

Cavalieri's Principle, 132
Census, 307
Center, 102
Central Angle
   defined, 102, 113
Chart of Standards, iii
Chord, 102
Circle
   arc length, 113, 115
   arcs, 102
   area, 105
   area of sector, 119
   center, 102
   central angle, 102, 113
   chord, 103
   circumference, 102, 104, 108
   circumscribed, 112
   circumscribed angle, 103
   constructions, 111, 112
   definitions, 102

diameter, 102, 104, 108
dissection argument, 105
equations, 120
inscribed, 111
inscribed angle, 102
inscribing a polygon, 40
major arc, 102
minor arc, 102
π, 104, 108
properties, 116
radian, 115
radius, 102, 104
secant, 103, 116
tangent, 102, 109, 116
Circumference
   defined, 102
   distance around a circle, 104
   ratio of circumference and diameter, 108
Circumscribed Angle, 102
Circumscribed Circle, 112
Compass, 28
Complement, 299
Complementary Angles
   defined, 13
   sum of the measures, 95
completing the square, 7
Complex Conjugate, 174
Complex Numbers
   absolute value of, 175
   adding and subtracting, 172
   dividing, 174
   multiplying, 173
   simplifying, 175
   written form, 170
Compound Events, 310
conclusion, 10
Conditional Probability
   defined, 307
   first and second events, 314
Conditional Statement, 10
Cone
   cross section, 127
   formation, 130

   properties, 125
Congruent Angles, 13
Congruent Complements Theorem, 16
Congruent Figures
   defined, 45
   transformations, 68–71
Congruent Segments, 13
Congruent Supplements Theorem, 16
Consecutive Exterior Angles, 13
Consecutive Interior Angles
   defined, 13
   theorem, 16
Consecutive Interior Angles Theorem, 16
Constructions
   angle bisector, 32
   circumscribed circle, 112
   copying a line segment, 29
   copying an angle, 30
   inscribed circle, 111
   parallel lines, 38
   perpendicular bisector, 35
   perpendicular lines, 37
   polygons inscribed in a circle, 40
   tangent line to a circle, 109
Contrapositive, 10
Converse, 10
Corollary
   angle-side theorem, 18
   base angles, 18
Corresponding Angles
   converse, 15
   defined, 13
Corresponding Sides
   triangles, 48
Cosine, 89
Counterexample, 10
CPCTC Theorem, 18
Cross Sections
   cones, 127
   cubes, 127
   cylinders, 128
   pyramids, 128
   rectangular prisms, 127

spheres, 128
Cube
   cross section, 127
   formation, 129
   properties, 125
Cube Roots, 163
Cylinder
   cross section, 128
   formation, 130
   properties, 125

Denominator, 187
Dependent Events
   defined, 307
   outcome, 311
Diameter
   defined, 102
   ratio of circumference and diameter, 108
   twice the length of radius, 104
Dilations
   defined, 76
   rules of, 64
   simlar figures, 78
Directrix, 270
Dissection Argument, 105
Division Property of Equality, 14

Elements, 299
Empty Set
   ∅, 299
Equally Likely Outcomes, 307
Equations
   linear systems
      solving by substitution, 224
   manipulating formulas and equations, 230
   of circles using the Pythagorean Theorem, 120
   solutions of, 272, 286
   solving polynomial equations analytically, 235
   solving polynomial equations graphically, 242
   solving with absolute values, 202
   solving with graphing calculator, 242

Even, 281
Event
   dependent, 307
   independent, 307
   mutually exclusive, 307
Expected Value, 307
Exponents
   dividing with, 153
   dividing with fractional exponents, 154
   exponents raised to an exponent, 149
   expressions, 149
   fractions raised to a power, 150
   multiplying fractional exponents, 148, 151
   multiplying polynomials, 185
   multiplying with, 148, 150
   multiplying with negative exponents, 152
   negative, 152
   of monomials with different variables, 185
   of polynomials, 183
   rational, 164
   understanding, 147
   when subtracting polynomials, 183
Expressions
   rewriting, 181

Factoring
   quadratic equations, 210
Flowcharts, 22
Focus, 270
Fractions
   raised to a power, 150
Functions
   combining functions 5 ways, 280
   comparing, 277
   finding common solutions, 274
   finding the rate of change, 275
   modeling data with quadratic functions, 287
   multiplicity of graphs of polynomial functions, 256
   real-world applications of polynomial functions, 261
   symmetry, 281

symmetry of graphs, 283
Fundamental Theorem of Algebra, 209

Geometric Constructions, 28
Geometric Relationships of Plane Figures, 51
Geometric Relationships of Solids, 135
Geometric Solids, 126
Graphing
   systems of inequalities, 226
   systems of three linear inequalities, 228
Graphing Calculator
   using to solve polynomial equations, 242

Hexagon
   inscribed in a circle, 40
HL Congruence Theorem, 18
Horizontal Shrink, 74
Horizontal Stretch, 74
Horizontal Stretch or Compression, 253
Hypotenuse, 82
Hypothesis, 10

Imaginary Numbers, 171
Independent Events
   defined, 307
   outcome, 311
Inequalities
   graphing systems of, 226
   graphing systems of three linear inequalities, 228
   multi-step, 200
   solving polynomial inequalities analytically, 247
   solving with absolute values, 202
   word problems, 205
Inscribed Angle
   defined, 102
   measure of, 113
   quadrilaterals, 114
Inscribed Circle, 111
Intersection
   ∩, 301
Inverse, 10
Irrational Numbers, 165

Lead Coefficients
   positive and negative, 256
Line of Reflection, 56
Line Segment
   copying, 29
   perpendicular bisector
      constructing, 35
      proof, 36

Major Arc
   defined, 102
   measure of, 113
Midpoint of a Segment, 13
Minor Arc
   defined, 102
   measure of, 113
Monomials
   defined, 182
   multiplying, 185
   multiplying by polynomials, 186
Multi-Step Algebra Problems, 195
Multiplication Property of Equality, 14
Multiplication Rule
   for finding the probability of two events, 316
   for independant events, 311
Mutually Exclusive Events
   defined, 307
   no common outcome, 313

Negative Lead Coefficient, 256
$P'(A)$
   probability of event A not occuring, 307

Odd, 281
Order of Operations, 155

Parabola, 270
Parallel Lines
   constructing, 38
Parentheses
   removing and simplifying in polynomials, 189
Parts of an Expression

simplifying, 179
Perfect Squares, 214
Perpendicular Bisector
  proof, 36
Perpendicular Lines
  constructing, 37
  theorem, 16
Perpendicular Lines Theorem, 16
Perpendicular Transversal Theorem, 17
π, pi
  ratio of a circumference of a circle, 104
  similarity of circles, 108
Plane Figures
  geometric relationships, 51
  similar, 53
Points on a Perpendicular Bisector, 36
Polygons
  inscribed in a circle, 40
Polynomial Equations
  solving polynomial equations analytically, 235
Polynomials
  adding, 183
  defined, 182
  dividing by monomials, 187
  multiplying by monomials, 186
  subtracting, 183
Population, 307
Positive Lead Coefficient, 256
Postulate, 12
Preface, xi
Prism
  defining names of, 126
  properties, 125
$P(A)$
  probability of event A occuring, 307
Probability, 307
  addition rule, 313
  chance, 308
  complement, 307
  compound events, 310
  conditional, 314
  defined, 307

  dependent events, 311
  expected value, 307
  independent events, 311
  multiplication rule, 311, 316
  mutually exclusive events, 313
  notation, 307
  two-way table, 317
Proofs
  flowchart, 22
  two-column, 20
  used for, 12
Properties of Congruent Triangles Theorem, 17
Proportions, 46
Pyramid
  cross section, 128
  defining names of, 126
  formation, 130
  properties, 125
Pythagorean Theorem
  applications, 85
  definition, 82
  missing leg, 84
  proof, 83
  similar triangles, 83

Quadratic Equation
  $ax^2 + bx + c = 0$, 211
  polynomials, 209
  solving or completing the square, 215
Quadratic Equations
  complex roots, 212
  real-world, 216
Quadratic Formula
  $\dfrac{-b \pm \sqrt{b^2 - 4ac}}{2a}$, 211
Quadratic Functions
  modeling data with, 287
Quadrilateral
  angle properties, 114

Radian, 115
Radical Equations, 197
Radius

defined, 102
half the length of diameter, 104
Rational Numbers, 165
Ratios, 46
Real Numbers
computing with, 166
definition, 165
Reciprocal, 152
Recognizing Errors in Problems, 206
Rectangular Prism
cross section, 127
formation, 129
properties, 125
Rectangular Pyramid, 125
Reflection
comparing transformations, 74
line of, 56
rules of, 64
transformations, 64
Reflexive Property of Congruence, 14
Reflexive Property of Equality, 14
Removing Parentheses and Simplifying, 189
Right Angle Theorem, 16
Right Triangle
30-60-90, 87
45-45-90, 87
finding the missing leg of, 84
Pythagorean Theorem, 82
special, 87
Rigid Motion, 68–71
Rotation
around a point, 62
comparing transformations, 74
rules of, 64
transformations, 64

Sample, 307
Sample Space, 307
SAS, 15
Scale Factor
dilations, 76
triangles, 48
Secant

defined, 102
properties of circles, 116
Sector
area of, 119
Segment
congruent, 13
midpoint, 13
Segment Addition Postulate, 15
Set, 299
Similar Figures
defined, 45
dilation, 78
plane figures, 50
solid figures, 137
Similar Triangles, 94
AA theorem, 47
defined, 47
Pythagorean theorem, 83
scale factor, 48
Sine, 89
Solid Geometry Word Problems, 134
Special Right Triangles, 87
Sphere
cross section, 128
formation, 130
properties, 125
Square
inscribed in a circle, 40
Square Pyramid, 125
Square Roots
adding, subtracting, and simplifying, 160
definition, 159
dividing and simplifying, 162
multiplying and simplifying, 161
simplifying using factors, 159
Squares
solving perfect squares, 214
sovling the difference of two squares, 213
SSS, 15
Straight Edge, 28
Subset
$\subseteq$, 300
Subtitution Property of Equality, 14

Subtraction
    of polynomials, 183
Subtraction Property of Equality, 14
Subtrahend, 183
Supplementary Angles
    defined, 13
    postulate, 15
Symmetric Property of Congruence, 14
Symmetric Property of Equality, 14
Symmetry
    functions, 281
    graphs of functions, 283
Synthetic division
    removing and simplifying in polynomials, 188

Table of Contents, x
Tangent, 89
    constructing, 109
    defined, 102
    intersects a circle, 109
    properties of circles, 116
Theorem, 12
    AA, 47
    triangles, 47
Three-Dimensional Figures
    cone, 125
    cube, 125
    cylinder, 125
    rectangular prism, 125
    rectangular pyramid, 125
    sphere, 125
    square pyramid, 125
    triangular prism, 125
    triangular pyramid, 125
Transformations
    algebraic notation, 64
    Congruency, 69–71
    congruency, 68
    dilation, 64, 74, 76
    graphing, 253
    horizontal shrink, 74
    horizontal stretch, 74
    reflection, 56, 64, 74
    rotation, 62, 64, 74
    rules, 64
    shapes onto itself, 66
    translation, 60, 64, 74
    vertical shrink, 74
    vertical stretch, 74
Transitive Property of Congruence, 14
Transitive Property of Equality, 14
Transitivity of Parallel Lines Theorem, 16
Translations, 253
    comparing, 74
    of a geometric figure, 60
    rules of, 64
Transversal, 13
Triangle Sum Theorem, 17
Triangles
    AA theorem, 47
    ASA, 15
    corresponding sides, 48
    proportional sides, 48
    SAS, 15
    scale factor, 48
    similar, 47
    $30 - 60 - 90$, 87
    $45 - 45 - 90$, 87
    SSS, 15
    trigonometric ratios, 89
Triangular Prism, 125
Triangular Pyramid, 125
Trigonometric Ratios
    arccos, 90
    arcsin, 90
    arctan, 90
    defined, 89
    other shapes, 97
    $\cos A = \frac{\text{adj.}}{\text{hyp.}}$, 89
    $\tan A = \frac{\text{opp.}}{\text{adj.}}$, 89
    similar triangles, 94
    $\sin A = \frac{\text{opp.}}{\text{hyp.}}$, 89
Trinomials, 182

Two Step Algebra Problems, 194
Two-Column Proofs, 20
Two-Way Table, 317

Union
  ∪, 302

Variable
  adding polynomials, 183
  multiplying monomials, 185
Venn Diagram
  intersection, 301
  union, 302
  visual way to see two or more variables, 303
Vertical Angles
  defined, 13
  theorem, 16
Vertical Angles Theorem, 16
Vertical Shrink, 74
Vertical Stretch
  comparing transformations, 74
  graphing transformations, 253
Volume
  spheres, cones, cylinders, and pyramids, 132

Photographs by W. Lee Youngblood

Merrill Publishing Co.
well Co.
Ohio

1974 by Bell & Howell Co. All rights reserved. No part may be reproduced in any form, electronic or mechanical, photocopy, recording or any information storage and re-n, without permission in writing from the publisher.

Standard Book Number: 0-675-08846-1

Congress Catalog Card Number: 73-93272

1 2 3 4 5 6 7 8 9 10—77 76 75 74

Printed in the United States of America

Student-Ce...

in the Secon...

Publishe...
Charles...
A *Bell &*...
Columb...

Copyrig...
of this...
includir...
trieval s...

Interna...

Librar...

Charl...